METHODS IN MOLECULAR BIOLOGY

For further volumes:
http://www.springer.com/series/7651

For over 35 years, biological scientists have come to rely on the research protocols and methodologies in the critically acclaimed *Methods in Molecular Biology* series. The series was the first to introduce the step-by-step protocols approach that has become the standard in all biomedical protocol publishing. Each protocol is provided in readily-reproducible step-by step fashion, opening with an introductory overview, a list of the materials and reagents needed to complete the experiment, and followed by a detailed procedure that is supported with a helpful notes section offering tips and tricks of the trade as well as troubleshooting advice. These hallmark features were introduced by series editor Dr. John Walker and constitute the key ingredient in each and every volume of the *Methods in Molecular Biology* series. Tested and trusted, comprehensive and reliable, all protocols from the series are indexed in PubMed.

Hepatic Stellate Cells

Methods and Protocols

Edited by

Ralf Weiskirchen

Institut für Molekulare Pathobiochemie, Experimentelle Gentherapie und Klinische Chemie (IFMPEGKC), Universitätsklinikum Aachen AöR, Aachen, Germany

Scott L. Friedman

Division of Liver Diseases, Icahn School of Medicine at Mount Sinai, New York, NY, USA

Humana Press

Editors
Ralf Weiskirchen
Institut für Molekulare
Pathobiochemie, Experimentelle
Gentherapie und Klinische Chemie
(IFMPEGKC)
Universitätsklinikum Aachen AöR
Aachen, Germany

Scott L. Friedman
Division of Liver Diseases
Icahn School of Medicine at Mount Sinai
New York, NY, USA

ISSN 1064-3745 ISSN 1940-6029 (electronic)
Methods in Molecular Biology
ISBN 978-1-0716-3206-2 ISBN 978-1-0716-3207-9 (eBook)
https://doi.org/10.1007/978-1-0716-3207-9

This Humana imprint is published by the registered company Springer Science+Business Media, LLC, part of Springer Nature.
The registered company address is: 1 New York Plaza, New York, NY 10004, U.S.A.

Preface

Hepatic stellate cells have moved into the spotlight of liver cell biology because of their pleiotropic functions that extend well beyond extracellular matrix production to pivotal roles in hepatic homeostasis, immunity, and metabolism. Once viewed as an obscure cell type burdened by multiple terminologies that sowed confusion, stellate cells are now a focal point of study across dozens of laboratories throughout the world, and the development of methods to isolate, evaluate, and manipulate this cell type have vaulted them into the very center of efforts to elucidate hepatic biology and establish prospects for treating disease. In doing so, interest has coalesced across the world among a community of collaborative and talented scientists.

In the spirit of this collegiality, it is timely to standardize and share methods for stellate cell isolation, characterization, and modulation. Herein is our effort to accelerate progress by disseminating validated methods detailed by experts in the field. We are extremely grateful to our colleagues for their contributions, and hopeful that, by outlining these techniques, we will further accelerate the outstanding science that has steadily unveiled the mysteries of stellate cell biology and their role in disease.

Aachen, Germany *Ralf Weiskirchen*
New York, NY, USA *Scott L. Friedman*

Contents

Contributors

RUCHI BANSAL • *Translational Liver Research, Department of Medical Cell BioPhysics, Faculty of Science and Technology, University of Twente, Enschede, The Netherlands; Translational Liver Research, Department of Medical Cell BioPhysics, Technical Medical Centre, Faculty of Science and Technology, University of Twente, Enschede, The Netherlands*

RICHELL BOOIJINK • *Translational Liver Research, Department of Medical Cell BioPhysics, Faculty of Science and Technology, University of Twente, Enschede, The Netherlands*

DAVID A. BRENNER • *Department of Medicine, University of California, San Diego School of Medicine, San Diego, CA, USA*

NATHALIE EYSACKERS • *Liver Cell Biology Research Group, Vrije Universiteit Brussel, Brussels, Belgium*

SCOTT L. FRIEDMAN • *Division of Liver Diseases, Icahn School of Medicine at Mount Sinai, New York, NY, USA*

LEO A. VAN GRUNSVEN • *Liver Cell Biology Research Group, Vrije Universiteit Brussel, Brussels, Belgium*

ADRIEN GUILLOT • *Charité Universitätsmedizin Berlin, Department of Hepatology & Gastroenterology, Berlin, Germany*

MAYA W. HAAKER • *Division of Cell Biology, Metabolism & Cancer, Faculty of Veterinary Medicine, Department of Biomolecular Health Sciences, Utrecht University, Utrecht, The Netherlands*

LIONEL HEBBARD • *Department of Molecular and Cell Biology, College of Public Health, Medical and Veterinary Sciences, Centre for Molecular Therapeutics, Centre for Tropical Bioinformatics, Australian Institute of Tropical Medicine and Health, James Cook University, Townsville, QLD, Australia*

J. BERND HELMS • *Division of Cell Biology, Metabolism & Cancer, Faculty of Veterinary Medicine, Department of Biomolecular Health Sciences, Utrecht University, Utrecht, The Netherlands*

SONG HWA HONG • *Laboratory of Liver Research, Graduate School of Medical Science and Engineering, Korea Advanced Institute of Science and Technology (KAIST), Daejeon, Republic of Korea*

JEROEN W. A. JANSEN • *Division of Cell Biology, Metabolism & Cancer, Faculty of Veterinary Medicine, Department of Biomolecular Health Sciences, Utrecht University, Utrecht, The Netherlands*

WON-IL JEONG • *Laboratory of Liver Research, Graduate School of Medical Science and Engineering, Korea Advanced Institute of Science and Technology (KAIST), Daejeon, Republic of Korea*

HEE-HOON KIM • *Laboratory of Liver Research, Graduate School of Medical Science and Engineering, Korea Advanced Institute of Science and Technology (KAIST), Daejeon, Republic of Korea*

KYURAE KIM • *Laboratory of Liver Research, Graduate School of Medical Science and Engineering, Korea Advanced Institute of Science and Technology (KAIST), Daejeon, Republic of Korea*

TATIANA KISSELEVA • *Department of Surgery, University of California, San Diego School of Medicine, San Diego, CA, USA*

MYRTHE KLAVER • *European Research Institute for the Biology of Aging (ERIBA), University Medical Center Groningen, University of Groningen, Groningen, The Netherlands*

MARLENE SOPHIA KOHLHEPP • *Charité Universitätsmedizin Berlin, Department of Hepatology & Gastroenterology, Berlin, Germany*

PATRICIA F. LALOR • *Centre for Liver and Gastroenterology Research and National Institute for Health Research (NIHR) Birmingham Biomedical Research Centre, Institute of Immunology and Immunotherapy, University of Birmingham, Birmingham, UK*

CHRISTIAN LIEDTKE • *Department of Medicine III, RWTH University Hospital Aachen, Aachen, Germany*

XIAO LIU • *Department of Medicine, University of California, San Diego School of Medicine, San Diego, CA, USA; Department of Surgery, University of California, San Diego School of Medicine, San Diego, CA, USA*

INGE MANNAERTS • *Liver Cell Biology Research Group, Vrije Universiteit Brussel, Brussels, Belgium*

MICHITAKA MATSUDA • *Karsh Division of Gastroenterology and Hepatology, Department of Medicine, Cedars-Sinai Medical Center, Los Angeles, CA, USA*

STEFFEN K. MEURER • *Institute of Molecular Pathobiochemistry, Experimental Gene Therapy and Clinical Chemistry (IFMPEGKC), RWTH University Hospital Aachen, Aachen, Germany*

HAN MOSHAGE • *Department of Gastroenterology and Hepatology, University Medical Center Groningen, University of Groningen, Groningen, The Netherlands*

YULIA A. NEVZOROVA • *Department of Immunology, Ophthalmology and Otolaryngology, School of Medicine, Complutense University Madrid, Madrid, Spain*

ELLIE NORTHALL • *Centre for Liver and Gastroenterology Research and National Institute for Health Research (NIHR) Birmingham Biomedical Research Centre, Institute of Immunology and Immunotherapy, University of Birmingham, Birmingham, UK*

PANTELITSA PAPAKYRIACOU • *Centre for Liver and Gastroenterology Research and National Institute for Health Research (NIHR) Birmingham Biomedical Research Centre, Institute of Immunology and Immunotherapy, University of Birmingham, Birmingham, UK*

KLAAS POELSTRA • *Department of Nanomedicine and Drug Targeting, Groningen Research Institute of Pharmacy, University of Groningen, Groningen, The Netherlands*

LIEN REOLIZO • *Karsh Division of Gastroenterology and Hepatology, Department of Medicine, Cedars-Sinai Medical Center, Los Angeles, CA, USA*

KAROLINA SAFRANSKA • *Vascular Biology Research Group, Department of Medical Biology, UiT The Arctic University of Norway., Tromso, Norway*

PAU SANCHO-BRU • *Liver Cell Plasticity and Tissue Repair Lab at Institut d'Investigacions Biomèdiques August Pi i Sunyer (IDIBAPS), Barcelona, Spain; Centro de Investigación Biomédica en Red de Enfermedades Hepáticas y Digestivas (CIBERehd), Barcelona, Spain; University of Barcelona, Barcelona, Spain*

SARAH K. SCHRÖDER • *Institute of Molecular Pathobiochemistry, Experimental Gene Therapy and Clinical Chemistry (IFMPEGKC), RWTH University Hospital Aachen, Aachen, Germany*

ROBERT F. SCHWABE • *Department of Medicine, Columbia University, New York, NY, USA*

EKIHIRO SEKI • *Karsh Division of Gastroenterology and Hepatology, Department of Medicine, Cedars-Sinai Medical Center, Los Angeles, CA, USA; Department of Biomedical Sciences, Cedars-Sinai Medical Center, Los Angeles, CA, USA*

SANDRA A. SERNA-SALAS • *Department of Gastroenterology and Hepatology, University Medical Center Groningen, University of Groningen, Groningen, The Netherlands*
EMMA L. SHEPHERD • *Centre for Liver and Gastroenterology Research and National Institute for Health Research (NIHR) Birmingham Biomedical Research Centre, Institute of Immunology and Immunotherapy, University of Birmingham, Birmingham, UK*
KAREN K. SORENSEN • *Vascular Biology Research Group, Department of Medical Biology, UiT The Arctic University of Norway., Tromso, Norway*
ABEL A. SOTO-GÁMEZ • *Department of Radiation Oncology, University Medical Center Groningen, University of Groningen, Groningen, The Netherlands; Department of Biomedical Sciences of Cells and Systems, University Medical Center Groningen, University of Groningen, Groningen, The Netherlands*
QIUYAN SUN • *Department of Medicine, Columbia University, New York, NY, USA*
FRANK TACKE • *Charité Universitätsmedizin Berlin, Department of Hepatology & Gastroenterology, Berlin, Germany*
CARMEN G. TAG • *Institute of Molecular Pathobiochemistry, Experimental Gene Therapy and Clinical Chemistry (IFMPEGKC), RWTH University Hospital Aachen, Aachen, Germany*
LEON TERSTAPPEN • *Department of Medical Cell BioPhysics, Faculty of Science and Technology, University of Twente, Enschede, The Netherlands*
RAQUEL A. MARTÍNEZ GARCÍA DE LA TORRE • *Liver Cell Plasticity and Tissue Repair Lab at Institut d'Investigacions Biomèdiques August Pi i Sunyer (IDIBAPS), Barcelona, Spain*
MIRIAM WANKELL • *Department of Molecular and Cell Biology, College of Public Health, Medical and Veterinary Sciences, Centre for Molecular Therapeutics, Centre for Tropical Bioinformatics, Australian Institute of Tropical Medicine and Health, James Cook University, Townsville, QLD, Australia*
RALF WEISKIRCHEN • *Institut für Molekulare Pathobiochemie, Experimentelle Gentherapie und Klinische Chemie (IFMPEGKC), Universitätsklinikum Aachen AöR, Aachen, Germany*
SABINE WEISKIRCHEN • *Institute of Molecular Pathobiochemistry, Experimental Gene Therapy and Clinical Chemistry (IFMPEGKC), RWTH University Hospital Aachen, Aachen, Germany*
ZONGMEI WU • *Department of Gastroenterology and Hepatology, University Medical Center Groningen, University of Groningen, Groningen, The Netherlands*
ESTHER A. ZAAL • *Division of Cell Biology, Metabolism & Cancer, Faculty of Veterinary Medicine, Department of Biomolecular Health Sciences, Utrecht University, Utrecht, The Netherlands*

About the Editors

Ralf Weiskirchen PhD was born on February 2, 1964 in Bergisch Gladbach, North Rhine-Westphalia, Germany. After his school education, he studied biology and made his PhD with distinction at the University of Cologne in Germany. Thereafter, he worked as a Research Associate at the Institute of Biochemistry, University of Innsbruck, Austria. Back in Germany, he habilitated at the RWTH University Hospital Aachen and became a Professor in 2007. Currently, he is Head of the Institute of Molecular Pathobiochemistry, Experimental Gene Therapy and Clinical Chemistry (IFMPEGKC) at the RWTH University Hospital Aachen. His major research interest is the analysis of TGF-β/BMP and PDGF signaling pathway in the pathogenesis of liver disease. Professor Weiskirchen maintains a variety of national and international cooperations that are focused on molecular aspects of hepatic disease formation and therapy. Moreover, his work is concentrated on the identification and evaluation of novel biomarkers.

Scott L. Friedman MD was born June 13, 1955 in Brooklyn, New York. He received his undergraduate degree *cum laude* at Rensselaer Polytechnic Institute and medical degree at the Mount Sinai School of Medicine where he was President of the lambda chapter of *Alpha Omega Alpha* honorary medical society. He pursued an internal medicine residency at Beth Israel Hospital, Harvard Medical School, and then a gastroenterology/hepatology fellowship at University of California San Francisco, where he remained on the faculty for 10 years. In 1997, he joined the faculty at Mount Sinai where he is currently the Dean for Therapeutic Discovery and Chief of Liver Diseases at the Icahn School of Medicine at Mount Sinai. He has performed pioneering research into the underlying causes of fibrosis associated with chronic liver disease. His work has spawned an entire field that is now realizing its translational and therapeutic potential. In 1995–1996, he was a Senior Fulbright Scholar at the Weizmann Institute in Israel. He has directly mentored over 90 fellows and students. Dr. Friedman has been awarded the *EASL International Recognition Award* and *Distinguished Achievement Awards* from both the AASLD and the American Liver Foundation. He is widely respected as a key opinion leader working closely with the biotech and pharmaceutical industry in developing new therapies for liver disease.

Chapter 1

Isolation, Purification, and Culture of Primary Murine Hepatic Stellate Cells: An Update

Steffen K. Meurer, Sabine Weiskirchen, Carmen G. Tag, and Ralf Weiskirchen

Abstract

In the healthy liver, quiescent hepatic stellate cells (HSCs) are found in the perisinusoidal space (i.e., the space of Dissé) in close proximity to endothelial cells and hepatocytes. HSCs represent 5–8% of the total number of liver cells and are characterized by numerous fat vacuoles that store vitamin A in the form of retinyl esters. Upon liver injury caused by different etiologies, HSCs become activated and acquire a myofibroblast (MFB) phenotype in a process called transdifferentiation. In contrast to quiescent HSC, MFB become highly proliferative and are characterized by an imbalance in extracellular matrix (ECM) homeostasis, by producing an excess of collagen and blocking its turnover by synthesis of protease inhibitors. This leads to a net accumulation of ECM during fibrosis. In addition to HSC, there are fibroblasts in the portal fields (pF), which also have the potency to acquire a myofibroblastic phenotype (pMF). The contributions of these two fibrogenic cell types (i.e., MFB and pMF) vary based on the etiology of liver damage (parenchymal *vs.* cholestatic). Based on their importance to hepatic fibrosis, the isolation and purification protocols of these primary cells are in great demand. Moreover, established cell lines may offer only limited information about the in vivo behavior of HSC/MFB and pF/pMF.

Here we describe a method for high-purity isolation of HSC from mice. In the first step, the liver is digested with pronase and collagenase, and the cells are dissociated from the tissue. In the second step, HSCs are enriched by density gradient centrifugation of the crude cell suspension using a Nycodenz gradient. The resulting cell fraction can be further optionally purified by flow cytometric enrichment to generate ultrapure HSC.

Key words Hepatic stellate cells, Portal fibroblasts, Myofibroblasts, Nycodenz, FACS-sorting, Density gradient, Desmin, Phalloidin

1 Introduction

Although HSCs are a relatively uncommon cell type in the liver with a cell count below 10% of the total liver cells, these cells play a central role in the pathogenesis of chronic liver disease. They are located in the space of Disse in close proximity to hepatocytes and sinusoidal endothelial cells, to which they stay in direct contact by

Ralf Weiskirchen and Scott L. Friedman (eds.), *Hepatic Stellate Cells: Methods and Protocols*, Methods in Molecular Biology, vol. 2669, https://doi.org/10.1007/978-1-0716-3207-9_1,

means of their protrusions [1, 2]. There is a plethora of diseases of different etiologies affecting the liver and leading to fibrosis. Whereas cholestatic diseases cause a ductular reaction relying primarily on the activation of portal fibroblasts, toxic insults to hepatocytes tend to engage HSC activation. Therefore, HSCs are considered to be (at least one of) the primary cell type(s) in the liver responsible for the disruption of ECM homeostasis, leading to scar tissue formation during hepatic injury. As a result, these cells have been in the focus of many studies during the last decades [3]. Understanding HSC biology is therefore considered a promising basis to identify novel therapeutic targets designed to attenuate or even reverse extracellular matrix production. A hallmark of these cells, when kept in primary culture on uncoated ordinary culture dishes, is that they become activated and undergo the typical gradual phenotypic transition from a "quiescent," retinoid-containing cell type to an "activated," highly proliferative, fat- and retinoid-losing phenotype. Additionally, the "transdifferentiation" process is associated with increased proliferation and migratory behavior. Key events are the tremendous increase in matrix molecule synthesis and the expression of inhibitors of matrix proteases, which prevent ECM degradation. Importantly, this in vitro transdifferentiation process mimics the characteristics of this cell type that also occur during hepatic fibrosis (Fig. 1).

First protocols for the isolation of HSC were established in 1982 [4]. Due to the high order of the collagenous scaffold serving to shape the liver, in the respective isolation protocols, HSCs were first released from rat liver by digesting the tissue with pronase E and collagenase. In the second step, enrichment of HSC was facilitated by performing density gradient centrifugation using Metrizamide and centrifugal elutriation. The resulting final cell fraction contained more than 70% of HSC [4]. Subsequently, protocols were established that introduced a few other compounds for preparing continuous or discontinuous density gradients (Table 1).

The key features of HSC which discriminate them from other liver cells are as follows: (i) the average volume of parenchymal cells (i.e., hepatocytes) is an order of magnitude higher than from HSC, and (ii) the retinoid content of HSC results in a significantly lower cell density (i.e., higher buoyancy).

The latter property enables HSC to float on several types of gradient materials. This is the reason why most of the original, pioneering protocols used livers from old, fatty animals as cell donors since the cellular retinyl ester content was greatly increased.

Nevertheless, although these "conventional" HSC isolation procedures have long been the "gold standard" methodologies to purify large numbers of HSC with an acceptable purity and a typical "HSC phenotype," some drawbacks have become apparent. One problem is the contamination of purified HSC with other cell types, which may lead to misinterpretation of experimental in vitro data.

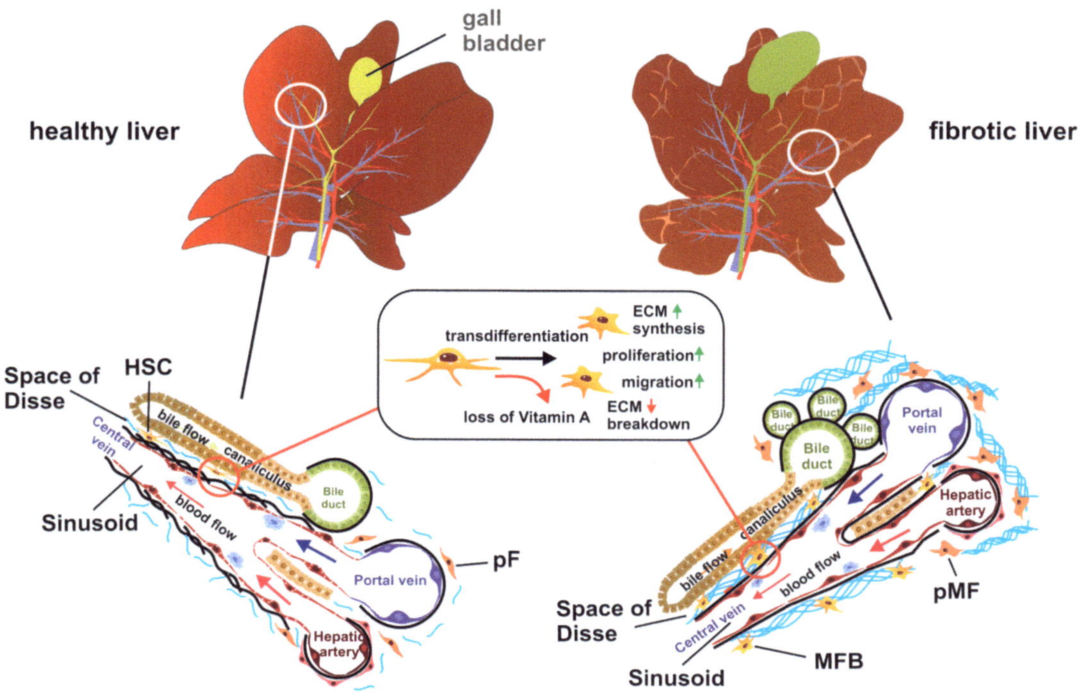

Fig. 1 HSCs are central in the development of fibrosis. Insult to the liver leads to a shift of a healthy to a fibrotic liver (*upper part*). This process is associated with characteristic changes in the functional module, i.e., the liver lobe (*lower part*). These changes include increased expression of interstitial collagens (*blue lines*), increased thickening of the basal membrane (*black lines*), and, depending on the insult, bile duct proliferation (*green circles*). A key event in the initiation/development of fibrosis is the activation and transdifferentiation of hepatic stellate cells (HSC) to myofibroblast-like cells (MFB) (*central part*). *Brown* hepatocytes; *green* biliary epithelial cells; *red* endothelial cells covering the inside of sinusoids, portal vein, and hepatic artery; *orange* HSC and portal myofibroblasts (pMF); *pink* smooth muscle cells covering the outside of the hepatic artery

Table 1
Compounds for preparing continuous or discontinuous density gradients

Compound	CAS registry number	References
Arabinogalactan (Stractan)	9036-66-2	[5, 6]
Metrizamide (Amipaque)	31112-62-6	[7]
Iohexol (Nycodenz)	66108-95-0	[8–10]
Iodixanol (OptiPrep)	92339-11-2	[11–13]
Polyvinylpyrrolidone-coated colloidal silica (Percoll) particles	65455-52-9	[14]

These contaminations become especially prominent when highly sensitive techniques are applied (e.g., RT-qPCR) where a small contaminating cell type can confound interpretation of data. Another hurdle is that the HSC population in a healthy liver is functionally heterogeneous with regard to its capacity for retinoid/ lipid storage and expression of HSC activation markers; single cell technologies have further reinforced their extensive heterogeneity [15].

Therefore, two decades after initial HSC isolation protocols were developed, isolation of rat HSCs was also achieved using side scatter-activated cell sorting. This method sorted on average $1.4 \pm 0.4 \times 10^6$ cells with feasible sorting runs for up to 4 hours, resulting in more than 5.0×10^6 highly purified viable cells [16].

Today, there is an urgent need to isolate highly pure HSC from a wide range of transgenic mouse models. Therefore, we and others have developed novel fluorescence-activated cell sorting (FACS) or magnetic cell sorting (MACS) protocols that are suitable to isolate ultrapure HSC from mouse livers [17–19]. In principle, FACS-based purification strategies take advantage of the endogenous retinoid fluorescence of HSC for their separation by FACS machines that are equipped with a UV laser, allowing the identification of these cells by their vitamin A-specific autofluorescence from other cell subpopulations (Fig. 2). However, those HSCs with little or no retinoid will not be identified by this approach.

In the following, we provide an update on a standard operation procedure (SOP) for the isolation of primary mouse HSC that was originally published in year 2017 in the Springer Protocols *Fibrosis: Methods and Protocols* [20]. The protocol includes the classical enzymatic digestion of the liver tissue, cellular enrichment by centrifugation through a discontinuous Nycodenz density gradient, and an optional UV light flow cytometric sorting step allowing recovery of ultrapure HSC.

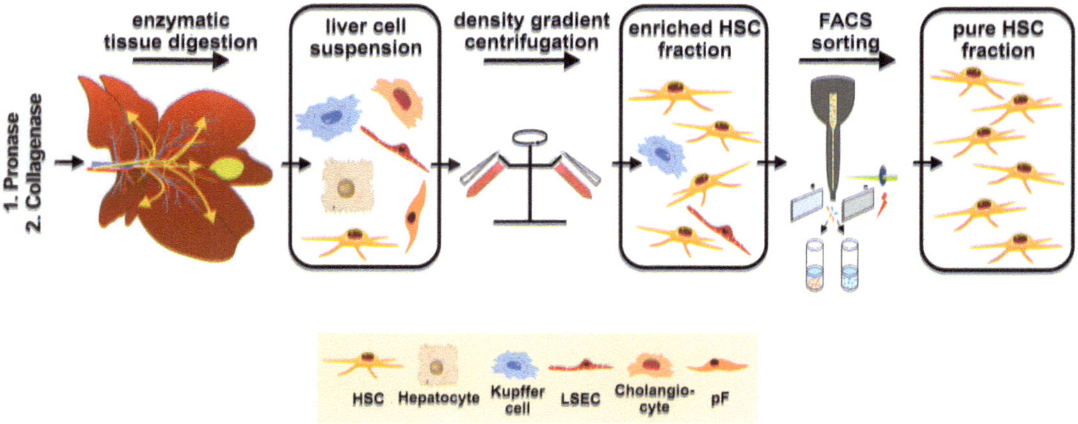

Fig. 2 Flow chart of the individual steps for isolation of murine HSC. (*From left to right*) In the first step, enzymatic treatment of the liver results in a cell suspension comprising all individual cells of the liver including blood cells. In the next step, HSCs, which contain a high amount of lipid droplets, are enriched by density gradient centrifugation. In the final step, a pure HSC fraction is generated by FACS sorting using cell size/ granularity and vitamin A autofluorescence as selection markers

2 Materials for Isolation and Culturing of Cells

2.1 General Notes

The outlined protocol allows the purification of approximately $0.5–2 \times 10^6$ HSC from one mouse. In our laboratory, we routinely prepare 3–5 mice in a batch to obtain sufficient amounts of cells for subsequent in vitro experiments. The optimal workflow when preparing more than one mouse requires two well-trained persons working in parallel. One person performs the in situ perfusion of the liver and the in vitro digestion, while the other person oversees the FACS sorter setup, FACS sorting, and seeding of the cells. The process takes about 4 h for preparation of a single cell suspension, 1 h for separating HSC from parenchymal cells by gradient centrifugation, and 1–2 h for FACS sorting. Additional times may be necessary for setup and calibration of the FACS machine.

All experiments depicted here were approved by the official State animal care and use committee in Germany (LANUV, Recklinghausen, Germany). The implementation of this protocol at other locations may require approval by relevant local authorities.

2.2 Animals

1. The cellular yield may be strongly dependent on the genetic background of the mice used as cell donors. The mice used in the outlined experiments had a C57BL/6 genetic background and were generated by in-house crossings. However, this protocol is also suitable for mice from other genetic backgrounds (e.g., BALB/c). The optimal weight of animals for the isolation procedure is 20–25 g, which is typically reached at an age of 20–50 weeks. The animals are housed at 4–5 mice per cage in a room maintained at constant temperature of 20 °C with a relative humidity of 50% and a 12-h on/off light cycle. The animals have free access to a regular standard mouse chow and tap water. The animals are housed under specific-pathogen-free (SPF) conditions, according to the guidelines of the Federation for Laboratory Animal Science Associations (FELASA).
2. Animals that are subject to animal experimentation need to be housed and treated according to the local animal welfare requirements. The method described here requires in situ perfusion, a procedure which requires permission from the animal welfare authorities in most countries.

2.3 Anesthesia and Antiseptics

1. Prior to starting the surgery, the mouse is anesthetized with appropriate anesthetics. In our laboratory, the animals are anaesthetized using 6–8 mg/kg body weight of xylazine hydrochloride and 90–120 mg/kg body weight of ketamine hydrochloride.

2. For sterilization of the skin, we use a colorless polyalcohol skin antiseptic. This antiseptic contains 70% (v/v) 2-propanol, water, butan-1,3-diol, perfume oil, and traces of quinoline yellow. The usage of this antiseptic is in full accordance with the German federal law regarding the protection of animals and the guidelines of FELASA.

2.4 Reagents, Consumables, and Instrumentation

Standard laboratory equipment (e.g., heated water bath, magnetic stirrer, peristaltic pump including sterile hoses, precooled centrifuges, electric fur shaver, standard light microscope, fluorescence microscope equipped with a red filter) is required. In addition, several routine materials including surgical instruments, sterile preparation tools (scissors, forceps), and plastic ware (50 mL tubes, disposable syringes, Pasteur pipettes, etc.) are necessary to carry out the protocol. Moreover, cell culturing is done in a standard cell culture room equipped with a sterile hood and a heated-humidified incubator and carbon dioxide control in which 5% CO_2 can be adjusted.

2.5 Buffers

Prepare all solutions using ultrapure water, and sterilize all materials by filtering with a 0.20 μM filter or autoclaving at 121 °C for 20 min. All solution should be kept at 4 °C until use. The formulation of perfusion buffer 1 and perfusion buffer 2 was slightly modified from an established protocol for isolation of rat HSC. In the respective protocol, these buffers were named SC-1 and SC-2 [21].

1. Perfusion buffer 1 (SC-1): 8 g/L NaCl, 400 mg/L KCl, 78 mg/L $NaH_2PO_4 \times H_2O$, 151 mg/L $Na_2HPO_4 \times 2\ H_2O$, 2380 mg/L HEPES, 350 mg/L $NaHCO_3$, 190 mg/L EGTA, 900 mg/L glucose, and 6 mg/L phenol red were adjusted to pH 7.3–7.4 with 10 N NaOH.
2. Perfusion buffer 2 (SC-2): 8 g/L NaCl, 400 mg/L KCl, 78 mg/L $NaH_2PO_4 \times H_2O$, 151 mg/L $Na_2HPO_4 \times 2\ H_2O$, 2380 mg/L HEPES, 350 mg/L $NaHCO_3$, 560 mg/L $CaCl_2 \times 2\ H_2O$, and 6 mg/L phenol red were adjusted to pH 7.3–7.4 using 10 N NaOH.
3. Gey's balanced salt solution (GBSS): 370 mg/L KCl, 210 mg/L $MgCl_2 \times 6\ H_2O$, 70 mg/L $MgSO_4 \times 7\ H_2O$, 75 mg/L $Na_2HPO_4 \times 2\ H_2O$, 30 mg/L KH_2PO_4, 991 mg/L glucose, 227 mg/L $NaHCO_3$, 225 mg/L $CaCl_2 \times 2\ H_2O$, 8 g/L NaCl, and 6 mg/L phenol red were adjusted to pH 7.3–7.4.
4. GBSS for Nycodenz solution: GBSS without 8 g/L NaCl.
5. Hank's balanced salt solution (HBSS) (without Ca^{2+}/Mg^{2+}): 0.4 g KCl, 0.06 g KH_2PO_4, 8 g NaCl, 0.35 g $NaHCO_3$, 0.048 g Na_2HPO_4, and 1 g glucose per liter water.

6. Phosphate-buffered saline (PBS): 8 g NaCl, 0.2 g KCl, 1.44 g Na_2HPO_4, and 0.24 g KH_2PO_4 are dissolved in water, adjusted with 1N HCl to pH 7.4, and filled to 1 L. For cellular characterization of sorted cells, you will need a 3× PBS solution.
7. Sheath fluid: HBSS (10×) is diluted with water to HBSS (1×) and according to manufacturer's instructions supplemented with 4.7 mL per liter of 7.5% $NaHCO_3$ solution.
8. FACS sorting buffer: HBSS is supplemented with 10 mM HEPES, 0.06% (w/v) BSA, and 0.3 mM EDTA and adjusted to pH 7.3–7.4.
9. FACS collection buffer: HBSS is supplemented with 10 mM HEPES and 20% FBS.

2.6 Enzyme Solutions

The activity of the necessary enzymes may vary significantly between different sources and lots. If you attempt to perform HSC isolation on a regular basis, we strongly recommend that you reserve larger amounts of successfully tested enzymes from a specific batch (i.e., with fixed lot numbers). This will increase the reproducibility of this protocol.

1. Pronase E solution: Pronase E (EC 3.4.24.4, 4.000.000 PU/g) from *Streptomyces griseus* is dissolved at 0.5 mg/mL in perfusion buffer 2 and sterile-filtered (0.20 μM). The solution is freshly prepared.
2. Collagenase P solution: Collagenase P (EC 3.4.24.3) from *Clostridium histolyticum* is dissolved at 0.75 U/mL in perfusion buffer 2 and sterile-filtered (0.20 μM). The solution is freshly prepared.
3. Pronase E/Collagenase P solution: 0.4 mg/L Pronase E, 1.5 U/mL Collagenase P are dissolved in perfusion buffer 2 and sterile-filtered (0.20 μM). The solution is freshly prepared.
4. DNase I solution: 100 mg DNase grade II from bovine pancreas (EC 3.1.21.1) is dissolved in 50 mL GBSS (final concentration 2 mg/mL), sterile-filtered (0.20 μM), and stored frozen in 1 mL aliquots at −20 °C.

2.7 Density Centrifugation Gradients

1. Nycodenz solution: Nycodenz is added at a final concentration of 28.6% (w/v) into GBSS without NaCl, sterile-filtered (0.20 μM), and stored at 4 °C. The final concentration during the density gradient centrifugation prepared from this stock solution is 8% (w/v) (*see* **Notes 1–2**).
2. Please note that other compounds for preparation of density gradients suitable for HSC preparation are available (*see* **Note 3**).

2.8 Culture Media and Cell Culture Plasticware

The individual components for HSC culture media are commercially available from different providers. We have no preference for any company. For HSC cultures, the following solutions are necessary:

1. Dulbecco's modified Eagle medium (DMEM): We obtain DMEM with 4.5 g/L glucose without L-glutamine.
2. L-glutamine (200 mM): 29.23 mg/mL L-glutamine in 0.85% (w/v) NaCl solution.
3. Penicillin-streptomycin solution (100×): This bacteriostatic and bactericidal stock solution contains 10,000 IU potassium penicillin and 10 mg streptomycin sulfate per mL.
4. Fetal bovine serum (FBS): In our experiments, we use FBS that is EU approved. According to the manufacturer's information, each lot of FBS is tested for sterility and for the ability to support the growth of several different cell lines, both sequential growth curves and plating efficiencies.
5. HSC culture medium: The final medium for HSC culturing is composed out of DMEM supplemented with 10% FBS, 4 mM L-glutamine, and 1% penicillin-streptomycin solution.

2.9 Consumables and Glassware

1. 50 mL plastic tubes.
2. 15 mL plastic tubes.
3. Tissue culture plates.
4. Hypodermic needle (26G × 0.5″, 0.45 × 13 mm) for cannulation.
5. Cannula (1.5 × 120 mM, chromed brass neck, stainless beveled steel tip) with Luer lock for preparation of Nycodenz gradient.
6. Standard syringes (10 mL) with Luer lock connection (2.0 × 70 mM).
7. Hydrophilic nylon net filter with pore size 40 μM.
8. Self-adhesive tapes.
9. Cotton swabs.
10. Erlenmeyer glass flask (250 mL).

2.10 Surgical Instruments

1. Standard straight forceps.
2. Strabismus scissors.
3. Surgical scissors (11.5 cm).
4. Dissecting forceps (145 mm).
5. Bulldog clamp (curved jaw length, 38 mm).

2.11 Additional Materials for Characterization of Isolated Cells

2.11.1 Antibodies

Primary HSCs express a number of typical intracellular markers. These include desmin, vimentin, glial fibrillary acidic protein (GFAP), cellular communication network factor 2/connective tissue growth factor (CCN2/CTGF), lecithin retinol acyltransferase (LRAT), synemin, platelet-derived growth factor receptor-β, cytoglobin, and many others [22]. The most reliable marker is α-smooth muscle actin (α-SMA), whose expression is increased in HSC during culture-induced activation and transdifferentiation. We recommend testing for vimentin, fibronectin, CCN2/CTGF, and/or α-SMA. Expression analysis can be done by Western blot or immunocytochemistry.

2.11.2 Cellular Characterization

1. Trypan blue solution (0.4%): The ready-to-use, sterile filtered solution is stored at room temperature (*see* **Note 4**).
2. Rhodamine phalloidin conjugate: This reagent is reconstituted in methanol to a final concentration of 200 U/L (equivalent to ~6.6 μM). The stock solution is stable when stored frozen at −20 °C (*see* **Note 5**).
3. 4% Paraformaldehyde solution: To prepare 100 mL of this solution, 4 g of extra pure paraformaldehyde is added to 60 mL sterile water and heated to 60 °C under stirring on a heated magnetic stirrer. Thereafter, two or three drops of 2N NaOH are slowly added until the solution becomes almost clear. The heater is turned off, and 33.3 mL 3× PBS (0.457 M NaCl, 9 mM KCl, 14.3 mM $Na_2HPO_4 \times 7\ H_2O$, 4.67 mM KH_2PO_4) is added. The solution is adjusted to pH 7.2 with 1N HCl and stored in aliquots at −20 °C until use.
4. DAPI solution (5 mg/mL): The stock solution is prepared by dissolving 10 mg DAPI in 2 mL deionized, sterile water. The solution is stable for at least 6 months when stored at −20 °C protected from light.
5. Nail polish (clear).

2.12 Additional Equipment for FACS Sorting

1. FACS sorter equipped with a UV laser.
2. FACS collection tubes.
3. Bright-line hemacytometer.
4. Falcon cell strainer (with pore size 70 μm).

3 Methods

All surgical procedures are carried out under clean conditions. Therefore, all surgical forceps and scissors that are needed during the procedure are properly sterilized prior usage according to the general guidelines necessary to perform aseptic surgeries in animals. In addition, animal preparations are done in separate labor spaces.

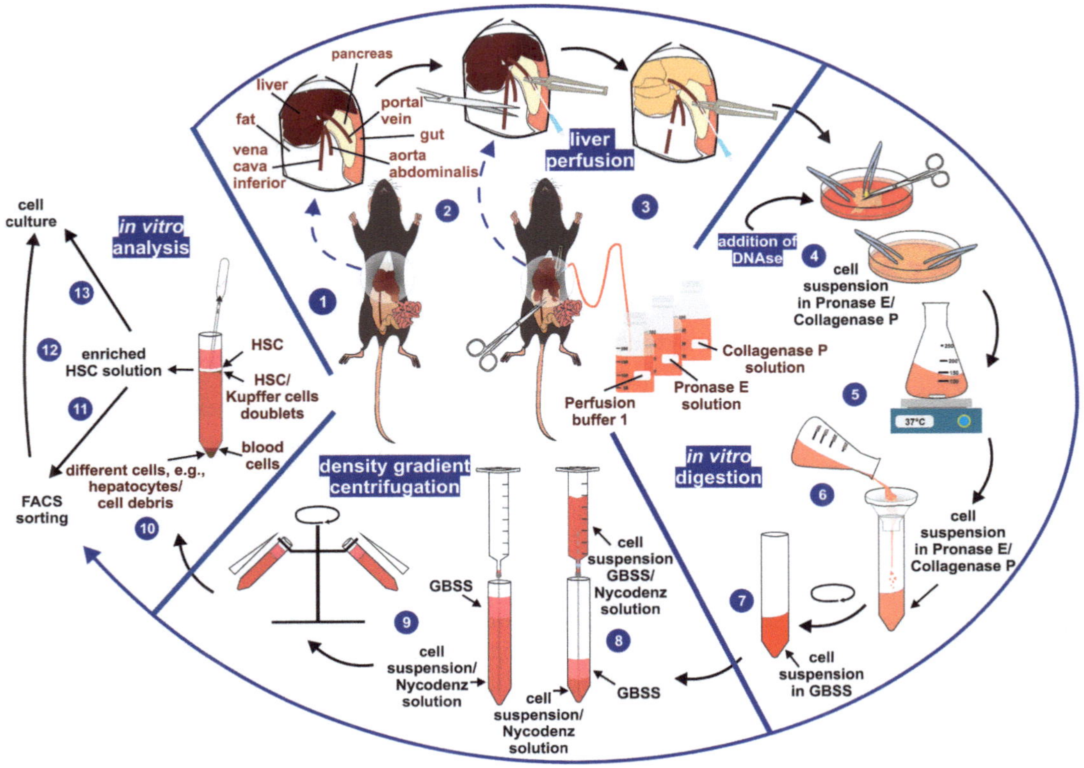

Fig. 3 Detailed schematic overview of experimental steps for isolation of murine HSC. In the first step, the mouse is anesthetized, and the abdominal cavity is opened. The liver is lifted and shifted to the diaphragm, while the intestine is positioned to the right side. Thereby, the *vena porta* and *vena cava* are exposed ❶. Then the perfusion system is connected to the portal vein, fixed, and the *vena cava* is cut to permit perfusion buffer outflow ❷. The progress of the enzymatic perfusion is displayed by swelling and color change of the liver ❸. In the next step, the liver is excised and decapsulated, and cells are dispersed in a sterile plastic culture dish in the presence of Pronase E, collagenase P, and DNase I ❹. Remaining cell clumps are dissolved by stirring the cell suspension at 37 °C on a magnetic stirrer ❺. Subsequently, the suspension is filtered through a nylon mesh ❻ and prepared for density gradient centrifugation ❼. Then, the cell suspension is supplemented with Nycodenz solution and layered under GBSS ❽. After centrifugation ❾, HSCs are enriched in the white shining interphase, while hepatocytes and blood cells are found at the bottom of the gradient ❿. Thereafter, the enriched cell fraction can either be further purified by FACS analysis ⓫ and cultured ⓬ or, alternatively, directly brought into culture after density gradient centrifugation ⓭. For details of each step, refer to the text

The method described here is optimized for the isolation of HSC from the livers of five mice. Other hepatic cell fractions are discarded during the purification (*see* **Note 6**).

The individual steps performed during the protocol are schematically summarized in Fig. 3.

In principal, there are three major work steps:

(i) Perfusion: A perfusion device is first connected to the *vena portae*. Thereby, the liver is first perfused with solutions containing Pronase E and Collagenase P. To drain the perfusion solution after the liver passage, the *vena cava* inferior is cut with scissors.

(ii) Liver removal and cell release: In the second step, the liver is excised and removed from the animal, the tunica broken, and the liver cells dispersed in a Petri dish.

(iii) Density gradient centrifugation: In the last step, the resultant liver cell suspension from the digested liver tissue is purified by centrifugation on a discontinuous Nycodenz density gradient. Thereafter, cells can be directly cultured or alternatively further purified by FACS sorting.

3.1 Presurgical Preparation

1. Anesthetize the mouse by intraperitoneal injection with 6–8 mg/kg body weight of xylazine and 90–120 mg/kg body weight of ketamine. Sufficient anesthetization is obtained when the following vital criteria are reached: regular spontaneous breathing, no reflex after setting of pain stimuli on toes, and no response to pain.
2. Shave the operation area with an electric fur shaver.
3. Sterilize the shaved abdominal skin with 70% ethanol (or an approved polyalcohol skin disinfectant).

3.2 Surgical Procedures

1. Place the mouse on the working surface, and fix the limbs with tapes so that the animal stays as straight as possible. Optional: Cover the operation area with a fluid-impermeable, self-adhesive drape.
2. Open the abdomen with a midline laparotomy up to the sternum by cutting the cutis plus fascia at the same time with 11.5 cm surgical scissors (Fig. 4a).
3. Cut the peritoneum along the *linea alba* to prevent bleeding, and open the peritoneal cavity (Fig. 4b). Expose the abdominal cavity as wide as possible.
4. Shift the stomach and the intestine sideways to the right, and lift the liver with a cotton swab so that the ventral side sticks to the diaphragm and the *hilum* is clearly visible. The portal vein and the *vena cava inferior* (plus *aorta abdominalis*) should now be exposed (Fig. 4c, d).

3.3 Liver Perfusion

1. Set up and actuate the liver perfusion system by equilibrating it with sterile, pre-warmed perfusion buffer 1 (*see* **Note 7**).
2. Allow the water bath to attain the correct temperature (37 °C), place a hypodermic needle (26G cannula) at the end of the hose system, and remove all air bubbles within the system. Set the flow rate of the perfusion pump to 6.5 mL/min.
3. Stop the perfusion pump, and insert the cannula of the perfusion set in the portal vein. It should be positioned parallel to

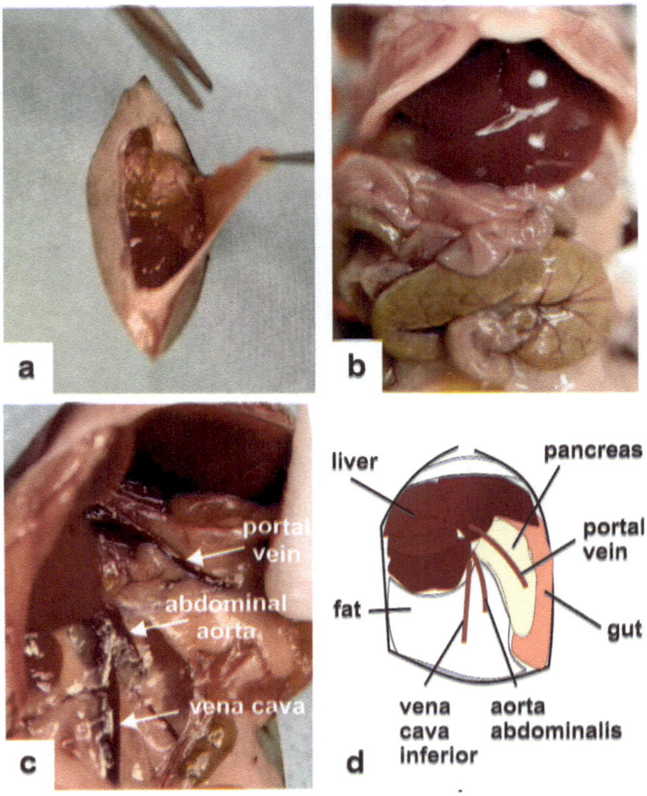

Fig. 4 Experimental setup for isolation of murine HSC. Prior starting, the workspace is prepared, and the sterilized instrumentation is arranged. The anesthetized and shaved mouse is placed onto the working area and prepared for surgery. (**a, b**) The abdomen is opened. (**b**) The liver is uncovered, and (**c, d**) the ventral side of the liver is lifted and attached to the diaphragm with a cotton swab so that the portal vein is freely exposed

the vein, close to where the splenic vein is entering the portal vein in direction to the liver (Fig. 5a).

4. Fix the cannula at the portal vein with a bulldog clamp in the middle of the injection site and the cannula tip (Fig. 5b) (*see* **Note 8**).
5. Start the perfusion pump, and immediately cut the inferior *vena cava* with sharp scissors (Fig. 5c). This drops the pressure within the liver and drains the perfusion solution after the liver passage. Perfusion buffer 1 now flows through the liver.
6. Perfuse the liver with a flow rate of 6.5 mL/min for 4 min and 30 s.
7. Stop the perfusion pump, and carefully change perfusion buffer 1 to pre-warmed Pronase E solution. (*see* **Notes 9**)

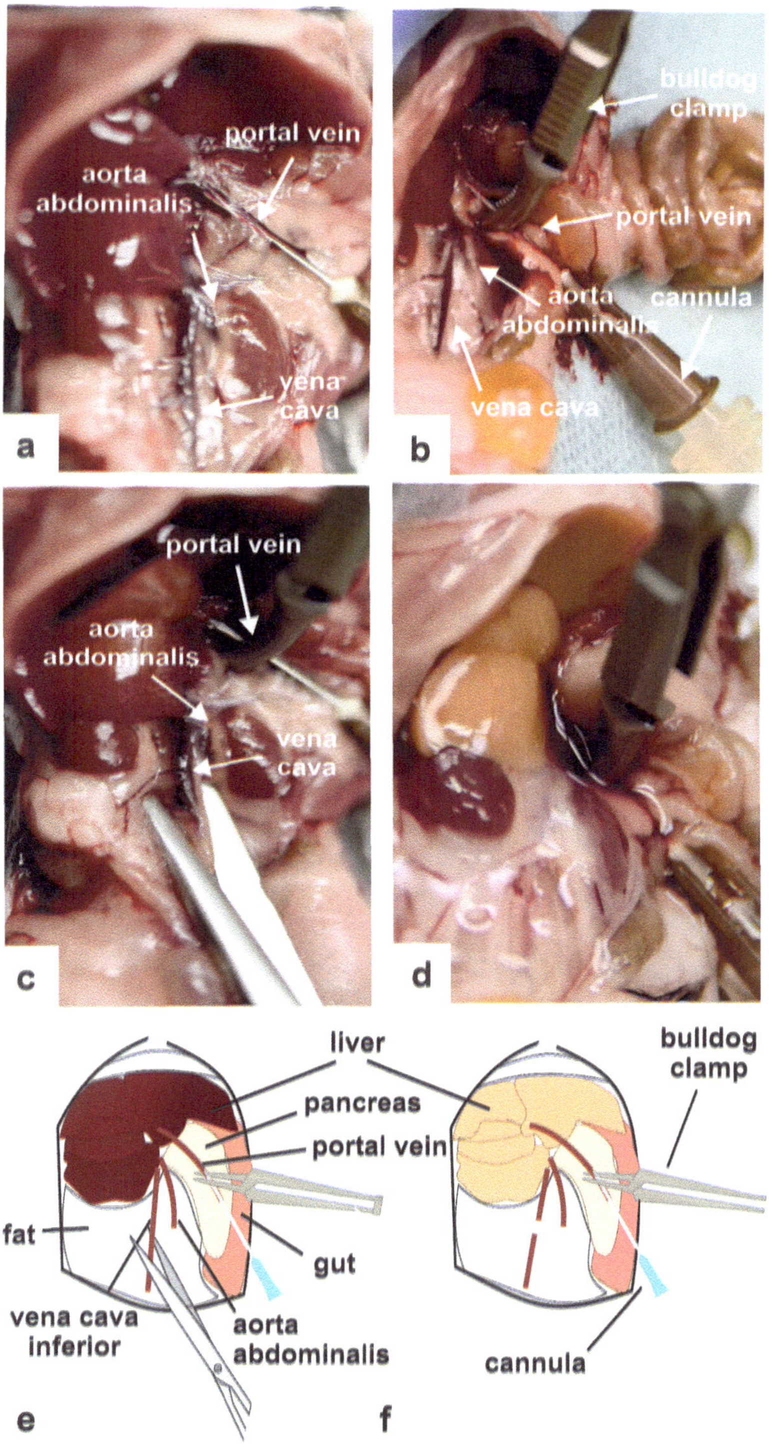

Fig. 5 Liver perfusion. (**a**) A cannula is carefully plugged into the portal vein without rupturing the vessel wall and (**b**) positioned into a stable position by fixing it with a bulldog clamp. (**c**) The *inferior vena cava* is cut with scissors and (**d**) the peristaltic pump started, allowing the perfusion buffer to flow freely through the liver.

8. Perfuse the liver for 4 min and 30 s with a flow rate of 6.5 mL/min.
9. Stop the perfusion pump, and change buffer to Collagenase P solution. Start the pump, and again perfuse the liver for 4 min and 30 s with a flow rate of 6.5 mL/min. Ongoing enzymatic digestion is reflected by swelling of the liver lobules and loss of typical liver shape. The color of the liver turns from dark red to light brown, while residual red areas reflect incomplete perfusion (Fig. 5d).
10. When the liver is digested, immediately cut the thoracic part of the *vena cava interior*, and disconnect the perfusion system to prevent air entering the liver.
11. Separate the liver from the gastric band, and cut the *arteria hepatica* with small scissors.

3.4 Excision of the Liver and In Vitro Digestion

1. Gently remove the liver from the body cavity with sterile forceps (Fig. 6a) (*see* **Note 10**).
2. Place the liver into a 50 mL plastic tube on ice with 20 mL Pronase E/Collagenase P solution, and start to prepare the next mouse following the advice given above. Collect all livers in this tube.

   Note: The following steps are optimized for a total amount of five livers. In case of different numbers of livers, the protocol can be adapted accordingly.
3. Transfer all digested livers into a sterile Petri dish. The typical appearance of the liver is depicted in Fig. 6b. Add 1 mL DNase I solution.
4. Remove the gallbladder (that is normally colored yellow if filled with bile) using scissors and tweezers (Fig. 6c). Remove the Glisson's capsule of each liver, and disperse the liver cells with the help of tweezers (Fig. 6d). At the end of this process, the liver tissue is completely resolved in individual cells, and only a few cell clumps remain.
5. Transfer this cell suspension into a sterile Erlenmeyer flask, in which a magnetic stir bar is placed. Fill up to a volume of 120 mL with pre-warmed, sterile Pronase E/Collagenase P solution (Fig. 7a).

Fig. 5 (continued) During the complete liver perfusion, the perfusion buffers are pre-warmed to 37 °C by storing them in a heated water bath. The perfusion is successively continued with different perfusion solutions. During this process, the liver is cleaned from blood (visible by a change from dark red to brown), and the ongoing enzymatic digestion is associated with a swelling of the liver lobules and a loss of the typical liver shape. (**e, f**) For better explanation of the localities of individual anatomical structures, schemes of figures depicted in (**c**) and (**d**) are shown

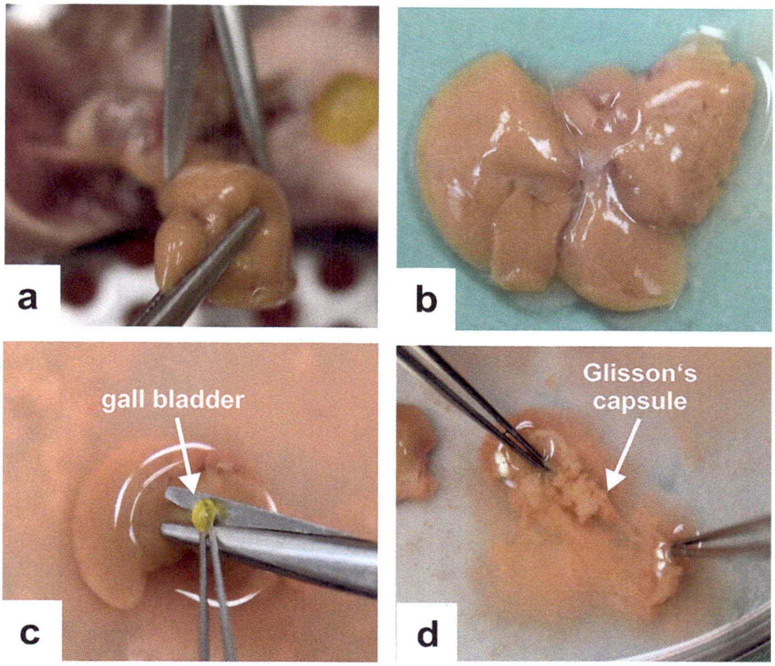

Fig. 6 Removal of digested liver and releasing of dispersed cells. (**a**) After successful digestion, the liver architecture is amorphous and has a very light color. The liver is grabbed with anatomic tweezers and removed from the abdominal cavity and put into a Petri dish filled with Pronase E/Collagenase P solution including DNase I. (**b**) The typical appearance of a digested liver is shown. (**c**) The gallbladder is carefully removed. (**d**) The Glisson's capsule is removed to release the dispersed cells, while the digested liver is grabbed with forceps and intensively shaken to release the dispersed cells

6. Try to resolve the remaining cell clumps by carefully stirring the cell suspension at 37 °C for 20 min in a water bath (Fig. 7a) (*see* **Note 11**)
7. Filter the cell suspension with the dispersed cells through a 70 μm cell strainer (Fig. 7b) into six 50 mL plastic tubes (~20 mL per tube). Thereafter, fill each tube up to 50 mL with cold perfusion buffer 2 (Fig. 7c). Centrifuge the suspensions for 10 min at 4 °C and 600 × g.
8. Remove the supernatants with a sterile Pasteur pipette (Fig. 7d), and resuspend each pellet in 10 mL cold GBSS, and add 150 μL DNase I solution to the cell suspensions (Fig. 7e). Combine them into four 50 mL plastic tubes (Fig. 7f). Fill up the tubes with cold GBSS, and centrifuge again for 10 min at 4 °C and 600 × g.
9. After centrifugation, aspirate the supernatants completely with a sterile Pasteur pipette. Resuspend each pellet in 10 mL cold GBSS, and add 150 μL DNase I. Pool the suspensions into two 50 mL plastic tubes.

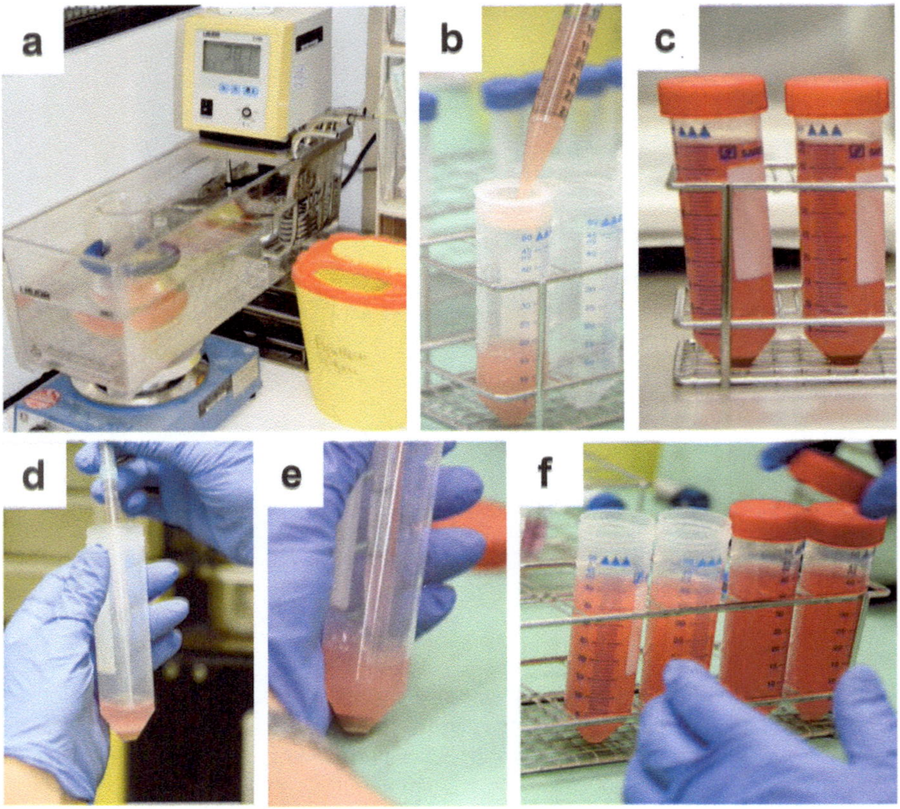

Fig. 7 Destruction of cell clumps and enrichment of hepatic stellate cells by centrifugation. (**a**) The tissue cell suspension is transferred into a sterile 250 mL glass Erlenmeyer flask, in which a magnetic stir bar is placed. Remaining cell clumps in the cell suspension are removed by stirring at low speed at 37 °C. (**b, c**) The cell suspension is filtered into six 50 mL plastic tubes. (**d**) After centrifugation, the supernatant is removed. (**e, f**) The cell pellets are then resuspended in GBSS, DNase I is added, and the suspension is combined into four 50 mL plastic tubes and centrifuged

3.5 Nycodenz Density Gradient Centrifugation

1. Fill up each tube to 36.0 mL with cold GBSS, and add 14.0 mL Nycodenz solution to each tube to reach a final Nycodenz concentration of 8% (w/v).
2. Place ten 15 mL plastic tubes into a rack, and pre-fill each tube with 1.5 mL cold GBSS (~ 2 tubes/animal). Place a sterile syringe with a special Luer lock cannula (120 mm, autoclaved) into each tube (Fig. 8a).
3. Fill 10 mL of the cell/Nycodenz solution into each syringe. The solution flows through the cannula under the pre-filled GBSS (Fig. 8b).
4. Centrifuge the tubes for 22 min at 4 °C and 1500 × g (*see* **Note 12**).

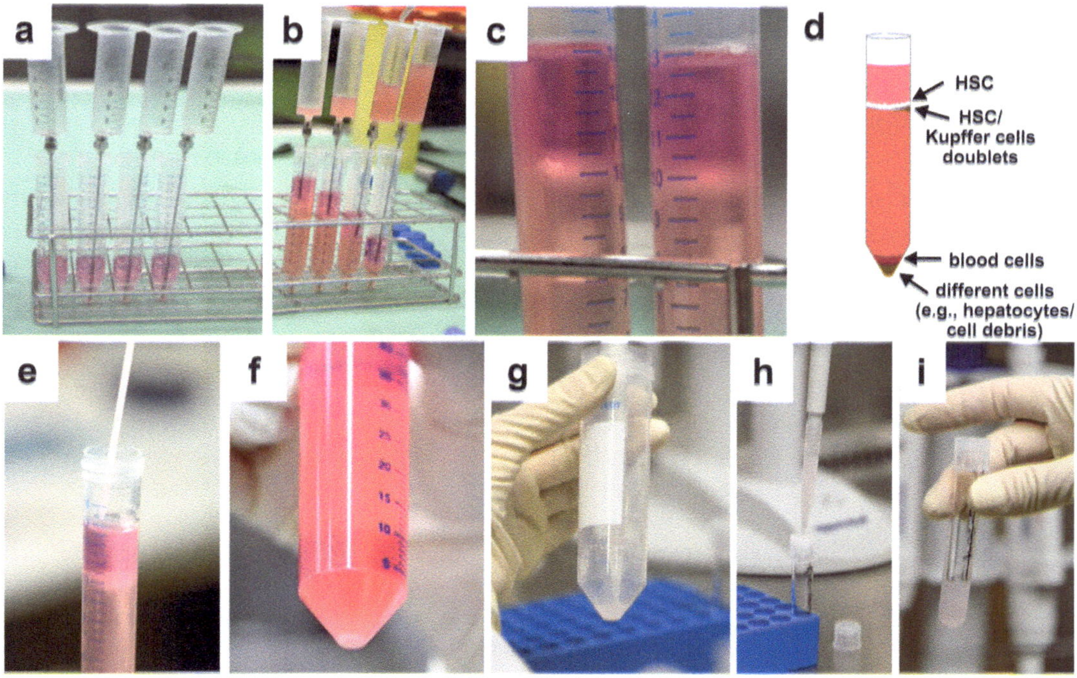

Fig. 8 Nycodenz gradient centrifugation. (**a, b**) Preparation of 15 mL tubes for Nycodenz density gradient centrifugation (~ 2 tubes for each animal). (**c, d**) After density gradient centrifugation, HSC can be identified as a white cell layer that is floating on the surface of the gradient. (**e**) HSC-enriched cell layers are collected, pooled in 50 mL Falcons, filled with GBSS, and centrifuged again. (**f**) After centrifugation, a white pellet is visible. (**g**) The supernatant is removed, resuspended in FACS sorting buffer, and (**h**) filtered through a 40 μM filter. (**i**) The slightly cloudy cell solution is now ready for FACS sorting

5. After centrifugation, a pellet containing hepatocytes (brown) and blood cells (red) is visible in each tube. HSCs are found in the interphase of both solutions. They appear as a white condensed band (Fig. 8c, d). The cells enriched in the white bands are pooled and collected in a 50 mL plastic tube using a blunt cannula made out of plastic and a smooth operating syringe, in which the plunger is equipped with a manual rubber stopper (Fig. 8e).
6. Fill the cell suspension to 50 mL with cold GBSS (or FACS sort buffer, if the cells are intended to be further sorted by FACS), and centrifuge in a sterile 50 mL tube for 10 min at 4 °C and 600 × g. After centrifugation, a clear white pellet containing the enriched fraction of HSC should be visible at the bottom of the tube (Fig. 8f). The supernatant is removed (Fig. 8g). If the cells are intended to be cultured without subsequent FACS sorting, the cell pellet is now resuspended in culture medium, and the cells are plated on culture dishes with a density of $2.0 \times 10^5 - 3.0 \times 10^5$ cells/well of a 6-well plate.

7. If the cells should be further purified by FACS sorting, discard the supernatant, resuspend the cells in FACS sorting buffer, and filter them carefully through a 40 μm filter for FACS sorting (Fig. 8h). After this step, the solution should have a white opaque appearance (Fig. 8i).

3.6 Fluorescence-Activated Cell Sorting (FACS)

Although flow cytometry and FACS have become standard methods for separating individual subsets of cells from heterogeneous cell populations, there are several potential pitfalls and safety concerns, which should be considered when using this technique (*see* **Note 13**).

For FACS purification of primary HSC, we use a cell sorter that is equipped with a 355 nm UV laser and a 488 nm blue laser. The respective protocol uses a three-step gating strategy, in which the cells are first gated based on their forward and sideward scattering. Thereby, smaller and granulated cells were framed. In the subsequent gating hierarchy, we perform a doublet discrimination gating, which is recommended by the manufacturer. In the last step, we select HSC from the generated single cell population by their emission of autofluorescence after excitation of vitamin A with a UV laser. This excitation light is measured in the "Indo-1" channel of the corresponding detector. A short overview of the FACS sorting workflow is given in Fig. 9.

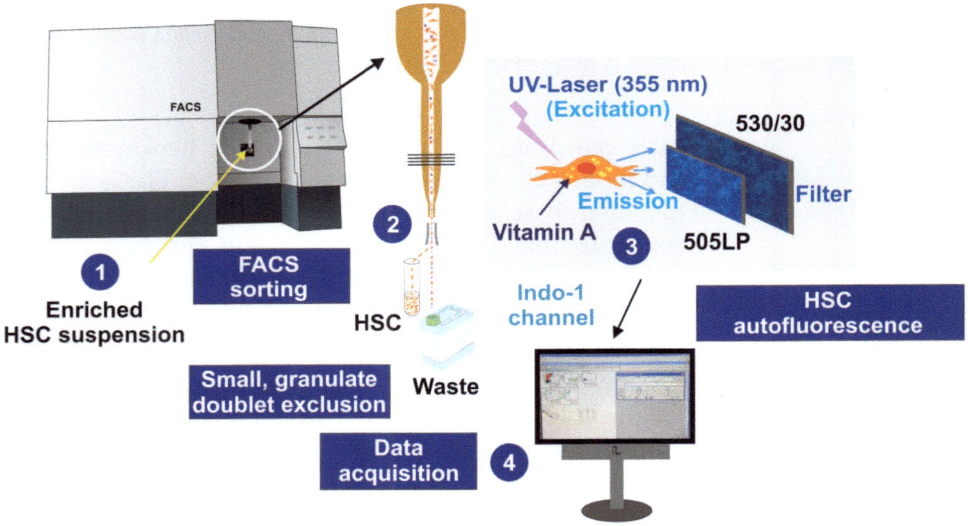

Fig. 9 Schematic overview of the FACS sorting protocol for purification of ultrapure murine HSC. The enriched HSC suspension is placed into the FACS machine ❶. Sorting of HSC is done first by side- and forward scatter discriminating size and granularity ❷ and then by excitation with UV light from a 355 nm laser and measuring the emission resulting from the cellular retinol content in the "Indo-1" channel that uses a 505-nm-long pass and a 530 ± 30 nm band pass filter ❸. The sorting is monitored and documented using appropriate hard- and software ❹. For details of each step refer to the text

1. Set up and actuate the FACS machine. For sorting of HSC, use the following parameters: sample loading port to 4 °C at 300 rpm, 100 μm nozzle, pressure 20 psi, threshold rate ~5000 events/s, threshold 5000, precision mode purity (yield mask 32, purity mask 32, phase mask 0), and collection device set to 4 °C.
2. Verify that the suitable cytometer configuration for HSC sorting is chosen, and perform the calibration steps recommended by the manufacturer. Verify that the side stream formation (Fig. 10a, b) and the drop break-off are properly adjusted (Fig. 10c, d).
3. Fill a collection tube with 1 mL FACS collection buffer, and put it into the collection device at the position of the left-side stream (Fig. 11a).
4. Insert the enriched HSC suspension into the sample injection chamber of the FACS sorter, and acquire a representative amount of cells (Fig. 11b).

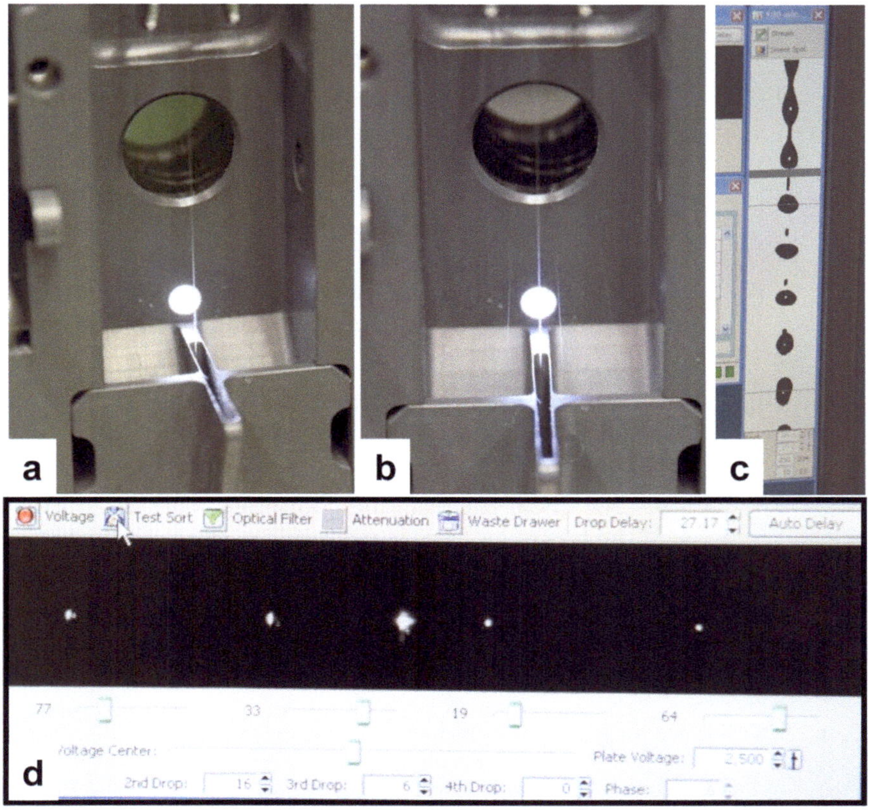

Fig. 10 FACS setup. The sorting condition, drop formation, and deflection are crucial in FACS sorting. (**a–d**) The side stream formation and the drop break-off that are shown as live images are depicted

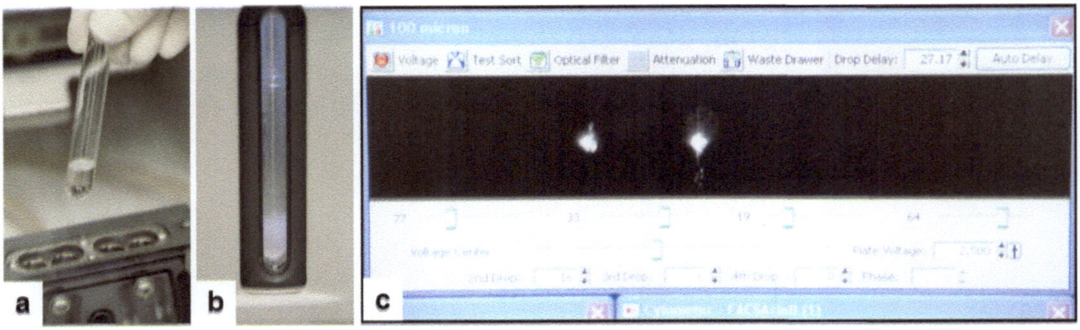

Fig. 11 FACS sorting. (**a**) A collection tube is filled with 1 mL FACS collection buffer; put it into the collection device. (**b**) During the sort, the cells are sorted into the collection device positioned in the collection chamber. (**c**) A representative image showing the flow of the desired cells (i.e., HSC) and waste cells (i.e., non-HSC) is depicted

5. Dilute the enriched HSC suspension with FACS sorting buffer so that the intended threshold rate during the measurement will not be exceeded.
6. Start the cell sorting process. During sorting, you can follow the correct sorting into the HSC and waste fraction by inspection of the different streams (Fig. 11c).
7. Based on the chosen gating strategy, sorting of HSC is done by exciting the cell suspension with a UV laser light and measuring the emission in the "Indo-1" channel passing 505-nm-long pass (505LP) and a 530 ± 30 nm (530/30) band pass filter. The emission measured in the "DAPI" channel (450/50) serves as a control (Fig. 12a). The "Indo-1"-positive population with higher size contains an amount of contaminating cell doublets, which should be excluded by adjusting the gate [17, 23].
8. During the cell sorting, the collection tube located in the collection device will fill up with sorted HSC and buffer (*see* **Note 14**).
9. Document the resulting counting events and purity of cells in the hierarchy (% parent) during sorting (Fig. 12b).
10. To confirm successful cell sorting and to determine the purity of the sorted HSC preparation, an aliquot of the sorted suspension is again analyzed by FACS sorting (Fig. 13). Again, document the resulting counting events and purity of cells in the hierarchy (% parent) during the sorting process. (*see* **Note 15**).
11. Centrifuge the collecting tubes at 4 °C for 10 min at 750 × g (*see* **Note 16**).

a

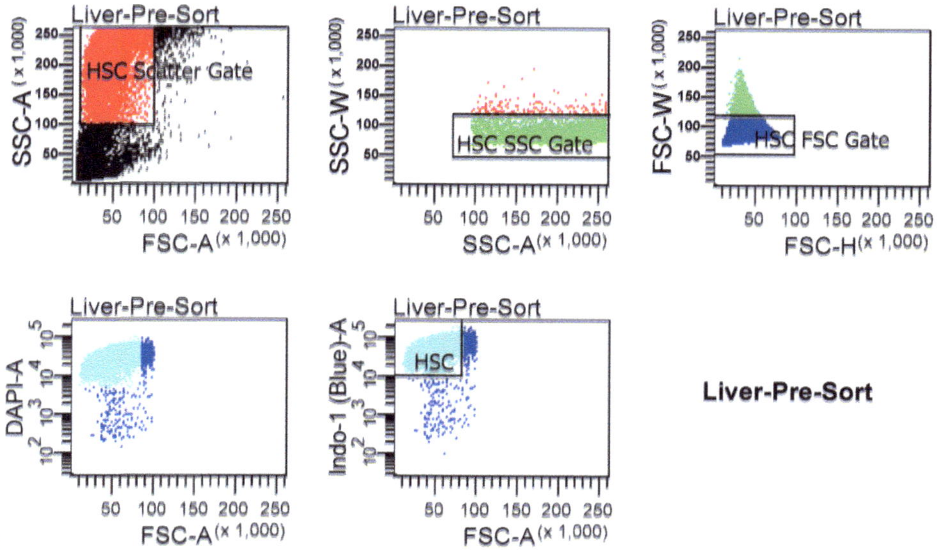

b

Population	#Events	%Parent	%Total
All Events	50,000	####	100.0
HSC Scatter Gate	26,130	52.3	52.3
HSC SSC Gate	25,903	99.1	51.8
HSC FSC Gate	23,990	92.6	48.0
HSC	22,842	95.2	45.7

Fig. 12 Example of a FACS experiment. (**a**) Sorting of cell suspension is conducted using the granularity and autofluorescence of HSC as sorting criteria. (**b**) In each FACS run, the number of events and information about the number of cells collected are documented

12. Remove the supernatants, carefully resuspend the cell pellets, and pool them into one tube in a total of 1 mL HSC culture medium (Fig. 14a–c) (*see* **Note 17**).

3.7 Determination of Cell Yield and Viability

1. Determine the final yield of the cells by use of an automated cell counter or a hemocytometer (*see* **Note 18**).
2. Determine cell viability. Therefore, dilute an aliquot of the resuspended cell suspension, e.g., 20 μL with 80 μL of FACS sorting buffer. Mix 20 μL of this dilution with 20 μL of trypan blue solution (Fig. 15a, b).
3. Place the cover glass on the top of the Neubauer chamber so that it covers the central counting area. In this step, it is important to achieve the correct depth between the cover glass and the central counting area (*see* **Note 19**).
4. Load the hemocytometer with 10 μL from the solution prepared in **step 2** (Fig. 15c). Please avoid introducing air bubbles into the counting chamber.

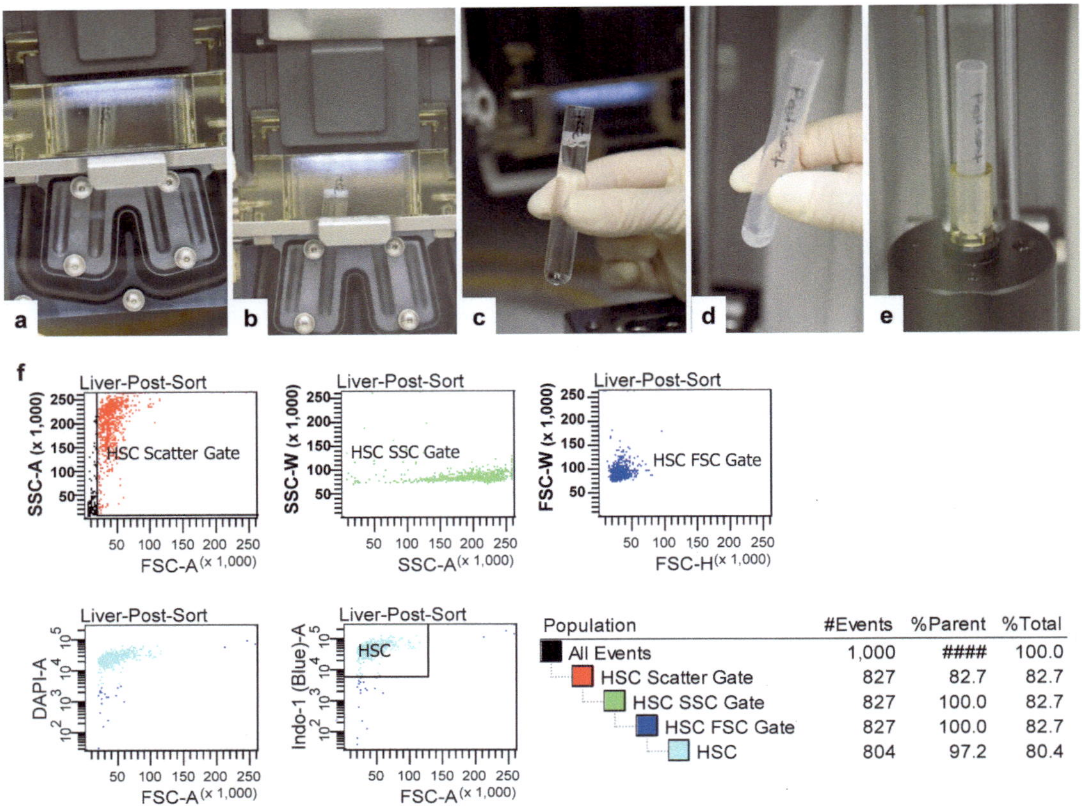

Fig. 13 Determination of cell purity by a second round of FACS sorting. (**a**–**e**) For the determination of the purity of sorted HSC, an aliquot of cells (at least 1000 events) is once analyzed under the same condition as during the initial sort. (**f**) The results of such a typical "post-sort" purity control run are depicted

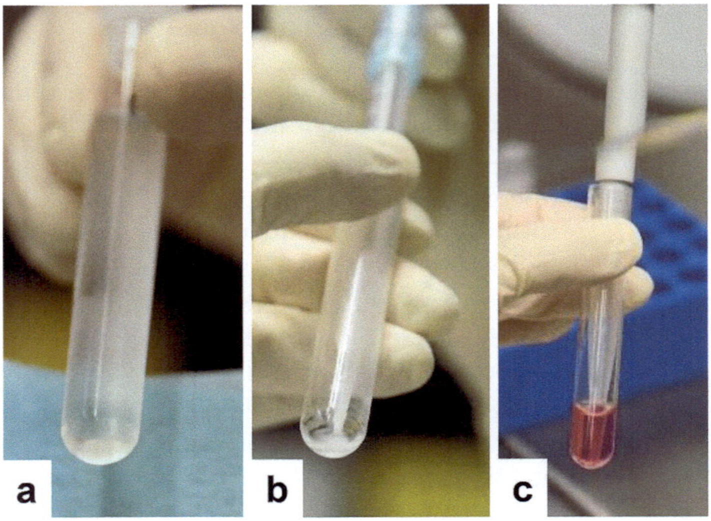

Fig. 14 Concentration of purified HSC. (**a, b**) The FACS-separated HSC are concentrated by centrifugation and (**c**) resuspended in cell culture medium. Please note the typical white pellet at the bottom of the centrifugation tube

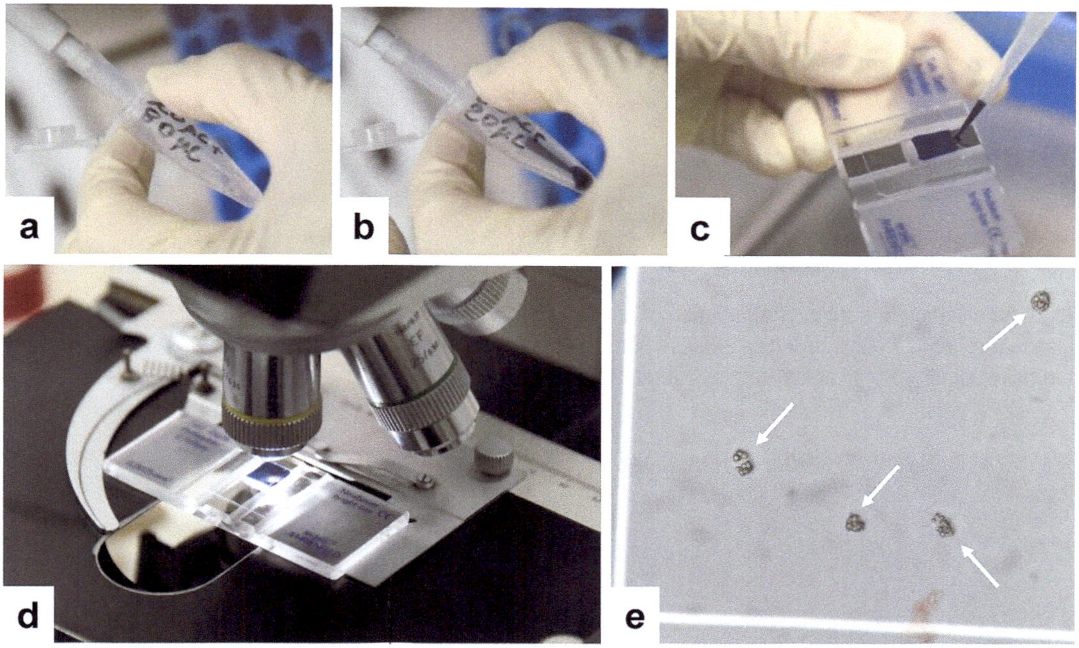

Fig. 15 Viability assay. (**a, b**) An aliquot of the cell fraction is mixed with the trypan blue solution. (**c**) The mixture is pipetted into a counting chamber. (**d**) The counting chamber is placed under a standard light microscope, and viable and dead cells are counted. (**e**) The typical appearance of freshly isolated HSC (marked with white arrows) is shown

5. Place the Neubauer chamber on the microscope stage (Fig. 15d). Count the unstained and stained cells in four big squares in the corners of the central counting area (which is equal to 16 small squares). The staining with trypan blue is a dye exclusion method used to selectively mark dead cells. Vital cells with intact cell membranes do not incorporate the dye, while dead cells appear in blue color (Fig. 15e) (*see* **Note 20**).
6. Calculate the total cell number of the cell suspension using the following formula: total cell number in the cell suspension = (number of counted cells/4) × dilution factor × mL suspension × 10,000.
7. Calculate the viability (%) in the cell suspension using the following formula: viability (%) = (total number of unstained cells/total number of cells) × 100.

3.8 Determination of Cell Purity

Primary HSCs show some typical characteristics and express some typical markers that are not found in other hepatic cell types. The most typical feature of these cells is the occurrence of large cytoplasmic fat droplets that can be easily detected in microscopy by their high refraction under UV light, exhibiting a striking rapidly fading blue-green autofluorescence when excited with 328 nm light. These droplets are well represented in Nomarski differential

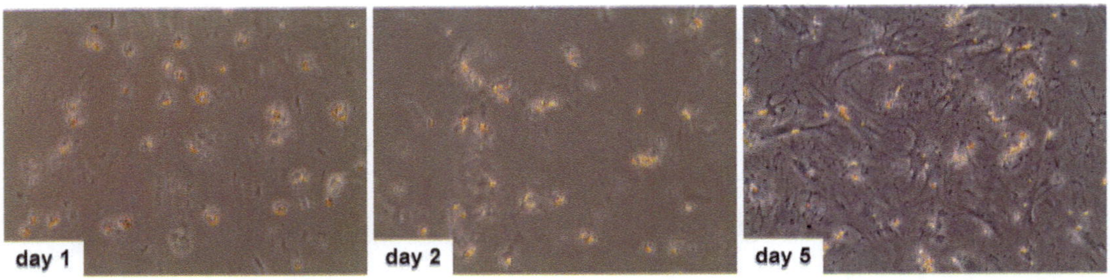

Fig. 16 Activation and transdifferentiation of primary HSC in culture. Isolated HSCs were cultured on uncoated plastic for 1 day, 2 days, or 5 days. Please note that during culturing, the content of vitamin A decreases, and the cells acquire a typical star-shaped fibroblastic phenotype

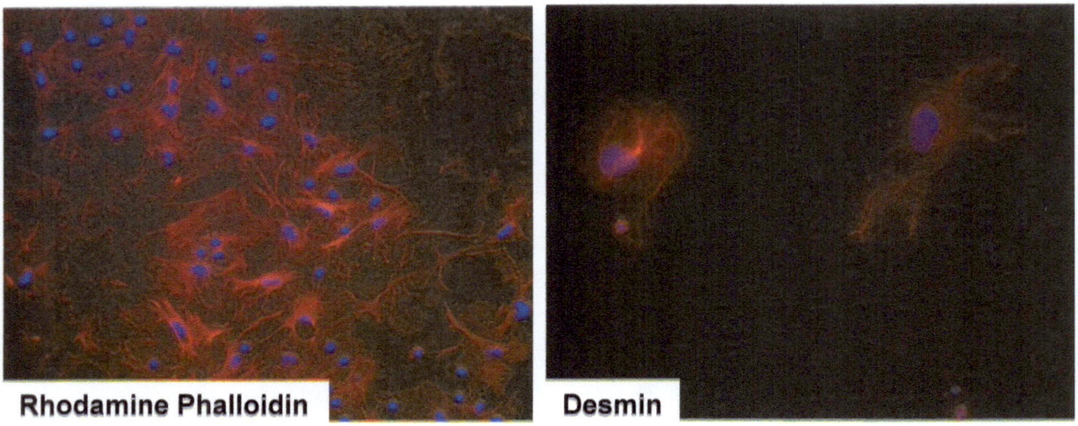

Fig. 17 Rhodamine phalloidin stain and immunocytochemical stain for desmin. (*left panel*) HSCs were stained with a rhodamine phalloidin conjugate (*red*) and DAPI (*blue*). (*right panel*) Cultured HSCs were immunostained with an antibody directed against desmin (*red*), and nuclei were counterstained with DAPI (*blue*). Original magnifications are 100× (*left panel*) and 400× (*right panel*)

interference-contrast microscopy (Fig. 16). Moreover, the fat droplets in early cultures of HSC can be identified in Oil Red O or Nile Red stain [24–26].

In addition, the expression of GFAP and desmin together is indicative of early quiescent HSC. Upon culturing, also the appearance of α-SMA is a reliable marker to positively identify activated HSC. Conversely, contaminating cells can be identified by staining for the surface markers F4/80, CD68, or CD11b that are present on resident macrophages (i.e., Kupffer cells) and RECA-1 (endothelial cell marker) or by determination of appropriate hepatocyte markers.

As discussed above, the cell purity after FACS sorting can be performed using an aliquot (~1000 events) of the purified cells. To determine cell purity in a culture, the cells can be further stained with a rhodamine phalloidin complex (Fig. 17, left panel) that specifically stains F-actin or with an antibody specific for desmin (Fig. 17, right panel). Protocols for respective stains are given under 3.10.

For both procedures, the cells are plated and grown on poly-L-lysine-pre-coated glass slides layered into 12-well plates with HSC culture medium. In the following, protocols for both the rhodamine phalloidin stain and the desmin stain are given.

3.9 Cell Culturing

1. Dilute the purified cells (unsorted or sorted) in HSC culture medium, and plate the cells at a density of 2×10^4 cells/cm^2 in suitable tissue culture plates (*see* **Note 21**).
2. Incubate the cells in a humidified incubator with 5% CO_2 at 37 °C.
3. Immediately, after cell attachment (around 2 h or overnight), replace the medium, and put the plates back into the incubator (*see* **Note 22**).
4. The culture process on the rigid uncoated plastic will lead to "transdifferentiation" due to the mechanical force. During prolonged culturing, HSC lose their quiescence, resulting in an activated phenotype, in which the cells slowly lose their vitamin A droplets and acquire a typical star-shaped fibroblastic morphology. The activation of the cells is further accompanied by elevated expression of α-SMA and collagen type I. After the cells reach confluence, they can be enzymatically detached from the plates by trypsin or Accutase and subcultured for up to four passages. Each proteolytic detachment further promotes the activation process that finally results in a myofibroblast phenotype. The typical phenotypic alterations occurring during transdifferentiation are characteristic for primary HSC and not observed in immortalized HSC cell lines (*see* **Note 23**).
5. The cells that are prepared by this protocol are very homogeneous in regard to their phenotypic appearance (*see* **Note 24**).

3.10 Useful Stains to Verify HSC Identity

HSCs kept in culture show intracellular expression of vimentin and increased expression of α-SMA during prolonged culture times. In addition, the cells show increased expression and secretion of fibronectin and CTGF, which can be easily monitored by Western blot analysis (Fig. 18). In addition, there are many ways to demonstrate the identity of purified HSC. One simple possibility is the staining with a rhodamine phalloidin reagent, which stains the actin filaments.

Similarly, the expression of specific markers of HSC such as desmin can be efficiently proven by immunocytochemistry (*see* **Note 25**).

In the following, protocols are given for both methods.

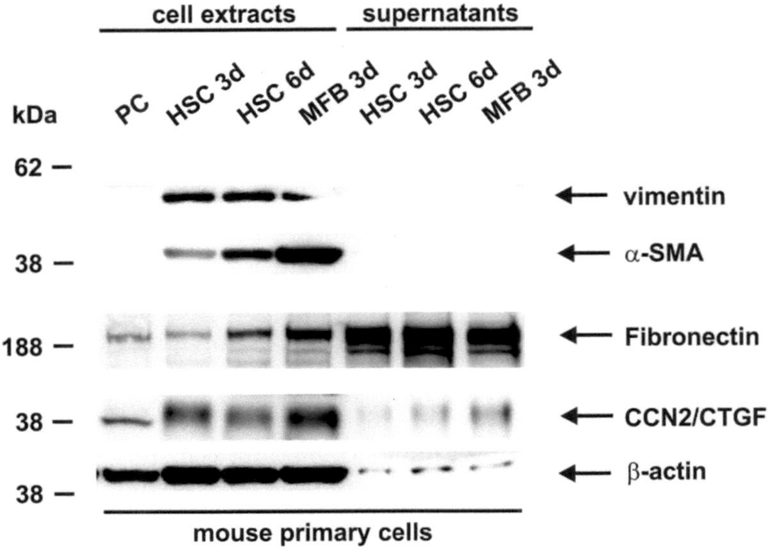

Fig. 18 Detection of typical HSC markers by Western blot analysis. Protein extracts of murine primary hepatocytes (PC), HSC cultured for 3 or 6 (HSC 3d, HSC 6d), and myofibroblasts established from primary HSC and cultured for additional 3 days (MFB 3d) were prepared. In addition, protein extracts of culture supernatants of respective cells were prepared. The proteins were separated by SDS-polyacrylamide electrophoreses, transferred on nylon membranes, and probed with antibodies directed against vimentin, α-SMA, Fibronectin, CCN2/CTGF, and β-actin.

3.10.1 Phalloidin Stain for Cultured HSC

1. One day after plating, remove the medium, and wash the cells once with ice-cold HBSS (without Ca^{2+}/Mg^{2+}).
2. Fix the cells in ice-cold 4% paraformaldehyde for 20 min on ice.
3. Remove the fixative, and wash the cells with ice-cold HBSS (without Ca^{2+}/Mg^{2+}).
4. Permeabilize the cells in ice-cold PBS supplemented with 0.2% Triton X-100 for 4 min on ice.
5. Remove the solution, and wash the cells in ice-cold HBSS (without Ca^{2+}/Mg^{2+}).
6. Block unspecific binding sites in PBS supplemented with 3% BSA for 5 min at room temperature.
7. Cover the cells with the rhodamine phalloidin conjugate (diluted 1:40) with PBS, and incubate at room temperature for 20 min.
8. Remove the staining solution, and wash the cells in HBSS (without Ca^{2+}/Mg^{2+}).
9. For counterstaining of nuclei, incubate the cells with DAPI (0.1 μg/mL) in HBSS (without Ca^{2+}/Mg^{2+}) for 5 min at room temperature.

10. Remove the staining solution, and again wash the cells in HBSS (without Ca^{2+}/Mg^{2+}).
11. Air-dry the stained cells, drop on an appropriate mounting medium, place upside down on a microscope slide, and seal with nail polish. Keep the cells protected from light to prevent photo-bleaching (*see* **Note 26**).
12. Document stained cells under a fluorescent microscope equipped with a red filter ("Rhodamine filter") and appropriate objectives and oculars.

3.10.2 Desmin Stain for Cultured HSC

1. Wash, fix, permeabilize, and block unspecific binding sites as described above (*see* Subheading 3.10.1, **steps 1–6**).
2. Prepare a suitable dilution of the primary desmin antibody in PBS supplemented with 3% BSA, and incubate the cells with that solution for 1.5 h at room temperature. As a negative control, an appropriate isotype control immunoglobulin should be used in parallel.
3. Remove the antibody solution, and wash the cells in HBSS (without Ca^{2+}/Mg^{2+}).
4. Block the cells in PBS supplemented with 3% BSA for 5 min at room temperature.
5. After blocking, add an appropriate dilution of a suitable fluorescently labeled secondary antibody directed against the chosen primary antibody for 30 min at 37 °C. The secondary antibody should also be prepared in PBS supplemented with 3% BSA.
6. Remove the antibody solution, and wash the cells in HBSS (without Ca^{2+}/Mg^{2+}).
7. For counterstaining of nuclei, incubate the cells in DAPI solution (0.1 μg/mL) in HBSS (without Ca^{2+}/Mg^{2+}) for 5 min at room temperature.
8. Remove the DAPI solution, and wash the cells in HBSS (without Ca^{2+}/Mg^{2+}).
9. Mount the cells, and document the results as described above (3.8.1; *see* **steps 11–12**).

4 Notes

1. Nycodenz is the trademark name for iohexol whose systematic name is 5-(N-2, 3-dihydroxypropylacetamido)-2, 4, 6-tri-iodo-N, N'-bis (2,3-dihydroxypropyl) isophthalamide. It is a non-ionic, tri-iodinated derivative of benzoic acids with three aliphatic hydrophilic side chains. The compound was originally developed as an X-ray contrast medium and has a

high solubility and low osmolality, which allows the generation of high-density solutions for the fractionation of diverse biomolecules and separating of living cells.

2. Nycodenz in solid form is stable for a period of 5 years when stored at room temperature and protected from light and humidity.
3. In other HSC isolation protocols, alternative compounds are used in density gradient centrifugation. Common are arabinogalactan (Stractan), metrizamide (Amipaque), iodixanol (OptiPrep), or polyvinylpyrrolidone-coated colloidal silica (Percoll) particles [5–14]. If you prefer to use one of these compounds, you have to adapt the concentration to form optimal gradients for HSC isolation. A summary of structures and properties of these compounds used for gradient density centrifugation for enrichment of HSCs can be found elsewhere [20].
4. Trypan blue ($C_{34}H_{24}N_6Na_4O_{14}S_4$, $M_r = 872.88$) is an acid azo dye used to distinguish between living and dead cells. This dye is absorbed only by cells with damaged cell membranes, but not by healthy viable cells. Thus, it is easy to count dead cells. This method is often referred to as the color exclusion method.
5. The rhodamine phalloidin conjugate binds to cellular actin filaments. Phalloidin is a bicyclic heptapeptide toxin (phallotoxin) isolated from the death cap mushroom, which binds and stabilizes filamentous actin (F-actin) and effectively prevents the depolarization of actin fibers. Rhodamine is a fluorescent fluorone dye. It can be easily detected with a fluorescent microscope at extinction/emission = 546/575 nm.
6. Some other protocols are established in which two or more hepatic cell subpopulations from murine liver are isolated simultaneously [27–29].
7. To avoid contaminations in the prepared cell suspension, it is essential that the hoses of the perfusion system are autoclaved.
8. Take care that you simultaneously clamp the silvery/translucend bile duct to guarantee that the perfusion buffers flow through the liver. Be sure that the bulldog clamp does not squeeze the *vena cava inferior.*
9. Take care not to introduce bubbles into the perfusion system. We recommend pre-warming of all buffers during the perfusion process in a water bath. For changing buffers, we only shift the hose of the perfusion apparatus into the next flask. Flasks that are not in use are closed with a bottle lid.
10. Take care not to damage the liver capsule. To do so, grab the liver carefully with large tweezers to avoid leakage.
11. Note: During the digest, adjust the pH with 1N NaOH. The perfusion buffer 2 should keep his red color (phenol red)

indicating pH 7.4. Usually 1 drop of 1N NaOH is required every other minute.

12. In this centrifugation step, the brake should be turned off.
13. It should be kept in mind that all sorters will produce biological/chemical hazards and aerosols that might have potential biosafety implications, which should be taken into account. Moreover, unfixed biological materials of animal origin can harbor known and unknown pathogens that may be transmitted through droplets and aerosols that are generated during the cell sorting process. A general introduction to this technique, its principles, sample preparation, FACS sorter setup, sort strategies, and a troubleshooting guide, is given elsewhere [30].
14. Take care to change the tubes in time to prevent overload.
15. In this process, you can modify the gating strategy to include all cells in the forward and sideward scatter ignoring doublet discrimination.
16. In this centrifugation step, the brake should be turned off.
17. The resulting pellet of ultrapure HSC shows a white color with diffuse shape and behaves very fragile.
18. We routinely determine the number of purified cells in a Neubauer bright-line chamber with a depth of 0.1 mm and a counting area of 0.0025 mm^2.
19. To guarantee the correct depth between cover glass and the central counting area, it is advisable to moisten the outer sections with breath and carefully adjust the slip correctly onto the counting chamber. The formation of multicolor Newton's rings confirms the proper attachment.
20. HSC have a typical appearance characterized by a high content of intracellular fat droplets with their specific bright shine and a typical rounded cell phenotype.
21. The cell plates should be uncoated and suitable for culturing of sensitive adherent primary cells.
22. As the quiescent HSCs are very fragile, carefully replace the medium with new pre-warmed medium without damaging the cell monolayer.
23. There are several immortalized HSC lines from mice available that differ in morphology, growth characteristics, and chromosomal endowment. Most of these cell lines represent an activated myofibroblast-like phenotype. Although most phenotypic features remain constant for many passages in these continuous lines, it is obvious that the results obtained in these cell lines may significantly differ from results obtained in experiments performed with primary cells. In particular,

they have limitations in mimicking the cellular dynamics of primary HSC that undergo a transdifferentiation process upon culturing to a myofibroblastic phenotype. Therefore, these immortalized cell lines might not be optimal to analyze fundamental biological processes such as cell cycle, loss of fat and cytoplasmic lipid droplets, apoptosis, necrosis, or pyroptosis. However, when trying to establish the purification of HSC in your lab, these lines may serve as good positive controls when testing your primary cells for expression of typical HSC markers. Details about several murine HSC lines such as GRX, Sv68c-IS, A640-IS, M1-4HSC, HSC Col-GFP, and A7 that were established by different procedures are summarized in detail elsewhere [31, 32].

24. Nowadays, it is well accepted that there exists significant HSC heterogeneity in a healthy liver. Individual subpopulations differ in fat content and metabolism, activation status, marker gene expression, or other biological features [15]. We acknowledge that our protocol might preferentially purify a specific HSC fraction that is particularly rich in fat and vitamin A. These cells share a similar activation status. However, continuous advancement in the field might allow the adaptation of the current protocol. Especially, the identification of new markers exposed to the surface of HSC could be used to purify specific HSC subsets by FACS sorting.
25. Desmin is an intermediate filament protein associated with both smooth and skeletal muscle differentiation. It has a species-dependent molecular mass of 50,000 to 55,000 Dalton. It is found in skeletal, cardiac, and smooth muscle cells. This protein is expressed in HSC regardless of their activation status [33].
26. At this stage, the cells can be stored at 4 °C for up to 6 months.

Acknowledgments

Authors' laboratory is supported by grants from the German Research Foundation (WE2554/13-1, WE2554/15-1, WE2554/17-1, and ME 3431/2-1) and a grant from the Interdisciplinary Centre for Clinical Research within the faculty of Medicine at the RWTH Aachen University (grant PTD 1-5).

References

1. Fujita H, Tamaru T, Miyagawa J (1980) Fine structural characteristics of the hepatic sinusoidal walls of the goldfish (*Carassius auratus*). Arch Histol Jpn 43(3):265–273. https://doi.org/10.1679/aohc1950.43.265
2. Häussinger D, Kordes C (2019) Space of Disse: a stem cell niche in the liver. Biol Chem 401(1):81–95. https://doi.org/10.1515/hsz-2019-0283
3. Tacke F, Weiskirchen R (2012) Update on hepatic stellate cells: pathogenic role in liver

fibrosis and novel isolation techniques. Expert Rev Gastroenterol Hepatol 6:67–80. https://doi.org/10.1586/egh.11.92

4. Knook DL, Seffelaar AM, de Leeuw AM (1982) Fat-storing cells of the rat liver. Their isolation and purification. Exp Cell Res 139: 468–471. https://doi.org/10.1016/0014-4827(82)90283-X
5. Friedman SL, Roll FJ (1987) Isolation and culture of hepatic lipocytes, Kupffer cells, and sinusoidal endothelial cells by density gradient centrifugation with Stractan. Anal Biochem 161:207–218. https://doi.org/10.1016/0003-2697(87)90673-7
6. Pinzani M, Gesualdo L, Sabbah GM, Abboud HE (1989) Effects of platelet-derived growth factor and other polypeptide mitogens on DNA synthesis and growth of cultured rat liver fat-storing cells. J Clin Invest 84:1786–1793. https://doi.org/10.1172/JCI114363
7. de Leeuw AM, McCarthy SP, Geerts A, Knook DL (1984) Purified rat liver fat-storing cells in culture divide and contain collagen. Hepatology 4:392–403. https://doi.org/10.1002/hep.1840040307
8. Schäfer S, Zerbe O, Gressner AM (1987) The synthesis of proteoglycans in fat-storing cells of rat liver. Hepatology 7:680–687. https://doi.org/10.1002/hep.1840070411
9. Weiskirchen R, Gressner AM (2005) Isolation and culture of hepatic stellate cells. Methods Mol Med. 117:99–113. https://doi.org/10.1385/1-59259-940-0:099
10. Maschmeyer P, Flach M, Winau F (2011) Seven steps to stellate cells. J Vis Exp 10(51). https://doi.org/10.3791/2710
11. Elsharkawy AM, Wright MC, Hay RT, Arthur MJ, Hughes T, Bahr MJ, Degitz K, Mann DA (1999) Persistent activation of nuclear factor-kappaB in cultured rat hepatic stellate cells involves the induction of potentially novel Rel-like factors and prolonged changes in the expression of IkappaB family proteins. Hepatology 30:761–769. https://doi.org/10.1002/hep.510300327
12. Graham JM (2002) Fractionation of hepatic nonparenchymal cells. ScientificWorldJournal 2:1347–1350. https://doi.org/10.1100/tsw.2002.283
13. Modak RV, Zaiss DM (2019) Isolation and culture of murine hepatic stellate cells. Bio Protoc. 9(21):e3422. https://doi.org/10.21769/BioProtoc.3422
14. Blomhoff R, Berg T (1990) Isolation and cultivation of rat liver stellate cells. Methods Enzymol 190:58–71. https://doi.org/10.1016/0076-6879(90)90009-P
15. D'Ambrosio DN, Walewski JL, Clugston RD, Berk PD, Rippe RA, Blaner WS (2011) Distinct populations of hepatic stellate cells in the mouse liver have different capacities for retinoid and lipid storage. PLoS One 6(9): e24993. https://doi.org/10.1371/journal.pone.0024993
16. Geerts A, Niki T, Hellemans K, De Craemer D, Van Den Berg K, Lazou JM, Stange G, Van De Winkel M, De Bleser P (1998) Purification of rat hepatic stellate cells by side scatter-activated cell sorting. Hepatology 27:590–598. https://doi.org/10.1002/hep.510270238
17. Bartneck M, Warzecha KT, Tag CG, Sauer-Lehnen S, Heymann F, Trautwein C, Weiskirchen R, Tacke F (2015) Isolation and time lapse microscopy of highly pure hepatic stellate cells. Anal Cell Pathol (Amst) 2015: 417023. https://doi.org/10.1155/2015/417023
18. Mederacke I, Dapito DH, Affò S, Uchinami H, Schwabe RF (2015) High-yield and high-purity isolation of hepatic stellate cells from normal and fibrotic mouse livers. Nat Protoc 10:305–315. https://doi.org/10.1038/nprot.2015.017
19. Paik YH, Iwaisako K, Seki E, Inokuchi S, Schnabl B, Osterreicher CH, Kisseleva T, Brenner DA (2011) The nicotinamide adenine dinucleotide phosphate oxidase (NOX) homologues NOX1 and NOX2/gp91(phox) mediate hepatic fibrosis in mice. Hepatology 53: 1730–1741. https://doi.org/10.1002/hep.24281
20. Weiskirchen S, Tag CG, Sauer-Lehnen S, Tacke F, Weiskirchen R (2017) Isolation and culture of primary murine hepatic stellate cells. In: Rittié L (ed) Fibrosis. methods in molecular biology, vol 1627. Humana Press, New York. https://doi.org/10.1007/978-1-4939-7113-8_11
21. Wang YQ, Ikeda K, Ikebe T, Hirakawa K, Sowa M, Nakatani K, Kawada N, Kaneda K (2000) Inhibition of hepatic stellate cell proliferation and activation by the semisynthetic analogue of fumagillin TNP-470 in rats. Hepatology 32:980–989. https://doi.org/10.1053/jhep.2000.18658
22. Acharya P, Chouhan K, Weiskirchen S, Weiskirchen R (2012) Cellular mechanisms of liver fibrosis. Front Pharmacol 12:671640. https://doi.org/10.3389/fphar.2021.671640
23. Donnenberg AD, Donnenberg VS (2007) Rare-event analysis in flow cytometry. Clin Lab Med 27:627–652. https://doi.org/10.1016/j.cll.2007.05.013

24. Neyzen S, Van de Leur E, Borkham-Kamphorst E, Herrmann J, Hollweg G, Gressner AM, Weiskirchen R (2006) Cryopreservation of hepatic stellate cells. J Hepatol 44:910–917. https://doi.org/10.1016/j.jhep.2005.07.008
25. Listenberger LL, Brown DA (2007) Fluorescent detection of lipid droplets and associated proteins. Curr Protoc Cell Biol. Chapter 24: Unit 24.2. https://doi.org/10.1002/0471143030.cb2402s35
26. Asimakopoulou A, Borkham-Kamphorst E, Henning M, Yagmur E, Gassler N, Liedtke C, Berger T, Mak TW, Weiskirchen R (2014) Lipocalin-2 (LCN2) regulates PLIN5 expression and intracellular lipid droplet formation in the liver. Biochim Biophys Acta 1842(10): 1513–1524. https://doi.org/10.1016/j.bbalip.2014.07.017
27. Riccalton-Banks L, Bhandari R, Fry J, Shakesheff KM (2003) A simple method for the simultaneous isolation of stellate cells and hepatocytes from rat liver tissue. Mol Cell Biochem 248:97–102. https://doi.org/10.1023/A:1024184826728
28. Mohar I, Brempelis KJ, Murray SA, Ebrahimkhani MR, Crispe IN (2015) Isolation of non-parenchymal cells from the mouse liver. Methods Mol Biol 1325:3–17. https://doi.org/10.1007/978-1-4939-2815-6_1
29. Werner M, Driftmann S, Kleinehr K, Kaiser GM, Mathé Z, Treckmann JW, Paul A, Skibbe K, Timm J, Canbay A, Gerken G, Schlaak JF, Broering R (2015) All-in-one: advanced preparation of human parenchymal and non-parenchymal liver cells. PLoS One 10(9):e0138655. https://doi.org/10.1371/journal.pone.0138655
30. Davies D (2012) Cell separations by flow cytometry. Methods Mol Biol 878:185–199. https://doi.org/10.1007/978-1-61779-854-2_12
31. Herrmann J, Gressner AM, Weiskirchen R (2007) Immortal hepatic stellate cell lines: useful tools to study hepatic stellate cell biology and function? J Cell Mol Med 11:704–722. https://doi.org/10.1111/j.1582-4934.2007.00060.x
32. Meurer SK, Alsamman M, Sahin H, Wasmuth HE, Kisseleva T, Brenner DA, Trautwein C, Weiskirchen R, Scholten D (2013) Overexpression of endoglin modulates TGF-β1-signalling pathways in a novel immortalized mouse hepatic stellate cell line. PLoS One 8:e56116. https://doi.org/10.1371/journal.pone.0056116
33. Van Rossen E, Vander Borght S, van Grunsven LA, Reynaert H, Bruggeman V, Blomhoff R, Roskams T, Geerts A (2009) Vinculin and cellular retinol-binding protein-1 are markers for quiescent and activated hepatic stellate cells in formalin-fixed paraffin embedded human liver. Histochem Cell Biol. 131(3):313–325. https://doi.org/10.1007/s00418-008-0544-2

Chapter 2

Differentiation of Hepatic Stellate Cells from Pluripotent Stem Cells

Raquel A. Martínez García de la Torre and Pau Sancho-Bru

Abstract

Hepatic stellate cells (HSCs) are non-parenchymal cells with a mesenchymal origin involved in vitamin A storage and extracellular matrix (ECM) homeostasis. In response to injury, HSCs activate and acquire myofibroblastic features, participating in the wound healing response. Upon chronic liver injury, HSCs become the main contributors to ECM deposition and to the progression of fibrosis. Due to their relevant roles in liver function and pathophysiology, it is of utmost importance to develop means to obtain HSCs for liver disease modeling and drug development. Here, we describe a directed differentiation protocol from human pluripotent stem cells (hPSCs) to obtain functional HSCs (PSC-HSCs). The procedure is based on the subsequent addition of growth factors during 12 days of differentiation. PSC-HSCs can be used for liver modeling and drug screening assays, hence emerging as a promising and reliable source of HSCs.

Key words Hepatic stellate cell differentiation, Pluripotent stem cells, Fibrosis, In vitro models, Disease modeling, Drug development

1 Introduction

Fibrosis is a common trait for almost all chronic liver diseases, independently of their etiology. It is characterized by an excessive accumulation of extracellular matrix (ECM) proteins mostly produced by activated hepatic stellate cells (HSCs). Hence, the activation of HSCs can be understood as the central driver of hepatic fibrosis in liver injury as they secrete ECM compounds such as collagen I, III, and fibronectin, which leads to scar formation [1, 2]. The limited availability of primary HSCs has hampered the study of cell physiology and the establishment of in vitro models of chronic liver diseases for drug development [3, 4].

Several antifibrogenic drugs are being investigated in clinical trials with promising results, but currently, there are no approved treatments in clinical practice. Likewise, current human-based in vitro models do not fully recapitulate chronic liver disease

Ralf Weiskirchen and Scott L. Friedman (eds.), *Hepatic Stellate Cells: Methods and Protocols*, Methods in Molecular Biology, vol. 2669, https://doi.org/10.1007/978-1-0716-3207-9_2,

features and fibrogenesis, thus suggesting the need for reliable disease models for drug development.

Human embryonic and induced pluripotent stem cells, from hereinafter PSCs, present the ability to differentiate into virtually any cell type. Specifically, iPSCs can be obtained from adult cells by the overexpression of the pluripotency core genes and, thus, can be generated from patient-specific cells. Moreover, PSC expansion can be almost unlimited, and they are easy to modify genetically. For all these reasons, they are valuable tools for in vitro modelling [5].

Thanks to the studies on liver embryology performed in animal models during recent decades, pathways involved in embryonic development have been elucidated [6, 7]. This knowledge has been used to differentiate liver cells from PSCs by using specific cytokines and growth factors to simulate biological pathways involved in cell commitment [8–10]. First efforts in the hepatology field were focused on obtaining hepatocytes using directed differentiation strategies [11–14]. However, the number of strategies and differentiation protocols to generate non-parenchymal cells is more limited. Although the embryonic development of HSCs is not completely understood [15, 16], lineage tracing studies and human evidences support their mesenchymal origin [17]. Following this knowledge, different strategies have been reported to differentiate PSCs into HSCs [18–22].

In this chapter, we describe a directed differentiation protocol to obtain functional HSCs from human PSCs for disease modelling and drug development [20, 21].

2 Materials

2.1 Biological and Coating Reagents

1. Human PSCs: Differentiation can be performed using embryonic and iPSC cell lines (*see* **Note 1**).
2. Cell culture plate for differentiation and expansion of PSC-HSCs:Differentiation and expansion of differentiated cells are performed in culture plates coated with 2% Matrigel® Growth Factor Reduced (GFR) Basement Membrane Matrix diluted in DMEM low glucose in sterile cold conditions (*see* **Note 2**). Place 0.63 mL of 2% Matrigel solution in DMEM per cm^2. Incubate the plate at room temperature (RT) for 1 h or 20–30 min at 37 °C. Wash with PBS before seeding the cells (*see* **Note 3**).

2.2 Differentiation Media Components: Stock and Working Solutions

1. MCDB 201 medium: Dissolve 8.85 g of MCDB in 500 mL of Milli-Q water. Stir with magnetic bar for 10–20 min and adjust the pH to 7.2, following manufacturers' instructions. Filter the solution with a 0.22 μM filter, and store at 4 °C.

2. Dexamethasone: Dissolve 5 mg in 1 mL of pure ethanol to obtain a 5000 ng/μL stock solution. Dilute the stock solution to a final working concentration of 98 ng/μL in Milli-Q water. Aliquots from both stock and working solutions should be stored at −20 °C.
3. L-ascorbic acid: Dissolve 1.45 g of L-ascorbic acid in 500 mL PBS 1× to obtain a final solution at 10 mM. Stir with magnetic bar for 10–20 min in dark conditions, and filter it with a 0.22 μM filter. Aliquots should be stored at −20 °C.
4. Bone morphogenic protein (BMP4): Reconstitute the vial in 4 mM HCl containing 0.1% BSA to obtain a final working solution of 10 ng/μL. Diluted BMP4 should be stored at −80 °C, while powder growth factor should be stored at −20 °C, according to manufacturer's instructions.
5. Fibroblast growth factor 1 (FGF1): Reconstitute the vial in DPBS (−/−) containing 0.1% BSA to obtain a working solution of 10 ng/μL. Diluted FGF1 should be stored at −80 °C, while powder growth factor should be stored at −20 °C, according to manufacturer's instructions.
6. Fibroblast growth factor 3 (FGF3): Reconstitute the vial in DPBS (−/−) containing 0.1% BSA to obtain a working solution of 10 ng/μL. Diluted FGF3 should be stored at −80 °C, while powder growth factor should be stored at −20 °C, according to manufacturer's instructions.
7. Retinol: Reconstitute a vial in pure ethanol to obtain a stock solution of 1 M. Dilute it in DPBS (−/−) to a final working solution of 5 mM. Aliquots from both stock and working solutions should be stored at −80 °C.
8. Palmitic acid: Dissolve it to obtain a working solution of 0.1 M. Aliquots should be stored at −20 °C.

2.3 Cell Culture Medium

1. PSC maintenance medium: Use your standard medium for the maintenance of undifferentiated PSCs and for platting cells before starting differentiation.
2. Liver differentiation medium (LdM): 57% DMEM, low glucose, pyruvate. 40% MCDB 201 medium, 1% L-ascorbic acid, 0.25% insulin-transferrin-selenium (100×) (ITS-G), 0.25% linoleic acid-albumin from bovine serum albumin (LA-BSA), 0.1% 2-mercaptoethanol (50 mM), 0.4% dexamethasone, 1% penicillin-streptomycin mixture (P/S). For every other day, add the corresponding cytokines following Table 1.
3. Expansion medium: DMEM GlutaMAX, high glucose, pyruvate, 10% fetal bovine serum (FBS), and 1% P/S (*see* **Note 4**).
4. Freezing medium: 80% FBS and 20% DMSO.

Table 1
PSC-HSC differentiation outline

Time	Medium composition
Day 0 and day 2	LdM supplemented with 2 μL of BMP4 per mL
Day 4	LdM supplemented with 2 μL of BMP4 + 2 μL of FGF1 + 2 μL of FGF3 per mL
Day 6	LdM supplemented with 2 μL of FGF1 + 2 μL of FGF3 + 1 μL retinol + 1 μL palmitic acid per mL
Day 8, day 10, and day 12	LdM supplemented with 1 μL retinol + 1 μL palmitic acid per mL

2.4 Other Reagents

1. 0.5 mM EDTA solution: Dissolve 50 μL of UltraPure 0.5 M EDTA, pH 8 in 50 mL of DPBS (−/−), and filter with a 0.22 μM filter. Diluted EDTA can be stored at 2–8 °C for up to 6 months.
2. DPBS (−/−): Dulbecco's phosphate-buffered saline (10×) DPBS without calcium/magnesium.
3. Trypsin-EDTA 0.25% solution: Trypsin, 0.25% (1×) with EDTA × 4 Na.
4. RevitaCell supplement (100×) (*see* **Note 5**).

3 Methods

Perform all cell culture procedures in sterile conditions. Cells should be cultivated at 37 °C, 5% CO_2 conditions in a cell incubator.

3.1 Seeding of Human PSCs for HSC Differentiation

Warm the coated plate before seeding cells at 37 °C for at least 30 min. Also, warm at RT, PSC maintenance medium and EDTA 0.5 mM solution before starting.

1. When the PSC culture reaches 70% confluence, remove medium, and wash cells with DPBS (−/−) during 1 min.
2. Then, detach cells using the EDTA solution. Wait for 2–3 min at RT, and check the cell colonies under the microscope. Cells should start to separate between them.
3. Aspirate EDTA, and using PSC maintenance medium, disrupt the cell colonies by gently pipetting it. Collect cell suspension.
4. Wash the coated plate with DPBS (−/−).
5. Count cells and plate 15,000 cells/cm^2 using PSC maintenance medium. Move the plate to distribute them homogeneously, and place it in the incubator.

6. Usually, it takes 1–2 days to start the differentiation. Change the medium to fresh PSC maintenance medium 24 h after seeding, if cells have not reached the starting confluence.

3.2 PSCs Differentiation toward HSCs

To start the differentiation, PSCs colonies should be approximately 70% confluent, usually after 1–2 days (Fig. 1).

1. Remove PSC maintenance medium, and wash cell culture with DPBS (−/−) (*see* **Note 6**).
2. Change PSC maintenance medium to LdM supplemented with 2 μL/mL of BMP4 working solution to induce mesoderm differentiation. This is considered day 0 (*see* **Note 7**).
3. At day 2, wash cells with DPBS (−/−), and change medium to LdM supplemented with 2 μL/mL of BMP4 working solution. At this time of the differentiation protocol, the culture should be 90% confluent.

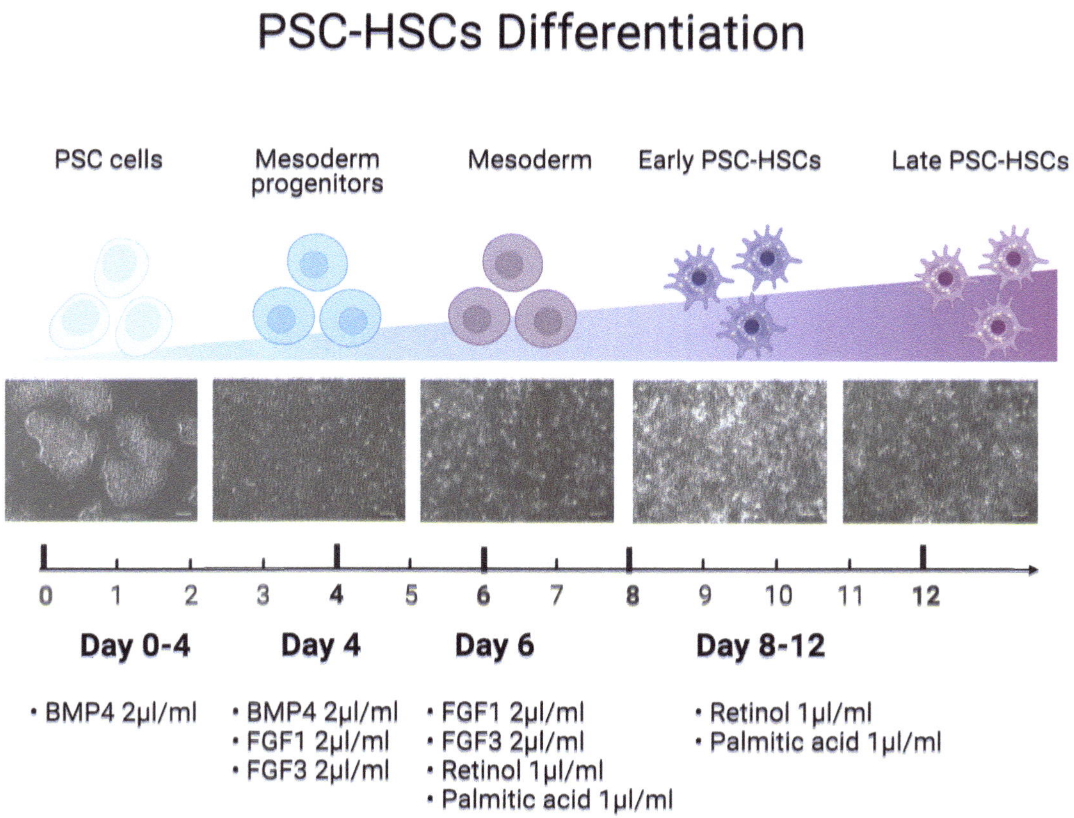

Fig. 1 Time course of the differentiation protocol: Microscopy images of cell culture morphology correspond to day 0, day 4, day 6, and day 12 of the differentiation. Scale bars = 100 μM. Schematic representation of the time course and the corresponding concentration of the cytokines added at each day of the differentiation protocol

4. At day 4, wash thoroughly cells as high cell death is observed at this point due to high proliferation rate of the culture. Then, add LdM supplemented with 2 μL/mL of each working solution (BMP4, FGF1 and FGF3), for the mesenchymal induction. At this time of the differentiation, the culture should be totally confluent.
5. At day 6, wash cells, and add to the culture 2 μL/mL of FGF1 and FGF3 working solution, 1 μL/mL of retinol, and palmitic acid working solution, to induce specification toward HSCs (*see* **Note 8**).
6. From day 8 until the end of the differentiation, LdM should be supplemented with 1 μL/mL retinol and palmitic acid of working solution to complete maturation of the differentiated cells. Nonetheless, the cleaning step before medium change is recommended.
7. At day 8, presence of lipid droplets in the differentiated cells can be noted under the microscope. Likewise, protein expression of PDGFRβ could be assessed by flow cytometry and should be around 40% at this point.
8. At the end of the differentiation, protein expression of PDFRβ and vitamin A content can be assessed by flow cytometry (*see* **Note 9**). Around 60% of PSC-HSCs should be positive for PDGFRβ. Close to 100% of cells should contain vitamin A. Also, typical markers of HSCs (PCDH7, LRAT, RELN, PDGFRB, ACTA2, and COL1A1) can be determined (Fig. 2).

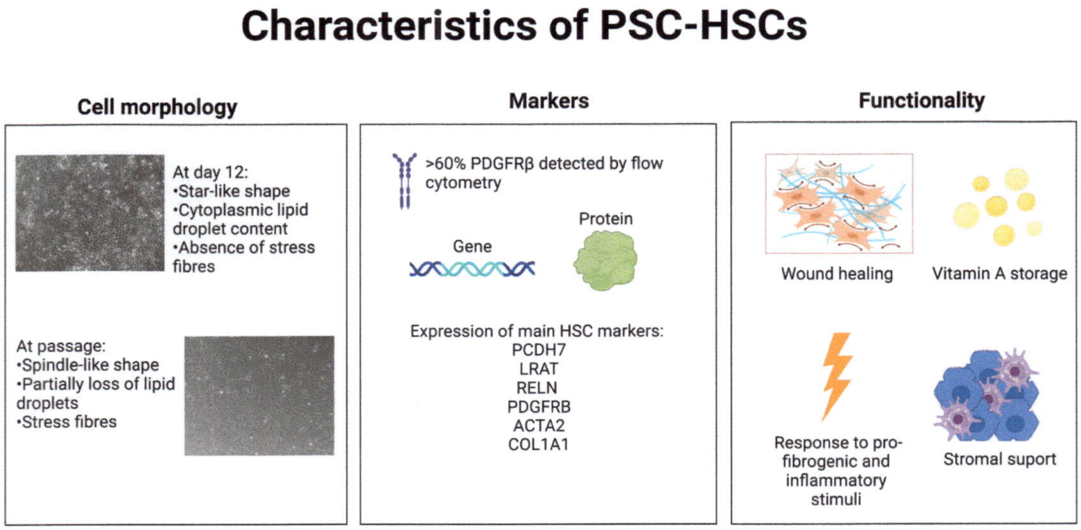

Fig. 2 Features of PSC-HSCs: Cell morphology of the differentiated cells at the end of the protocol and after passage. Main HSC markers that can be assessed after the differentiation. Functional analysis that can be performed using PSC-HSCs

3.3 Detaching Differentiated PSC-HSCs

Differentiated cells can be used directly at the end of the procedure as they display functional characteristics of HSCs. Also, they can be subcultured or cryopreserved to generate a cell bank of PSC-HSCs for future use. Next, we will tackle how to detach PSC-HSCs. Detachment of cells requires pre-warmed expansion medium and trypsin-EDTA 0.25% solution.

1. Wash cells for 1 min at RT with DPBS (–/–) before detaching them with trypsin-EDTA 0.25% during 2–3 min.
2. Add an equal amount of expansion medium as trypsin solution in order to stop the reaction. Collect cells by gently pipetting the medium (*see* **Notes 10** and **11**).
3. Centrifuge cell suspension at 300 × g for 5 min at RT.

3.4 Subculturing PSC-HSCs in 2D

Matrigel pre-coated plate is recommended to expand and perform further assays with PSC-HSCs and avoid activation due to plastic contact. Warm it at 37 °C 20–30 min before seeding them, and warm also the expansion medium.

1. Wash the new coated plates with DPBS (–/–).
2. After centrifugation, add an appropriate volume of expansion medium to plate 73,000 cells per cm^2. Cell suspension can be supplemented with RevitaCell (1:100) to improve cell viability after passage and cell filtration.
3. Seed the cells and change medium after 24 h. Medium should be changed every other day (*see* **Note 12**).
4. When cells achieved 70% confluence, they can be used for downstream assays previous starvation of serum (*see* **Notes 13** and **14**).

3.5 Freezing PSC-HSCs

1. After centrifugation, resuspend the pellet and count cells to freeze in each cryovial approximately 2 million of PSC-HSCs in 1 mL of cold freezing medium (80% FBS and 20% DMSO) (*see* **Note 15**).
2. Place the cryovials rapidly in a cryofreezing container, and keep it overnight at −80 °C.
3. After 24 h, transfer the cryovial to a liquid nitrogen tank for long-term storage.

3.6 Thawing of PSC-HSCs

Matrigel pre-coated plates are recommended to thaw and perform further assays with PSC-HSCs, which avoids activation through contact with the plastic dish. Warm the plate at 37 °C 30 min before seeding the cells and also warm the expansion medium. Ensure the water bath is at 37 °C for thawing the cryovial.

1. Thaw the cryovial at 37 °C in a water bath. When only an ice crystal remains, rinse with ethanol the cryovial to avoid contamination, and add 0.5 mL of expansion medium.
2. Next, transfer the cell content into a 15 mL tube containing expansion medium.
3. Centrifuge cells at 300 × g for 5 min at RT. Meanwhile, wash the new Matrigel plate with DPBS(−/−).
4. Aspirate the supernatant, and resuspend the cells with the corresponding volume of expansion medium to plate 100,000 cells per cm^2. Cell suspension can be supplemented with RevitaCell (1:100).
5. Move the plate gently before placing them in the incubator, to ensure a homogenous distribution.
6. After 24 h, discard the medium, and replace with fresh expansion medium. Medium changes should be performed every other day.

   When cells achieved 70% confluence, they can be used for downstream assays previous starvation of serum.

4 Notes

1. PSCs research should always be conducted in accordance with all relevant governmental and institutional guidelines and regulations.
2. Matrigel dilution should be performed under cold conditions to avoid polymerization. Cold pipette tips can be used.
3. Coated plates can be stored up to 1 week at 4 °C. Store them directly with the coating solution. Pre-warmed the plates 30 min before use, and wash them before seeding cells.
4. It is possible to subculture the PSC-HSCs in xeno-free conditions using 10% knockout serum as a replacement for the FBS.
5. Traditional ROCK inhibitors could be used instead of RevitaCell supplement. Use them at the concentration indicated in manufacturers' instructions.
6. The cleaning steps along the differentiation process are critical in order to remove possible leftovers of the medium and cytokines used previously.
7. It is recommended to add daily fresh cytokines to the LdM and warm it up at RT. Medium changes should be performed every 48 h following Table 1 and Fig. 1.
8. Palmitic acid is difficult to dilute, so the stock solution may precipitate. Vortex the stock solution before using it, and warm

the medium to help dilute the palmitic acid during medium preparation of days 6, 8, 10, and 12.

9. Until day 14, PSC-HSCs can be maintained in the same plate of the differentiation, but it is recommended to use them at day 12 for the downstream assays.
10. At the end of the protocol, PSC-HSCs are strongly attached as they secrete ECM, so it may take longer to detach them. Cell scrapper can be used.
11. When detaching cells, ECM and Matrigel can be released, so it is recommended to filter cell suspension using cell strainers (diameter = 100 μM).
12. The subculture of PSC-HSCs promotes activation phenotype. Therefore, it is recommended to use them at low passages.
13. Starvation of serum should be performed 12 h prior stimulation at 1% of serum. Stimuli should be also performed in the same medium.
14. It is possible to expand PSC-HSCs until passage three.
15. Maintain the freezing medium at 4 °C. It should be cold when added to the cell suspension.

Acknowledgments

P. S. - B. is supported by the Fondo de Investigación Sanitaria Carlos III, cofinanced by the Fondo Europeo de Desarrollo Regional (FEDER), Unión Europea, "Una manera de hacer Europa" (FIS PI20/00765, PI17/00673), DTS18/ 00088, COST Action H2020 PRO-EURO-DILI-NET CA17112, Ajudes d'Industria del Coneixement, Modalitat B: Ajudes Producte (PI046292-34024). R. M. is supported by PFIS (FI18/00215). Figures were created using BioRender.com.

References

1. Hernandez-Gea V, Friedman SL (2011) Pathogenesis of liver fibrosis. Rev Adv Annu Rev Pathol Mech Dis 6:425–456. https://doi.org/10.1146/annurev-pathol-011110-130246
2. Puche JE, Saiman Y, Friedman SL (2013) Hepatic stellate cells and liver fibrosis. Compr Physiol 3:1473–1492. https://doi.org/10.1002/CPHY.C120035
3. Perea L, Coll M, Sancho-Bru P (2015) Assessment of liver fibrotic insults in vitro. Methods Mol Biol 1250:391–401. https://doi.org/10.1007/978-1-4939-2074-7_30
4. Shang L, Hosseini M, Liu X et al (2018) Human hepatic stellate cell isolation and characterization. J Gastroenterol 53:6–17. https://doi.org/10.1007/s00535-017-1404-4
5. Shi Y, Inoue H, Wu JC, Yamanaka S (2017) Induced pluripotent stem cell technology: a decade of progress. Nat Rev Drug Discov 16: 115–130. https://doi.org/10.1038/nrd.2016.245
6. Zhao R, Duncan SA (2005) Embryonic development of the liver. Hepatology 41:956–967. https://doi.org/10.1002/hep.20691
7. Si-Tayeb K, Lemaigre FP, Duncan SA (2010) Organogenesis and development of the liver.

Dev Cell 18:175–189. https://doi.org/10.1016/J.DEVCEL.2010.01.011

8. Roelandt P, Vanhove J, Verfaillie C (2013) Directed differentiation of pluripotent stem cells to functional hepatocytes. Methods Mol Biol 997:141–147. https://doi.org/10.1007/978-1-62703-348-0_11
9. Sancho-Bru P, Roelandt P, Narain N et al (2011) Directed differentiation of murine-induced pluripotent stem cells to functional hepatocyte-like cells. J Hepatol 54:98–107. https://doi.org/10.1016/j.jhep.2010.06.014
10. Takebe T, Sekine K, Enomura M et al (2013) Vascularized and functional human liver from an iPSC-derived organ bud transplant. Nature 499:481–484. https://doi.org/10.1038/nature12271
11. Boon R, Kumar M, Tricot T et al (2020) Amino acid levels determine metabolism and CYP450 function of hepatocytes and hepatoma cell lines. Nat Commun 11:1393. https://doi.org/10.1038/s41467-020-15058-6
12. Collin de l'Hortet A, Takeishi K, Guzman-Lepe J et al (2019) Generation of human fatty livers using custom-engineered induced pluripotent stem cells with modifiable SIRT1 metabolism. Cell Metab 30:385–401.e9. https://doi.org/10.1016/j.cmet.2019.06.017
13. Baxter M, Withey S, Harrison S et al (2015) Phenotypic and functional analyses show stem cell-derived hepatocyte-like cells better mimic fetal rather than adult hepatocytes. J Hepatol 62:581–589. https://doi.org/10.1016/j.jhep.2014.10.016
14. Tricot T, Thibaut HJ, Abbasi K et al (2022) Metabolically improved stem cell derived hepatocyte-like cells support HBV life cycle and are a promising tool for HBV studies and antiviral drug screenings. Biomedicine 10:268. https://doi.org/10.3390/biomedicines10020268
15. Asahina K, Zhou B, Pu WT, Tsukamoto H (2011) Septum transversum-derived mesothelium gives rise to hepatic stellate cells and perivascular mesenchymal cells in developing mouse liver. Hepatology 53:983–995. https://doi.org/10.1002/hep.24119
16. Asahina K, Tsai SY, Li P et al (2009) Mesenchymal origin of hepatic stellate cells, submesothelial cells, and perivascular mesenchymal cells during mouse liver development. Hepatology 49:998–1011. https://doi.org/10.1002/hep.22721
17. Wesley BT, Ross ADB, Muraro D et al (2022) Single-cell atlas of human liver development reveals pathways directing hepatic cell fates. Nat Cell Biol. https://doi.org/10.1038/s41556-022-00989-7
18. Koui Y, Kido T, Ito T et al (2017) An in vitro human liver model by iPSC-derived parenchymal and non-parenchymal cells. Stem Cell Rep 9:490–498. https://doi.org/10.1016/j.stemcr.2017.06.010
19. Miyoshi M, Kakinuma S, Kamiya A et al (2019) LIM homeobox 2 promotes interaction between human iPS-derived hepatic progenitors and iPS-derived hepatic stellate-like cells. Sci Rep 9:2072. https://doi.org/10.1038/s41598-018-37430-9
20. Coll M, Perea L, Boon R et al (2018) Generation of hepatic stellate cells from human pluripotent stem cells enables in vitro modeling of liver fibrosis. Cell Stem Cell 23:101–113.e7. https://doi.org/10.1016/j.stem.2018.05.027
21. Vallverdú J, Martínez RA, de La Torre G et al (2021) Directed differentiation of human induced pluripotent stem cells to hepatic stellate cells. Nat Protoc 16:2542–2563. https://doi.org/10.1038/s41596-021-00509-1
22. Kumar M, Toprakhisar B, Van Haele M et al (2021) A fully defined matrix to support a pluripotent stem cell derived multi-cell-liver steatohepatitis and fibrosis model. Biomaterials 276:121006. https://doi.org/10.1016/j.biomaterials.2021.121006

Chapter 3

Testing Cell Migration, Invasion, Proliferation, and Apoptosis in Hepatic Stellate Cells

Miriam Wankell and Lionel Hebbard

Abstract

The hepatic wound repair process involves cell types including healthy and injured hepatocytes, Kupffer and inflammatory cells, sinusoidal endothelial cells (SECs), and hepatic stellate cells (HSCs). Normally, in their quiescent state, HSCs are a reservoir for vitamin A, but in response to hepatic injury, they become activated myofibroblasts that play a key role in the hepatic fibrotic response. Activated HSCs express extracellular matrix (ECM) proteins, elicit anti-apoptotic responses, and proliferate, migrate, and invade hepatic tissues to protect hepatic lobules from damage. Extended liver injury can lead to fibrosis and cirrhosis, the deposition of ECM that is driven by HSCs. Here we describe in vitro assays that quantify activated HSC responses in the presence of inhibitors targeting hepatic fibrosis.

Key words Liver fibrosis, HSCs, Proliferation, Invasion, Migration, Apoptosis

1 Introduction

HSCs (vitamin A-storing cells, lipocytes, interstitial cells, fat-storing cells, and Ito cells) are located in the space of Disse—the area between the sinusoids and hepatocytes. In response to sustained inflammatory stimuli such as viral hepatitis, aflatoxin, alcohol, and fatty liver disease leading to non-alcoholic fatty liver disease (NASH) and metabolic associated fatty liver disease (MAFLD), parenchymal cell damage occurs leading to the promotion of an inflammatory response to assist in liver repair [1, 2]. In acute liver injury, the repair process is rapid and short-lived. However, following sustained injury, the repair response supports the presence of active immune cells, which together alter the phenotype of the quiescent HSCs. The HSCs become activated, proliferate, and transdifferentiate to myofibroblast-like, contractile, proinflammatory, and fibrogenic cells [3]. Moreover, in this environment of sustained hepatic damage and inflammation, activated HSCs secrete ECM components such as collagen and laminin and

Ralf Weiskirchen and Scott L. Friedman (eds.), *Hepatic Stellate Cells: Methods and Protocols*, Methods in Molecular Biology, vol. 2669, https://doi.org/10.1007/978-1-0716-3207-9_3,

matrix metalloproteinases and tissue inhibitor of metalloproteinases (TIMPs), which can lead to the accumulation of scar tissue in the liver. Importantly prolonged liver disease due to the above etiologic factors maintains HSC activation leading to hepatic fibrosis and eventual progression to cirrhosis. In these instances, the HSCs produce abundant ECM that is not degraded, resulting in the appearance of ECM bundles surrounding clusters of hepatocytes, causing stiffening of the liver and hepatic cirrhosis that can promote portal hypertension and liver failure and also be a harbinger for hepatocellular carcinoma (HCC) [4–6]. A broad spectrum of hepatic diseases and high percentage of HCCs occur in the setting of fibrosis and cirrhosis. Hence, an area of intensive research is to find agents that can inhibit activated HSC activity to resolve hepatic scarring, leading to improved hepatic function and decreased HCC risk. Underlying these investigations are the use of standard HSC assays with immortalized or primary cells, enabling the evaluation of potential therapeutics targeted to this cell type. In this chapter, we describe essential assays for evaluating HSC migration, invasion, proliferation, and apoptosis.

2 Materials

Prepare all solutions at room temperature with deionized and sterile filtered water. Unless otherwise indicated, buffers are autoclaved, and those containing active enzymes can be prepared the day before the perfusion with a sterile filtered bottle-top 0.22 μM filter and stored at 4 °C until use. For liver perfusion studies, the reagents should be warmed to 37 °C to promote effective enzymatic dissociation. Tissues and cells at experiment end should be disposed of according to local health and safety regulations.

2.1 Hepatic Stellate Cells

For the studies described in this chapter, HSCs can be sourced as transformed cell lines, for example, human LX-2 [7], murine GRX [8], and rat HSC-T6 [9], or as primary murine, rat, or human HSCs isolated from perfused livers [10, 11].

1. Primary human, rat, and mouse HSCs are available from various commercial providers, such as Lonza and Sciencell. Cells should be cultured according to the protocols referenced above or as directed from each commercial provider (*see* **Note 1**). For the below procedures, rat primary HSCs are our cell of choice (*see* **Note 2**).
2. Sterility is required throughout, and the following reagents are required to isolate and grow primary rat HSCs: Dulbecco's modified Eagle media (DMEM), Pronase E, collagenase B, DNAse I (Roche), Histodenz (Merck), penicillin, streptomycin (Thermo Fisher Scientific), and fetal calf serum (Corning).

Supplement DMEM with heat inactivated fetal calf serum (FCS) (20%) and penicillin (100 U/mL).

3. 10× Gey's balanced salt solution (GBSS): For a liter solution, dissolve 2.2 g $CaCl_2{\cdot}2H_2O$, 3.7 g KCl, 0.3 g KH_2PO_4, 2.1 g $MgCl_2 \times 6H_2O$, 0.7 g $MgSO_4 \times 7H_2O$, 80 g NaCl, 2.27 g $NaHCO_3$, 1.2 g Na_2HPO_4, and 10 g D-Glucose. Sterile-filter and store at 4 °C, and dilute 1 in 10 in sterile water for final use. Remove $CaCl_2 \times 2H_2O$ and $MgCl_2 \times 6H_2O$ for Ca^{+2} and Mg^{2+} free GBSS.
4. Autoclaved surgical instruments are required, and include scissors, forceps, and scalpel blade and holder.
5. Trypan blue solution.
6. Rabbit anti-desmin and alpha smooth muscle actin antibodies.
7. For centrifugation, use laboratory-standard polypropylene conical bottom 15 and 50 mL plastic sterile centrifuge tubes.

2.2 Migration and Invasion Assays

1. Boyden chambers: Costar 3422, Transwell Permeable Supports 6.5 mM Insert, 8.0 polycarbonate membrane. They fit into a well of a standard 24-well cell culture plate.
2. Rat tail collagen, type 1.
3. Human monocyte chemotactic protein 1 (MCP-1).
4. Sterile phosphate-buffered saline.
5. Serum-free medium: DMEM without 20% FCS.
6. Mayer's hematoxylin solution.

2.3 Proliferation Assays

1. Proliferation ELISA, BrdU (colorimetric) kit.
2. Sterile cell culture clear 96-well plates.
3. Stop solution: 1 M H_2SO_4.

2.4 Apoptosis Assays

1. Caspase-3 activity assay kit.
2. Sterile Tris-buffered saline (TBS): Dissolve 6.05 g Tris and 8.76 g NaCl in 800 mL of H_2O. Adjust pH to 7.5 with 1 M HCl, and make volume up to 1 L with H_2O.
3. Tween 20.
4. TBST: TBS containing 0.05% Tween 20.
5. Protein block, serum-free.
6. 4′,6-Diamidino-2-phenylindole (DAPI).

3 Methods

3.1 Primary Rat HSC Isolation and Cell Culture

The isolation of rat HSCs from adult male Sprague-Dawley rat livers (weighing 400–500 g) by in situ perfusion and a single-step density gradient centrifugation are described below. Perform the perfusion and cell isolation under sterile conditions in a sterile laminar flow hood, and wipe down all centrifuges and surfaces with 70% ethanol beforehand.

1. Anesthetize the rat with ketamine (100 mg/kg body weight) and xylazine (20 mg/kg body weight), and shave the abdomen with electric grooming clippers.
2. Spray the abdomen with 70% ethanol; with surgical scissors, open the abdominal wall of the rat, and with the blunt side of the scissors, move the intestines to the side to expose the vena cava and portal vein.
3. To inhibit clotting, inject 1 mL of 1000 IU heparin into the inferior vena cava with a 25G needle.
4. Insert a fine 16–18-gauge cannula into the portal vein, and ligate into position with two silk ties (*see* **Note 3**).
5. Warm and maintain the enzymatic solutions at 37 °C in a clean water bath. Using a peristaltic pump, perfuse the liver firstly with 150 mL of Ca^{+2} and Mg^{2+} free GBSS, followed by 150 mL of Pronase solution (450 mg Pronase E) in GBSS and then 150 mL collagenase (40 mg Collagenase B) in GBSS at a flow rate of 20 mL/min.
6. Cut the vena cava to allow the liver to drain.
7. Remove the liver from the abdominal cavity to a sterile cell culture 100 mM dish.
8. Mince the liver with surgical scissors, and incubate with 0.2% Pronase E and 0.002% DNAse 1 in 50 mL of GBSS with agitation for 30 min at 37 °C.
9. To remove cell aggregates, filter the mixture through a nylon mesh filter (100 μM).
10. Centrifuge the liver cell suspension in a 50 mL plastic sterile centrifuge tube at 80 × g for 2 min to sediment the hepatocytes, and then remove the supernatant containing the residual cells.
11. Centrifuge the supernatant at 500 × g for 5 min to pellet the non-parenchymal cells.
12. Resuspend the non-parenchymal cells in 40 mL of DMEM culture medium, and centrifuge at 500 g for 5 min.
13. Resuspend the cells in 10 mL of GBSS, combined with 6 mL of GBSS containing 30% Histodenz (for an 11.25% final

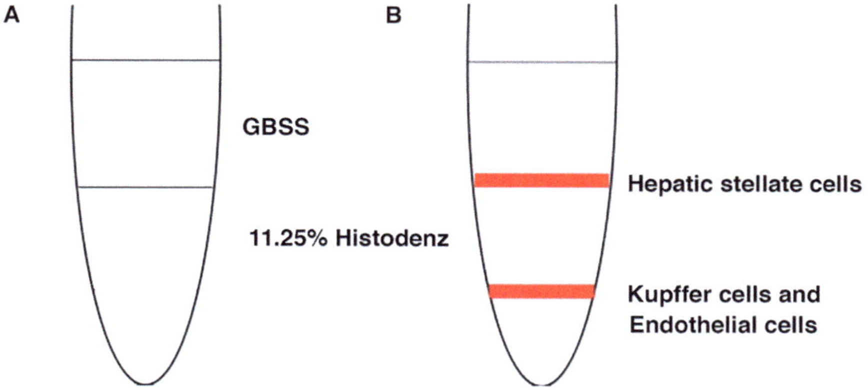

Fig. 1 Histodenz and cell suspension gradients before and after centrifugation. (**a**) The isolated cells are resuspended in GBSS containing Histodenz and layered with GBSS. (**b**) after centrifugation, the HSCs settle between the GBSS and 11.25% Histodenz, and the combination of Kupffer and Sinusoidal endothelial cells settles further down

gradient); transfer to a centrifuge tube and mix gently, and then layer 2 mL of GBSS on top of the cell suspension; centrifuge at 1400 × g for 21 min (low acceleration).

14. With a pipette, remove the thin cloudy layer of HSCs between the Histodenz and GBSS, wash twice with DMEM by centrifugation at 500 × g for 5 min, and seed in DMEM 20% FCS, containing penicillin and streptomycin (100 mg/mL; *see* Fig. 1 for detail).
15. Cell viability can be gauged by trypan blue exclusion and purity by anti-Desmin immunostaining and autofluorescence of the stored retinoids in the HSC lipid droplets.
16. Incubate the HSCs in 6-, 12-, or 96-well tissue culture plates at a density of 1.0×10^6, 0.5×10^6, or 3×10^3 cells per well, respectively, at 37 °C in a humidified atmosphere of air and 5% CO_2. Replenish the medium every second day. After being cultured for 5–7 days, the cells will become activated as evidenced by loss of vitamin A deposition and dendritic processes and positive staining for α-smooth muscle actin (*see* **Note 4**).

3.2 The HSC Migration Assay Using a Transwell Chamber

1. Prepare activated HSCs, and use Transwell chambers for the assay (*see* Fig. 2).
2. For siRNA or shRNA knockdown, treat the HSCs before undertaking the migration assay (*see* **Note 5**). Note that transfection efficiency for primary cells is typically lower than for cell lines: LX-2 are especially transfectable.
3. Fill the bottom of the 24-wells with 400 μL of 20% FCS DMEM.

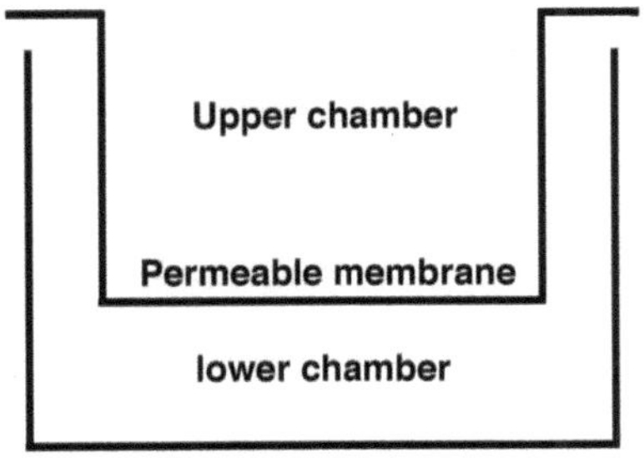

Fig. 2 Illustration of a Transwell chamber. The HSCs are placed in the inserts upper chamber on the membrane or collagen for migration and invasion assays, respectively. The inhibitors are supplied to the upper chamber to effect migration and invasion. A chemoattractant can be supplied in the lower chamber

4. Serum starve the HSCs for 24 h, and plate 200 μL of cells (2.5×10^4 cells per ml in serum-free medium) to the upper chamber (*see* **Note 6**).
5. Apply the inhibitory reagent to the upper chamber, and ensure a control well is run (*see* **Note 7**).
6. Incubate the chambers at 37 °C for 4–24 h to allow the migration of cells through the membrane into the lower chamber.
7. Fix the chamber with 4% paraformaldehyde in PBS, and stain with Mayer's hematoxylin (*see* **Note 8**).
8. Count six random fields using a phase-contrast microscope.

3.3 The HSC Migration Assay Using the Scratch-Wound Assay

1. Using 6-well plates, allow the primary rat HSCs to become activated and grow to confluency.
2. Draw a straight line with a thin permanent marker pen on the bottom of well.
3. Generate scratch wounds with a sterile 10 μL pipette tip.
4. Wash away the suspended cells with PBS.
5. Replace the media containing the respective inhibitors.
6. Measure wound closure after 0, 24, and 48 h with an inverted microscope equipped with a digital camera.
7. Use image software to measure the wounding differences.

3.4 HSC Invasion Assay

1. HSC invasion assays are a modification of a HSC Transwell migration assay and incorporate an ECM layer that sits on top of the permeable membrane, through which the HSCs in response to a chemoattractant must invade and migrate to the lower membrane side (*see* Fig. 2).

2. Coat the upper Transwell chamber with 50 μL of rat tail collagen I (2 mg/mL), and allow it to solidify for 1 h in a 37 °C cell incubator.
3. Use activated HSCs, and plate 5×10^4 cells in 100 μL of defined serum-free medium in the upper chamber of the Transwell chamber on top of the collagen (*see* **Note 6**).
4. The inhibitory reagents such as antibodies are supplied to the upper chamber containing the HSCs and collagen; ensure a control well is run (*see* **Note 7**).
5. To promote invasion and migration, add 400 μL of serum-free medium containing the chemoattractant Monocyte chemoattractant protein-1 (MCP-1) 200 ng/mL to the lower chamber of the 24-well plate.
6. Culture the HSCs for 48 h.
7. Remove the upper layer containing the collagen and unmigrated HSCs with a cotton swab.
8. Fixed the chamber with 4% paraformaldehyde in PBS, wash with sterile water, and stain with Mayer's hematoxylin (*see* **Note 8**).
9. Count six random fields using a phase-contrast microscope.

3.5 HSC Proliferation Assay

HSC proliferation can be evaluated with several commercial kits. We have previously used a Cell Proliferation ELISA, BrdU (colorimetric) kit from Roche (1164722901) that, depending on HSC cell number, can be applied in 6-, 12-, or 96-well format. The principle of the test is that BrdU (bromodeoxyuridine), a thymidine analogue during S phase of the cell cycle, is supplied to the cells and incorporates into the newly synthesized DNA. The now nuclear expressed BrdU is detected with a specific antibody with covalently bound peroxidase that in the presence of the chromogenic substrate 3,3′,5,5′-Tetramethylbenzidine (TMB) develops a permanent, insoluble, dark blue reaction product that can be quantified calorimetrically. siRNA or shRNA should be trialed beforehand in larger wells to qualify and quantify gene knockdown with supportive qPCR and Western blot analyses [12] (*see* **Note 9**).

1. Seed primary rat HSCs into 96-well plates at 3000 cells per well, and allow them to become activated (*see* **Note 10**).
2. Treat your cells with your compound of choice for a specific time, or use cells that have been previously treated with siRNA or shRNA.
3. At experiment end dilute the BrdU labelling reagent (1000×) at 1:100 in DMEM medium, and supply 10 μL to each well for 2 h.

4. Remove the labelling medium by gentle suction, and add 200 μL of the fixing solution (FixDenat) to the cells for 30 min.
5. Flick off the FixDenat solution, and tap the plate dry on tissue paper.
6. Dilute the anti-BrdU-peroxidase stock solution 1:100 in the supplied antibody dilution solution, add 100 μL to each well, and incubate for 30 min.
7. Remove the anti-BrdU-peroxidase working solution by suction, and wash the cells three times with PBS.
8. Add 100 μL of substrate solution to the cells for 30 min at room temperature.
9. Add 25 μL of the stop solution to each well, and measure the absorbance with an ELISA plate reader at 450 nM (reference wavelength, 690 nM).
10. For controls, use cells not incubated with the BrdU labelling reagent (*see* **Note 11**).

3.6 Cell Apoptosis Assay

A variety of apoptosis kits, through plate-based assays or visually for microscopy, can be used for gauging HSC apoptosis after a specific stimulus. For a plate-based assay, we have previously used a caspase-3 activity assay kit, which can be purchased from Merck (CASP3F) or Abcam (ab252897). The Caspase-3 activity assay kit (Fluorometric) employs a specific substrate, acetyl-Asp-Glu-Val-Asp-7-amino-4-trifluoromethyl coumarin (DEVD-AMC), which upon cleavage by active caspase-3 generates the highly fluorescent product release of 7-amino-4-methylcoumarin (AMC), which can be measured using excitation and emission wavelengths of 485 and 535 nM, respectively [13].

3.6.1 96-Well Assay

1. Prepare primary rat HSCs as described, and seed into 96-well plates at 3000 cells per well in quadruplicate; allow them to become activated.
2. Treat HSCs with an agent that induces cell death, and run controls, that is, untreated, or 1 mM hydrogen peroxide or staurosporine 1 μg/mL to induce apoptosis, for a specific time, 3–24 h.
3. By aspiration, remove the cell media, and wash the cells once with PBS.
4. The remaining cells are then lysed with the buffer provided in the kit, incubated on ice for 10 min, and then centrifuged at 10,000 g for 10 min.
5. Transfer equal volumes of the supernatant to a new 96-well plate, and add 50 μL of the reaction buffer and 10 μL of the Caspase 3 conjugate.

6. Mix by taping the plate, and incubate for 2 h at 37 °C.
7. With an advanced plate reader fluorimeter, measure the fluorescence immediately in kinetic mode every 10 min for up to 1 h at Ex/Em of 400/505 nM. Calculate the caspase 3 activity using a standard curve, as described below.
8. For fluorescence assays, the appropriate mode of measurement is required. Namely, if using white microplates with an opaque bottom, the plate reader should measure fluorescence by scanning the top. If using a white, clear bottom plate, the plate reader should measure fluorescence from underneath.

3.6.2 Standard Curve

1. To calculate caspase 3 activity, a standard curve is required to be prepared for each assay. This is best done as the assay is underway. Briefly, prepare an increasing series of AMC standards (25–250 nM) in lysis buffer, and measure the fluorescence (Ex/Em = 400/505 nM).
2. Draw a curve of the intensity values versus the concentration of the respective AMC solutions. For the actual samples, calculate the caspase 3 activity in nmole of AMC released per minute per ml of cell lysate or a positive control, based on the following formula:

$$\text{Activity, nmol AMC/ min/ml} = \text{nmole AMC} \times d/t \times v$$

v = sample volume in ml; d = dilution factor; t = reaction time in minutes.

3.6.3 Visual Apoptosis Assay

To observe apoptosis in primary rat HSCs, we have previously used the FAM-FLICA® caspase 3 and 7 staining kit (Immunochemistry Technologies) to see specific immunofluorescence of caspase 3 and 7 enzymes. The principle of each FLICA probe is that it contains a 3- or 4-amino acid sequence that is targeted by a different activated caspase. The recognition sequence sits between a green, fluorescent label, carboxyfluorescein (FAM) and a fluoromethylketone (FMK). The caspase enzyme cannot cleave the FLICA inhibitor probe but instead generates an irreversible bond with FMK at the specific target sequence and is also inhibited. We provide below a protocol adapted from the kit.

1. Plate 3×10^3 HSCs into each well of an 8-well glass chamber slides, and grow them until they are activated.
2. Expose the cells to an experimental treatment, and ensure that you have a control population of cells that receive a placebo and positive apoptosis control.
3. Treat the cells for the required time, and fix them in ice-cold acetone for 1 min.
4. Wash the cells with TBST three times for 5 min each.

5. Block with protein block, serum-free for 20 min.
6. Dilute the FLICA stock 1:50 in PBS, and apply to the cells for 1 h, protected from light.
7. Wash the cells three times with TBST, followed by a single wash with apoptosis wash buffer.
8. Incubate the sections with DAPI, dilute 1:1000 in PBS for 5 min, and wash three times with PBS.
9. Remove plastic chamber piece and sealer with strong forceps (*see* **Note 12**).
10. Mount a glass cover slip with a histology mounting media that limits fluorescence degradation.
11. Take images with a fluorescent microscope or laser scanning confocal microscope.
12. Quantify the number of apoptotic cells.

4 Notes

1. An important point to consider in the choice of transformed HSCs versus primary HSCs for your assays: for primary HSCs, their isolation is expensive and time-consuming and must be repeated many times to attain sufficient experimental replicates. One should also be mindful that the transformed cell lines are activated and hence differentiated and cannot replace the better functionality of primary HSCs.
2. For isolating your own primary HSCs, institutional human or animal ethics is required, and considerable experience in tissue perfusing and primary cell culture is necessary.
3. We recommend using surgical loups for this step.
4. For primary cells, they can be directly supplied to the plastic cell culture wells and become activated 5–7 days later prior to undertaking specific assays.
5. The use of siRNA or shRNA should be trialed beforehand in larger wells to qualify and quantify gene knockdown with supportive qPCR and Western blot analyses [12]. The cells are re-seeded into the Transwell chamber after the optimal time of gene knockdown has been established.
6. For migration and invasion assays, the primary rat HSCs are activated in 6-well plates over 5–7 days. On the assay day, the HSCs are removed by washing with a standard trypsin EDTA solution, counted, and resuspended in the media of choice.
7. Controls could include an untreated well or an appropriate control to ensure experimental integrity.

8. A simple way to fix the cells is to fill a well of a 24-well plate with 500 μL of 4% PFA and place the Transwell chamber in the well for 10 min, wash in water, and then stain with Mayer's hematoxylin in another well for 2–5 min; wash with water a few times. For microscopy, leave the Transwell in the empty 24-well plate, and focus on the underside of the membrane to observe the stained HSCs.
9. As per **Note 5**, siRNA or shRNA should be trialed beforehand, but the optimal time of knockdown should be found so that the BrdU assay best interprets the relevance of your gene of interest.
10. With proliferation assays, the number and viability of the primary rat HSCs will determine how many cells will be present after 5–7 days. Primary rat HSCs can also be grown in larger wells and, after activation is reached, transferred to 96 wells.
11. Controls can include inhibitors of proliferation. A blanking background cell-free well must also be used.
12. A scalpel can be used to remove the plastic glue so that the cover slip seals properly.

Acknowledgments

This work was supported by a grant from Tropical Australian Academic Health Centre (SF0000121) that is supported by the National Health and Medical Research Council (NHMRC) to LH.

References

1. Subramanian P, Hampe J, Tacke F, Chavakis T (2022) Fibrogenic pathways in metabolic dysfunction associated fatty liver disease (MAFLD). Int J Mol Sci 23(13). https://doi.org/10.3390/ijms23136996
2. Peiseler M, Schwabe R, Hampe J, Kubes P, Heikenwalder M, Tacke F (2022) Immune mechanisms linking metabolic injury to inflammation and fibrosis in fatty liver disease - novel insights into cellular communication circuits. J Hepatol 77(4):1136–1160. https://doi.org/10.1016/j.jhep.2022.06.012
3. Tsuchida T, Friedman SL (2017) Mechanisms of hepatic stellate cell activation. Nat Rev Gastroenterol Hepatol 14(7):397–411. https://doi.org/10.1038/nrgastro.2017.38
4. Singal AG, El-Serag HB (2022) Rational HCC screening approaches for patients with NAFLD. J Hepatol 76(1):195–201. https://doi.org/10.1016/j.jhep.2021.08.028
5. Younossi ZM (2019) Non-alcoholic fatty liver disease – a global public health perspective. J Hepatol 70(3):531–544. https://doi.org/10.1016/j.jhep.2018.10.033
6. Senzolo M, Garcia-Tsao G, Garcia-Pagan JC (2021) Current knowledge and management of portal vein thrombosis in cirrhosis. J Hepatol 75(2):442–453. https://doi.org/10.1016/j.jhep.2021.04.029
7. Xu L, Hui AY, Albanis E, Arthur MJ, O'Byrne SM, Blaner WS et al (2005) Human hepatic stellate cell lines, LX-1 and LX-2: new tools for analysis of hepatic fibrosis. Gut 54(1):142–151. https://doi.org/10.1136/gut.2004.042127
8. Borojevic R, Monteiro AN, Vinhas SA, Domont GB, Mourao PA, Emonard H et al (1985) Establishment of a continuous cell line from fibrotic schistosomal granulomas in mice livers. In Vitro Cell Dev Biol 21(7):382–390. https://doi.org/10.1007/BF02623469

9. Nanda I, Steinlein C, Haaf T, Buhl EM, Grimm DG, Friedman SL et al (2022) Genetic characterization of rat hepatic stellate cell line HSC-T6 for in vitro cell line authentication. Cell 11(11). https://doi.org/10.3390/cells11111783
10. Friedman SL, Roll FJ (1987) Isolation and culture of hepatic lipocytes, Kupffer cells, and sinusoidal endothelial cells by density gradient centrifugation with Stractan. Anal Biochem 161(1):207–218. https://doi.org/10.1016/0003-2697(87)90673-7
11. Wang J, Leclercq I, Brymora JM, Xu N, Ramezani-Moghadam M, London RM et al (2009) Kupffer cells mediate leptin-induced liver fibrosis. Gastroenterology 137(2): 713–723. https://doi.org/10.1053/j.gastro.2009.04.011
12. Ramezani-Moghadam M, Wang J, Ho V, Iseli TJ, Alzahrani B, Xu A et al (2015) Adiponectin reduces hepatic stellate cell migration by promoting tissue inhibitor of metalloproteinase-1 (TIMP-1) secretion. J Biol Chem 290(9): 5533–5542. https://doi.org/10.1074/jbc.M114.598011
13. Alzahrani B, Iseli T, Ramezani-Moghadam M, Ho V, Wankell M, Sun EJ et al (2018) The role of AdipoR1 and AdipoR2 in liver fibrosis. Biochim Biophys Acta 1864(3):700–708. https://doi.org/10.1016/j.bbadis.2017.12.012

Chapter 4

Phalloidin Staining for F-Actin in Hepatic Stellate Cells

Sarah K. Schröder, Carmen G. Tag, Sabine Weiskirchen, and Ralf Weiskirchen

Abstract

During the development of liver fibrosis, hepatic stellate cells undergo a transition from a quiescent phenotype into a proliferative, fibrogenic, and contractile, α-smooth muscle actin-positive myofibroblast. These cells acquire properties that are strongly associated with the reorganization of the actin cytoskeleton. Actin possesses a unique ability to polymerize into filamentous actin (F-actin) form its monomeric globular state (G-actin). F-actin can form robust actin bundles and cytoskeletal networks by interacting with a number of actin-binding proteins that provide important mechanical and structural support for a multitude of cellular processes including intracellular transport, cell motility, polarity, cell shape, gene regulation, and signal transduction. Therefore, stains with actin-specific antibodies and phalloidin conjugates for actin staining are widely used to visualize actin structures in myofibroblasts. Here we present an optimized protocol for F-actin staining for hepatic stellate cells using a fluorescent phalloidin.

Key words Hepatic stellate cells, Phalloidin, Staining, F-actin, Liver fibrosis

1 Introduction

Actin is one of the most abundant proteins in eukaryotic cells, and its amino acid sequence is evolutionarily conserved [1]. One characteristic of actin is its highly dynamic turnover between a monomeric (G-actin) and a polymeric filamentous state (F-actin), which forms robust filaments. These are composed of two twisted helices with a diameter of ~5 to 9 nM [2]. In mammals, there exist at least six actin isoforms that play essential roles in many cellular processes, including cell proliferation, differentiation, migration, contraction, focal adhesion maturation, extracellular matrix reorganization, gene regulation, signal transduction, and many others [1, 3]. The respective genes, *Acta1*, *Acta2*, *Actb*, *Actc1*, *Actg1*, and *Actg2*, encode isoforms that possess very similar amino acid sequence, with no isoform sharing less than 93% identity with any other isoform [4]. However, these isoforms cannot completely substitute for each another in vivo, indicating that the highly conserved actins

Ralf Weiskirchen and Scott L. Friedman (eds.), *Hepatic Stellate Cells: Methods and Protocols*, Methods in Molecular Biology, vol. 2669, https://doi.org/10.1007/978-1-0716-3207-9_4,

are functionally specialized for the tissues in which they predominate [5]. Globular actin will polymerize into filamentous actin whenever its concentration is greater than its critical concentration, a process that can be modified by numerous solution conditions [6].

During hepatic fibrosis, α-smooth muscle actin (α-SMA) encoded by the *Acta2* gene is significantly upregulated in hepatic stellate cells (HSCs), promoting myofibroblast contraction and migration [7–9]. Although *Acta2* null mice are born at Mendelian ratios and are viable, they show significant defects in vascular contractility and blood pressure regulation [10]. α-SMA transmits mechanical signals to the nucleus to regulate collagen type I expression in HSC, while the lack of *Acta2* leads to decreased liver fibrosis in vivo [9]. Importantly, α-SMA expression is a reliable marker of HSC activation, which precedes fibrous tissue deposition, rendering this cytoskeletal protein as a reliable marker to identify the earliest stages of hepatic fibrosis and to monitor the efficacy of potential antifibrotic therapies [11, 12].

There are many possibilities to quantitate α-SMA in HSC or in liver extracts. Most common is the semi-quantitative measurement by Western blot analysis that is done in many laboratories to document the transdifferentiation of HSC in culture or the increased expression in fibrotic liver tissue [13, 14]. In addition, species-specific α-SMA ELISA kits for mouse, rat, human, and many other species are commercially available. Nevertheless, the distribution of actin within cells can be best visualized by fluorescent phalloidin conjugates.

Phalloidin (i.e., bicyclic(Ala-DThr-Cys-*cis*-4-hydroxy-Pro-Ala-2-mercapto-Trp-4,5-dihydroxy-Leu)(S-3 → 6) is a bicyclic, water-soluble heptapeptide. It can be isolated from the poisonous green death cap mushroom (*Amanita phalloides*), which produces two families of toxic peptides, namely, the amatoxins and the phallotoxins (Fig. 1). Amatoxins are by far the more toxic toxins of the mushroom. They form a tight complex with eukaryotic RNA polymerases II, thereby inhibiting transcription, while phallotoxins bind to F-actin, disrupt plasma membranes, and cause massive efflux of calcium and potassium [16].

Phallotoxins possess a high binding affinity for the grooves between F-actin subunits over monomeric G-actin. Bound to F-actin, it prevents ATP hydrolysis, thereby shifting the equilibrium of monomers to filaments toward the filaments position. Based on these properties, phalloidin and its fluorescent derivatives have become useful compounds to stain and visualize F-actin filaments in different sample types including formaldehyde-fixed and permeabilized tissue sections, cell cultures, and cell-free experiments. In particular, the direct conjugation with fluorochrome to phalloidin makes it an excellent experimental tool for labeling filamentous actin and studying actin structures in eukaryotic cells [17]. In

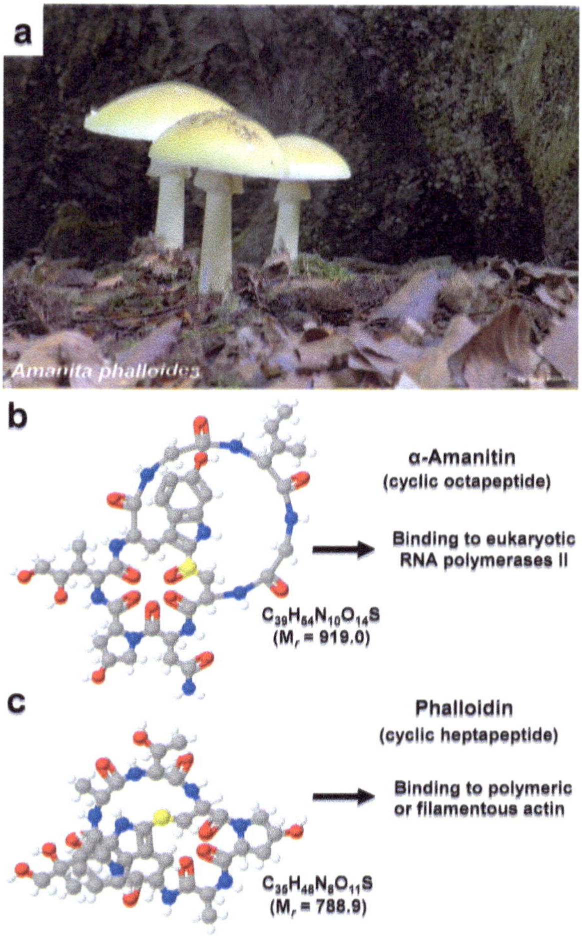

Fig. 1 α-Amanitin and phalloidin. (**a**) The green death cap, *Amanita phalloides*, is a poisonous basidiomycete fungus that is widespread in Europe. It produces two families of toxic peptides, the amatoxins and the phallotoxins (Image courtesy of Thomas Bauder). (**b**) The mycotoxin α-amanitin ($C_{39}H_{54}N_{10}014S$, $M_r = 919.0$, CAS-no.: 23109-05-9) is a cyclic octapeptide that can form a tight complex with eukaryotic RNA polymerases, thereby interfering with transcription. (**c**) On the contrary, phalloidin ($C_{35}H_{48}N_8O_{11}S$, $M_r = 788.9$, CAS-no.: 17466-45-4) is a cyclic heptapeptide that binds to filamentous actin, thereby preventing depolymerization and enhancing the stability of respective filaments. Both mycotoxins possess a distinctive tryptophan-cysteine crossbridge (i.e., a tryptathionine) that makes these compounds highly stable and resistant to proteases of the gastric tract and several hydroxylated amino acids [15]

particular, staining with phalloidin is favorable over α-SMA antibody stains, because antibodies recognize both monomer and polymer actin and hence tend to produce a high background during cell staining compared to probes that only have affinity for F-actin.

We here describe a simple but effective phalloidin staining protocol that is suitable to label actin filaments in cultured HSCs.

2 Materials

In the following protocol, we describe the staining procedure to visualize actin filaments (F-actin) by using fluorescence-labeled phalloidin probes. It is important for successful staining to use cells that grow adherent, such as hepatic stellate cells (HSC), which are important drivers of liver fibrosis. The procedure described herein for staining F-actin is adapted to the use of immortalized (hepatic stellate) cell lines but can also be successfully used in a modified form for primary rodent HSC [for isolation of primary HSC, see [18–20]. All volumes mentioned in this protocol are adapted to the use of 21 × 26 mm (1 mm corresponds to 0.0394 inches) coverslips in 6-well plates and can be adjusted by the user depending on the plate size. The materials belonging to the basic equipment of a laboratory (such as tips and pipettes) are assumed. Unless otherwise stated, all steps were performed in room temperature. The basic workflow for phalloidin staining is displayed in Fig. 2.

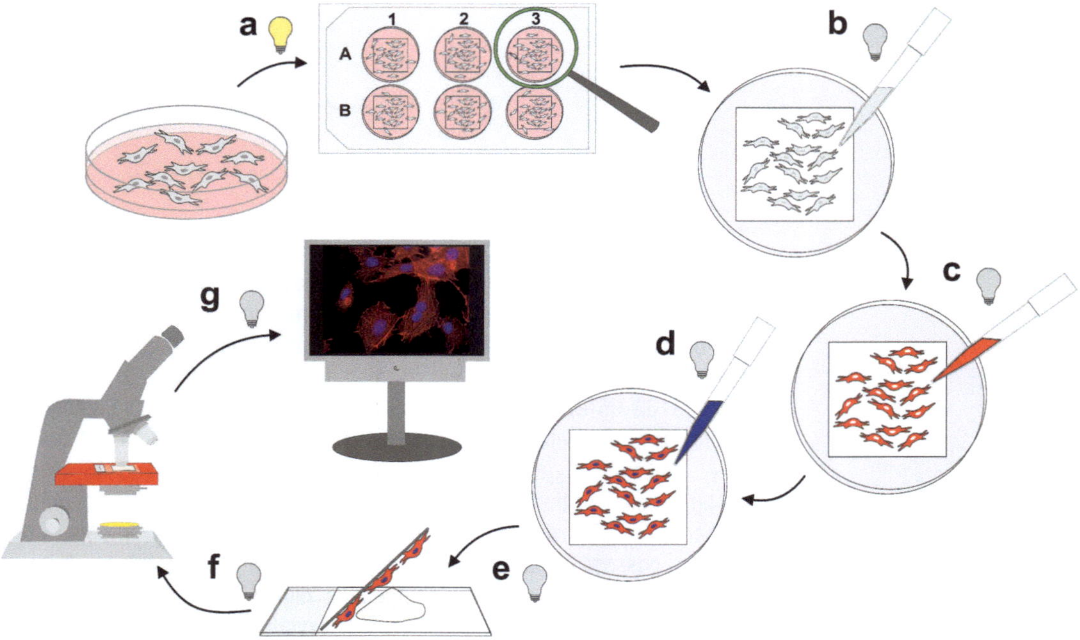

Fig. 2 Workflow for phalloidin stain of cultured cells. The phalloidin stain consists of seven different steps including (**a**) seeding of cells on glass coverslips, (**b**) fixing the cells in 3.7% paraformaldehyde and permeabilizing, (**c**) staining with a suitable fluorescently-labeled phalloidin conjugate, (**d**) nuclear counterstain with DAPI, (**e**) mounting the cells in an aqueous mounting medium on a microscopy slide, (**f**) analysis of stained cells with a fluorescence microscope, and finally (**g**) documentation of stained cells. To obtain optimal results and to avoid photobleaching of fluorescent dyes, steps (**b**)–(**g**) should be performed in the dark

2.1 Reagents and Equipment for Pretreatment of Glass Coverslips

1. Coverslips (e.g., 21 × 26 mm, when using 6-well plates).
2. Sealing film (e.g., Parafilm®).
3. Sterile distilled water (dH_2O).
4. 25% (vol/vol) HCl.
5. 70% Ethanol.
6. Sterile plastic box (to store cleaned coverslips).
7. Tweezers (sterile, cleaned with ethanol).
8. Horizontal shaker.

2.2 Cell Culture Reagents and Equipment

1. Adherent growing hepatic stellate cells.
2. Complete media (e.g., Dulbecco's modified Eagle medium with 4.5 g/L glucose supplemented with 10% fetal bovine serum (FBS), 1 mM sodium pyruvate, 4 mM L-glutamine, 100 μg/mL streptomycin sulfate, and 100 U/mL Penicillin G.
3. Solution to detach cells (e.g., Accutase®).
4. Sterile, cleaned glass coverslips (*see* Subheading 2.1).
5. Sterile PBS (1×).

2.3 Reagents and Equipment for Staining

1. Fixation solution: 3.7% paraformaldehyde (PFA) in phosphate-buffered saline PBS (buffered with phosphoric acid to pH 7.4).
2. Permeabilization solution: 0.1% sodium citrate/0.1% Triton-X-100 in PBS (precooled).
3. Fluorescence-conjugated phalloidin probes (e.g., rhodamine phalloidin R415, Thermo Fisher) or Alexa Fluor™ 488 Phalloidin A12379, Thermo Fisher) (*see* **Notes 1–5**).
4. 200 ng/mL (4′,6-diamidino-2-phenylindole) DAPI in PBS (for nuclear counterstaining).
5. Aqueous Mounting Medium (e.g., PermaFluor™).
6. Sterile PBS (1×).
7. PBS containing 1% bovine serum albumin (BSA).
8. Sterile dH_2O.
9. Microscope slides.
10. Fluorescence microscope (e.g., Nikon Eclipse E80i fluorescence microscope with NIS-Elements Vis software).
11. Optional: blocking solution—50% FBS/0.5% BSA in PBS.

3 Methods

3.1 Pretreatment of Glass Coverslips

1. Before glass coverslips can be used in cell culture, they need to be acid-washed and cleaned with ethanol. Therefore, place desired amount of glass coverslips in a glass dish, and cover with 25% HCl. Seal the dish (with Parafilm®), and incubate on a horizontal shaker overnight.
2. Remove the 25% HCl solution, and dispose it properly.
3. Rinse the coverslips under running tap water for 10 min.
4. Wash the coverslips in sterile water 3 times for 5 min.
5. Place the coverslips in 70% ethanol 3 times for 5 min, and store them in 70% ethanol (well-sealed with Parafilm®) in a box at 4 ° C until use.

3.2 Seeding Cells on Glass Coverslips

All reagents used for cell culture should be sterile or should be sterilized using the appropriate procedure (autoclave or sterile filter). In addition, a sterile working area is required for sterile handling (disinfection with 70% ethanol) and working under a cell culture hood to minimize the probability of contaminations. Unless otherwise stated, all steps were performed in room temperature. For seeding freshly isolated primary rodent HSC, *see* **Note 6**.

1. The cleaned coverslips (*see* Subheading 3.1, **steps 1–5**) were transferred to sterile 6-well plate and placed vertical for drying. After 10 min, coverslips were placed horizontally and washed with 1 mL sterile PBS.
2. Optional: For better cell adherence on the coverslips, they can be covered with cell culture medium (containing 10–20% FCS) or coated with poly-L-lysin (*see* **Note 7**) prior seeding the cells. If medium is used for covering the plate, this is stored in the incubator for 24 h. Aspirate the medium after 24 h, and continue with **step 3**.
3. Detach the cells to be stained as usual by using, e.g., Accutase® solution. In order to seed a defined number of cells, the cells should be counted using a hemocytometer or an automatized cell-counting device. Seed 250,000–400,000 cells/well (*see* **Note 8**) in appropriate standard cell culture medium in a 6-well plate, and let them grow to 60–70% confluence.

3.3 Fixation and Permeabilization

1. Carefully remove medium from the cells grown on coverslips in 6-well plates.
2. Wash the cells three times with 1 mL PBS (1×) for 3 min (*see* **Note 9**).

3. Fixate the cells in each well with 1 mL 3.7% PFA in PBS (buffered with phosphoric acid to pH 7.4) for 20 min in the dark (*see* **Notes 10** and **11**).
4. Wash the cells three times with 1 mL PBS (1×) for 3 min.
5. Cover the cells in each well with 1 mL precooled 0.1% sodium citrate/0.1% Triton-X-100 for 3 min on ice to permeabilize (*see* **Note 12**).
6. Wash the cells three times with 1 mL PBS (1×) for 3 min.
7. Optional: To minimize unspecific background staining, cells can be covered with 1 mL 50% FCS/0.5% BSA in PBS for 1 h (on a horizontal shaker, with low agitation, *see* **Note 13** for primary HSC).

3.4 Phalloidin Staining and Nuclear Counterstaining

All the following steps have to be performed in the dark as fluorescence-conjugated phalloidin probes are light sensitive.

1. Prepare the phalloidin stock solution according to the manufacturer's instructions. When using rhodamine phalloidin (No. R415) or Alexa Fluor™ 488 Phalloidin (No. A12379) from Invitrogen/Thermo Fisher, the final stock solution (40×) is made by adding 1.5 mL methanol to the vial containing 300 Units (*see* **Note 14**). As recommended by the manufacturer, this solution can be stored at −20 °C until the expiration date. Handle phalloidin probes with care as they are toxic.
2. Prepare the phalloidin working solution to stain the cells. Calculate 12.5 μL phalloidin stock solution (40×) in 500 μL PBS solution, containing 1% BSA per well. Incubate the cells with the phalloidin working solution for 20 min in the dark.
3. Wash the cells three times with 1 mL PBS (1×) for 3 min (optional: *see* **Notes 15** and **16**).
4. Prepare a 200 ng/mL DAPI solution in PBS. Apply 500 μL DAPI solution per well, and incubate for 30 min in the dark (*see* **Note 17**).
5. Wash the cells three times with 1 mL PBS (1×) for 3 min.
6. Rinse the cells once with 1 mL dH_2O.
7. Finally, place on drop of aqueous mounting medium (e.g., PermaFluor™) on a microscopy slide, and then carefully transfer the coverslips with tweezers from the 6-well plate to the slide (cell side down).
8. Let the slides dry for several hours, and document results with a fluorescence microscope. Typical results of phalloidin staining for immortalized rat HSC lines and primary mouse HSC are shown in Fig. 3. Based on the fluorochrome conjugated to the phallotoxin, appropriate filters must be used in microscopic evaluation. In our laboratory, we have made good experience

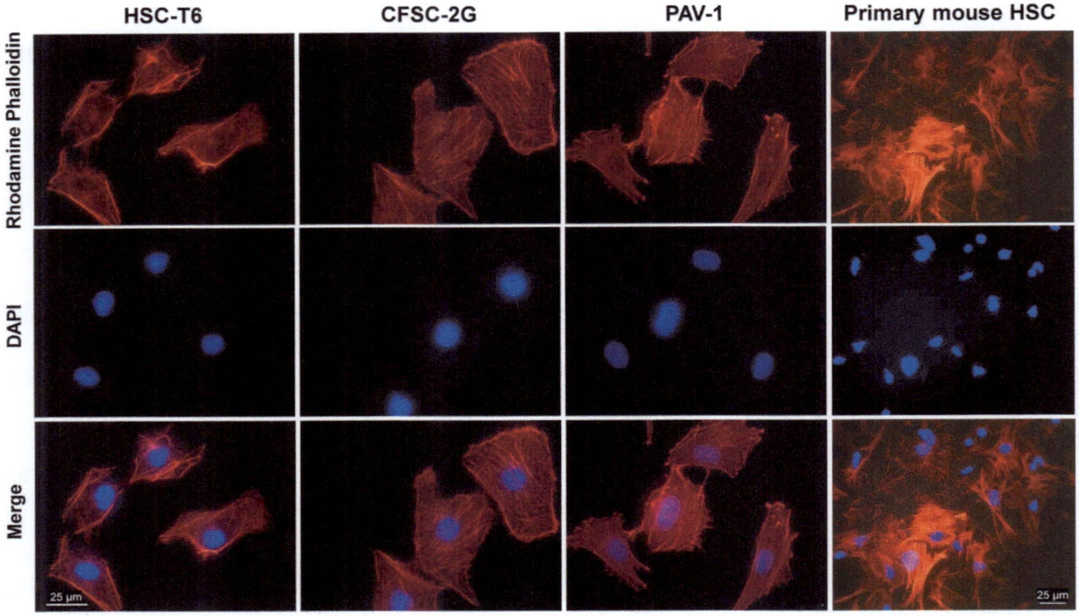

Fig. 3 Representative phalloidin stains in selected hepatic stellate cells. Different immortalized rat hepatic stellate cell lines (HSC-T6, CFSC-2G, and PAV-1) and primary mouse HSC cultured for 4 days were stained with rhodamine phalloidin (*red*). Nuclei were counterstained with DAPI (*blue*). Images were taken either with a Nikon Eclipse E80i fluorescence microscope at 600× magnification (HSC-T6, CFSC-2G, and PAV-1) or alternatively with a Leica DMLB fluorescence microscope at 400× magnification (primary mouse HSC). As previously shown, all these cells express large quantities of *Acta2* [21–24]

with the rhodamine phalloidin reagent that emits red fluorescence and Alexa Fluor™ 488 Phalloidin conjugate that emits green fluorescence (Fig. 4).

9. Slides can be stored at 4 °C in a dark chamber (*see* **Note 18**).

4 Notes

1. Phalloidin is toxic and should be handled with extreme caution. The LD_{50} of phalloidin (when administered via intraperitoneal injection) is approximately 1.9 mg/kg when injected into mouse [25]. Phalloidin is less toxic than α-amanitin but faster acting. Poisoning with phalloidin can be fatal in 1–2 h, while amanitin poisoning may be delayed for 12–24 h.
2. Several features of phalloidin are required to bind to F-actin. However, the side chain of amino acid 7 (g-d-dihyroxyleucine) is accessible for chemical modification without appreciable loss of affinity for actin.
3. Besides conventional fluorescent labels (rhodamine, Alexa Fluor, FITC, etc.), there are phalloidin conjugates available

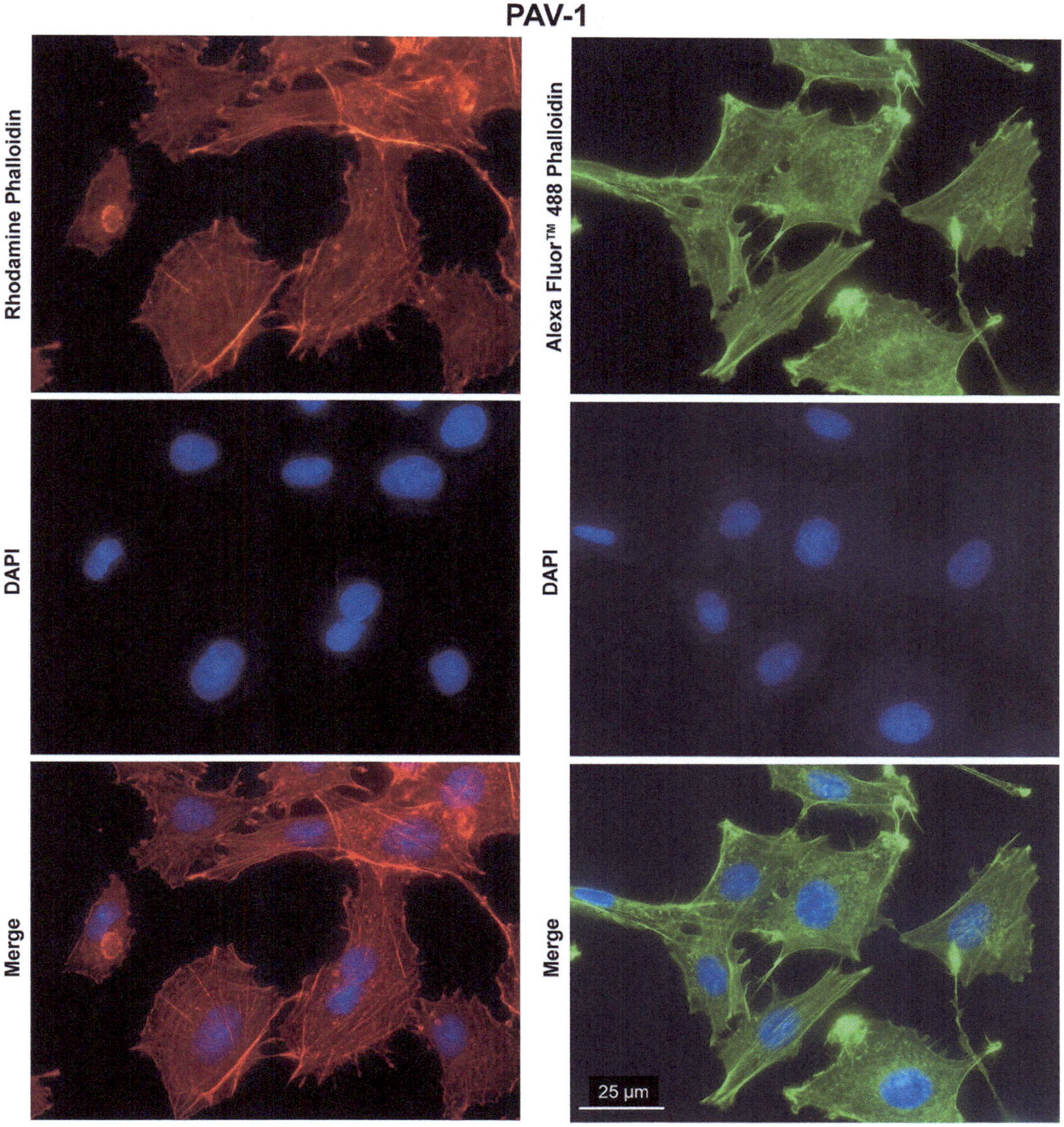

Fig. 4 Representative examples of phalloidin conjugates. PAV-1 cells were stained with rhodamine phalloidin (*left row*) or alternatively with Alexa Fluor™ 488 Phalloidin (*right row*). Nuclei were counterstained with DAPI and images taken with a Nikon Eclipse E80i fluorescence microscope at 600× magnification

for super-resolution microscopy and special imaging techniques. In these conjugates, phalloidin is coupled to cyanine-based fluorescent dyes (i.e., CF® dyes) that are superior in regard to brightness, photostability, and photochemical switching properties.

4. To prepare a stock solution of the phalloidin conjugate, dissolve the vial contents in methanol or DMSO (as recommend

by the manufacturer), and store the solution at −20 °C while protecting from light.

5. Phalloidin is pH sensitive. At elevated pH, the thioether bridge, which links the two cycles, is cleaved, and the affinity of phalloidin for actin is lost.
6. Seed around 200,000 primary rodent hepatic stellate cells in a 6-well plate on poly-L-lysine-coated glass coverslips.
7. For poly-L-lysine coating, flame coverslips with 70% ethanol to sterilize. Coat the coverslips with a 0.01% poly-L-lysine solution in Hanks' balanced salt solution (HBSS) w/o Ca^{2+}/Mg^{2+} (filtered sterile) for 15 min at room temperature. Afterwards, wash for 3 times with HBSS w/o Ca^{2+}/Mg^{2+}, let dry, and store at 4 °C until use.
8. It is necessary to find the optimal density for your cells. If cells are to confluent, F-actin stress fibers strongly overlap.
9. For primary HSC, the washing should be performed with cold HBSS without Ca^{2+}/Mg^{2+} instead of PBS in all steps.
10. Always use a methanol or acetone-free fixative, as these reagents can disrupt F-actin structure and prevent staining. It is best to use PFA as a fixative because it retains the quaternary protein structure of actin, which is necessary for high affinity binding.
11. The optimal condition to fixate primary HSC is to use 4% cold PFA (buffered with phosphoric acid to pH 7.4) as a fixative. Fixation should be done in the dark on ice.
12. The optimal condition to permeabilize primary HSC is to use ice-cold 0.2% Triton-X-100 in PBS for 4 min on ice.
13. For primary HSC, the blocking should be done with 3% BSA in PBS for 5 min.
14. The packing size of respective phalloidin conjugates is 300 units, in which 1 unit of fluorescently-labeled phalloidin is defined as the amount of material used to stain one microscope slide of fixed cells.
15. Phalloidin can be used in combination with antibodies suitable for immunofluorescence. In case you want to combine stainings, add phalloidin derivate to primary or secondary antibodies.
16. In addition, phalloidin staining can be used in combination with lipid droplet staining. Suitable dyes to detected lipid droplets such as BODIPY™ 493/503 or Nile Red [26] can be added (*see* Subheading 3.3; apply between **steps 3** and **4**).
17. Instead of using DAPI solution first, followed by mounting with aqueous mounting medium, there are anti-fade mounting media with already included DAPI. These can be used if

photobleaching has been previously excluded, which shortens the staining protocol.

18. Even robust fluorescent dyes can be photo-bleached ("fading") by intensive microscope light sources. This is caused by cleaving of covalent bonds or the occurrence of nonspecific reactions between the fluorophore and surrounding molecule. So care should be taken to prevent prolonged light exposure. It is best to store the phalloidin-stained cells in a suitable mounting medium at 4 °C in the dark.

Acknowledgments

Work from Dr. Weiskirchen's laboratory is supported by grants from the German Research Foundation (WE2554/13-1, WE2554/15-1, WE2554/17-1) and a grant from the Interdisciplinary Centre for Clinical Research within the faculty of Medicine at the RWTH Aachen University (grant PTD 1-5). The authors are grateful to Scott L. Friedman, Marcus Rojkind, and Patrick Sauvant for providing the established rat hepatic stellate cell lines HSC-T6, CFSC-2G, and PAV-1. We also thank Thomas Bauder for providing an image of the green death cap *Amanita phalloides*.

References

1. Dominguez R, Holmes KC (2011) Actin structure and function. Annu Rev Biophys 40: 169–186. https://doi.org/10.1146/annurev-biophys-042910-155359
2. Melak M, Plessner M, Grosse R (2017) Actin visualization at a glance. J Cell Sci 130(3): 525–530. https://doi.org/10.1242/jcs.189068
3. Sandbo N, Dulin N (2011) Actin cytoskeleton in myofibroblast differentiation: ultrastructure defining form and driving function. Transl Res 158(4):181–196. https://doi.org/10.1016/j.trsl.2011.05.004
4. Perrin BJ, Ervasti JM (2010) The actin gene family: function follows isoform. Cytoskeleton (Hoboken) 67(10):630–634. https://doi.org/10.1002/cm.20475
5. Khaitlina SY (2001) Functional specificity of actin isoforms. Int Rev Cytol 202:35–98. https://doi.org/10.1016/s0074-7696(01)02003-4
6. Doolittle LK, Rosen MK, Padrick SB (2013) Measurement and analysis of in vitro actin polymerization. Methods Mol Biol 1046: 273–293. https://doi.org/10.1007/978-1-62703-538-5_16
7. Ramadori G, Veit T, Schwögler S, Dienes HP, Knittel T, Rieder H, Meyer zum Büschenfelde KH. (1999) Expression of the gene of the alpha-smooth muscle-actin isoform in rat liver and in rat fat-storing (ITO) cells. Virchows Arch B Cell Pathol Incl Mol Pathol 59(6): 349–357. https://doi.org/10.1007/BF02899424
8. Rockey DC, Weymouth N, Shi Z. (2013) Smooth muscle α actin (Acta2) and myofibroblast function during hepatic wound healing. PLoS One 8(10):e77166. https://doi.org/10.1371/journal.pone.0077166
9. Rockey DC, Du Q, Weymouth ND, Shi Z (2019) Smooth muscle α-actin deficiency leads to decreased liver fibrosis via impaired cytoskeletal signaling in hepatic stellate cells. Am J Pathol 189(11):2209–2220. https://doi.org/10.1016/j.ajpath.2019.07.019
10. Schildmeyer LA, Braun R, Taffet G, Debiasi M, Burns AE, Bradley A, Schwartz RJ (2000) Impaired vascular contractility and blood pressure homeostasis in the smooth muscle alpha-actin null mouse. FASEB J 14(14):2213–2220. https://doi.org/10.1096/fj.99-0927com
11. Weiskirchen R, Weiskirchen S, Tacke F (2019) Organ and tissue fibrosis: molecular signals,

cellular mechanisms and translational implications. Mol Asp Med 65:2–15. https://doi.org/10.1016/j.mam.2018.06.003

12. Acharya P, Chouhan K, Weiskirchen S, Weiskirchen R (2021) Cellular mechanisms of liver fibrosis. Front Pharmacol 12:671640. https://doi.org/10.3389/fphar.2021.671640
13. Herrmann J, Borkham-Kamphorst E, Haas U, Van de Leur E, Fraga MF, Esteller M, Gressner AM, Weiskirchen R (2006) The expression of CSRP2 encoding the LIM domain protein CRP2 is mediated by TGF-beta in smooth muscle and hepatic stellate cells. Biochem Biophys Res Commun 345(4):1526–1535. https://doi.org/10.1016/j.bbrc.2006.05.076
14. Tag CG, Sauer-Lehnen S, Weiskirchen S, Borkham-Kamphorst E, Tolba RH, Tacke F, Weiskirchen R (2015) Bile duct ligation in mice: induction of inflammatory liver injury and fibrosis by obstructive cholestasis. J Vis Exp 96:52438. https://doi.org/10.3791/52438
15. Wong JH (2013) Chapter 25 – fungal toxins. In: Handbook of biological active peptides, 2nd edn. Elsevier, pp 166–168. https://doi.org/10.1016/B978-0-12-385095-9.00025-7
16. Schneider SM, Wiegang TJ (2017) Chapter 66: toxic mushroom ingestions. In: Auerbach's widlerness medicine, 7th edn. Elsevier, pp 1464–1490.e3. ISBN: 978-0323359429
17. Wulf E, Deboben A, Bautz FA, Faulstich H, Wieland T (1979) Fluorescent phallotoxin, a tool for the visualization of cellular actin. Proc Natl Acad Sci USA 76(9):4498–4502. https://doi.org/10.1073/pnas.76.9.4498
18. Weiskirchen R, Gressner AM (2005) Isolation and culture of hepatic stellate cells. Methods Mol Med 117:99–113. https://doi.org/10.1385/1-59259-940-0:099
19. Bartneck M, Warzecha KT, Tag CG, Sauer-Lehnen S, Heymann F, Trautwein C, Weiskirchen R, Tacke F (2015) Isolation and time lapse microscopy of highly pure hepatic stellate cells. Anal Cell Pathol (Amst) 2015:417023. https://doi.org/10.1155/2015/417023
20. Weiskirchen S, Tag CG, Sauer-Lehnen S, Tacke F, Weiskirchen R (2017) Isolation and culture of primary murine hepatic stellate cells. Methods Mol Biol 1627:165–191. https://doi.org/10.1007/978-1-4939-7113-8_11
21. Nanda I, Steinlein C, Haaf T, Buhl EM, Grimm DG, Friedman SL, Meurer SK, Schröder SK, Weiskirchen R (2022) Genetic characterization of rat hepatic stellate cell line HSC-T6 for in vitro cell line authentication. Cell 11(11):1783. https://doi.org/10.3390/cells11111783
22. Nanda I, Schröder SK, Steinlein C, Haaf T, Buhl EM, Grimm DG, Weiskirchen R (2022) Rat hepatic stellate cell line CFSC-2G: genetic markers and short tandem repeat profile useful for cell line authentication. Cell 11(18):2900. https://doi.org/10.3390/cells11182900
23. Sauvant P, Sapin V, Abergel A, Schmidt CK, Blanchon L, Alexandre-Gouabau MC, Rosenbaum J, Bommelaer G, Rock E, Dastugue B, Nau H, Azaïs-Braesco V (2002) PAV-1, a new rat hepatic stellate cell line converts retinol into retinoic acid, a process altered by ethanol. Int J Biochem Cell Biol 34(8):1017–1029. https://doi.org/10.1016/s1357-2725(02)00023-7
24. Krenkel O, Hundertmark J, Ritz TP, Weiskirchen R, Tacke F (2019) Single cell RNA sequencing identifies subsets of hepatic stellate cells and myofibroblasts in liver fibrosis. Cell 8(5):503. https://doi.org/10.3390/cells8050503
25. Listenberger LL, Brown DA (2007) Fluorescent detection of lipid droplets and associated proteins. Curr Protoc Cell Biol. Chapter 24: Unit 24.2. https://doi.org/10.1002/0471143030.cb2402s35
26. Wieland T (1963) Chemical and toxicological studies with cyclopeptides of *amanita phalloides*. Pure Appl Chem 6(3):339–350. https://doi.org/10.1351/pac196306030339

Chapter 5

Retinyl Ester Analysis by Orbitrap Mass Spectrometry

Jeroen W. A. Jansen, Maya W. Haaker, Esther A. Zaal, and J. Bernd Helms

Abstract

Retinoids are light-sensitive molecules that are normally detected by UV absorption techniques. Here we describe the identification and quantification of retinyl ester species by high-resolution mass spectrometry. Retinyl esters are extracted by the method of Bligh and Dyer and subsequently separated by HPLC in runs of 40 min. The retinyl esters are identified and quantified by mass spectrometry analysis. This procedure enables the highly sensitive detection and characterization of retinyl esters in biological samples such as hepatic stellate cells.

Key words Retinyl esters, Retinol, Vitamin A, Mass spectrometry, Orbitrap, Atmospheric pressure chemical ionization, High-performance liquid chromatography

1 Introduction

Vitamin A or retinol is an essential nutrient that must be acquired from the diet [1, 2]. Upon uptake of carotenoids like carotenes and β-cryptoxanthin, they can be converted to retinol via retinaldehyde [3]. Retinol and its derivatives are collectively called retinoids, with multiple isoforms of retinaldehyde and retinoic acid as the main biological active molecules. Retinoids are important for stem cell functions, reproduction, cell differentiation during embryogenesis and normal growth, cell metabolism in many different cell types, vision, and the immune system and play a role in the development of metabolic syndrome [4, 5].

Given the important roles of retinoids, it may not come as a surprise that unlike most other vitamins, their precursor molecule retinol can be stored within the body in relatively high levels to protect against the adverse effects of temporary insufficient dietary intake of vitamin A [6]. Retinol that is present in the body is stored as retinyl esters in specialized lipid droplets. The main storage places in the body are the liver and the lung, and low amounts of retinyl esters are also found in other tissues such as the eye, kidney, pancreas, adipose tissue, muscle, and brain [7–9].

Ralf Weiskirchen and Scott L. Friedman (eds.), *Hepatic Stellate Cells: Methods and Protocols*, Methods in Molecular Biology, vol. 2669, https://doi.org/10.1007/978-1-0716-3207-9_5,

Two different enzymatic activities have been implicated in the esterification of retinol to retinyl esters. The enzyme responsible for most of the retinol esterification is lecithin/retinol acyltransferase (LRAT), which transfers the acyl-chain from the *sn*-1 position of phosphatidylcholine to retinol via a trans-esterification reaction [10]. The other enzymatic activity is an acyl-coenzyme A (Acyl-CoA)/retinol acyltransferase (ARAT) reaction. Acyl-CoA/diacylglycerol acyltransferase 1 (DGAT1) has been shown to contain ARAT activity as well [11].

As a result of these different reaction mechanisms, the retinyl ester species that are synthesized by these enzymes are different. In the liver, LRAT is the main enzyme responsible for storage of retinyl esters in a subset of liver cells named hepatic stellate cells (HSCs) [8, 9, 12]. Retinyl palmitate is the main retinyl ester synthesized in the liver and in HSCs, with lower amounts of retinyl oleate and retinyl stearate also present [9, 13]. In the absence of LRAT, the total retinyl ester content decreases, and retinyl palmitate is no longer the main retinyl ester. Other retinyl ester species such as retinyl oleate are relatively more abundant in the absence of LRAT [9, 13]. Thus, the retinyl species composition is indicative of the enzymatic activity involved in the storage of retinyl esters in lipid droplets.

For the analysis of retinyl ester species in cells and tissues, several methods have been developed. These methods are complicated by the light-sensitive nature of retinoids [14], so appropriate measures must be taken to avoid light exposure. Most methods are based on reverse-phase HPLC, due to the hydrophobic nature of retinyl esters [15–17]. The introduction of mass spectrometry allowed refinement of these methods by combining HPLC with mass spectrometry (LC-MS), facilitating the identification of specific retinyl ester species and improving the sensitivity. Further refinement was achieved by high-resolution mass spectrometry, allowing the specific monitoring of multiple predefined retinyl ester species [13].

Here we describe a highly sensitive LC-MS method that has been optimized to analyze the retinyl palmitate and retinyl oleate species. The ratio of these two retinyl ester species identifies the enzymatic activity involved in their synthesis, providing insight in the molecular mechanism of retinyl ester synthesis and storage. This method can easily be expanded to include other retinyl ester species as well.

2 Materials

2.1 Laboratory Equipment

1. Workspace with yellow/red light (*see* **Notes 1** and **2**).
2. Chemical fume hood.

3. Vortex mixer.
4. Centrifuge with relative centrifugal force of 2000 RCF.
5. Nitrogen evaporator unit.
6. (U)HPLC system fitted with column oven and cooled autosampler (e.g., Vanquish, Thermo Fisher Scientific).
7. High-end Orbitrap mass spectrometer interfaced with APCI (e.g., Q-Exactive HF (Thermo Fisher Scientific).
8. HALO 90 Å C8 Column 2.7 μm, 3.0 × 150 mm (Advanced Technologies).
9. Kinetex SecurityGuard Ultra C8, 3.0 mm with holder.
10. Gas: high-purity N_2 (for drying samples and atmospheric pressure ionization).

2.2 Basic Consumables and Chemicals

1. Borosilicate amber glass tubes (16 × 150 mm, 15 mL) with Teflon-lined caps (*see* **Notes 3** and **4**).
2. Borosilicate glass Pasteur pipettes (long).
3. Amber autosampler vials and caps with Teflon-liner.
4. Pipette tips.
5. Vial racks.
6. Vial trays.
7. Milli-Q-grade deionized water (18 MΩ) or bottled HPLC-grade water.
8. Formic acid (100%) (LC-MS Grade) (*see* **Note 5**).
9. Chloroform (LC-MS Grade).
10. Methanol (LC-MS Grade).
11. Ethanol (LC-MS Grade).
12. Acetonitrile (LC-MS Grade).
13. Acetone (LC-MS Grade).
14. Butylated-hydroxytoluene.

3 Methods

To avoid photodecomposition [14], the extraction of retinoids from serum, cells, and tissues must be carried out rapidly "in the dark" (*see* **Note 1**). Laboratory windows should be covered with appropriate materials such as aluminum foil. A room with no windows is the ideal setting to perform this procedure. Artificial lighting should be provided by preferably a red or alternatively yellow light bulb. During the extraction procedures, butylated-hydroxytoluene is added to avoid free radical-mediated oxidation and photoisomerization. After extraction, retinoids must be stored in amber glass tubes.

In the liver, retinyl esters are stored in hepatic stellate cells. Procedures for the isolation of primary hepatic stellate cells from mouse liver have been described elsewhere [18]. In short, after liver perfusion, primary hepatic stellate cells were isolated from livers of mice by pronase/collagenase digestion followed by Nycodenz density gradient centrifugation. Due to the abundant presence of retinyl ester-containing lipid droplets, primary hepatic stellate cells have a lower buoyant density than other liver cells and float on the Nycodenz layer. The isolated hepatic stellate cells were cultured for 1 day on plastic culture dishes as described [19], before retinyl ester analysis was performed.

3.1 Preparation of Solutions (see Note 6)

1. Chloroform/methanol (1:2, v/v): In a clean 0.5-liter borosilicate bottle with Teflon-lined cap, add 100 mL chloroform to 200 mL methanol, and mix.
2. Methanol/acetonitrile/chloroform/water (46:20:17:17, v/v/v/v): In a clean 0.5-liter borosilicate bottle with Teflon-lined cap, add 230 mL methanol to 100 mL acetonitrile, 85 mL chloroform, and 85 mL Milli-Q water, and mix.
3. Solvent A (95% acetonitrile, 5% Milli-Q water, 0.1% formic Acid): To 950 mL of acetonitrile, add 50 mL Milli-Q water and 1 mL formic acid (100%). De-gas by sonication in a sonifier bath for 5 min.
4. Solvent B (85% acetone, 15% chloroform, 0.1% formic acid): To 850 mL of acetone, add 150 mL chloroform and 1 mL formic acid (100%). De-gas by sonication in a sonifier bath for 5 min.

3.2 Preparation of Standards

Prepare stock solutions of 1 mg/mL by dissolving 1 mg of the desired retinyl ester species in 1 mL of ethanol or chloroform.

3.3 Sample Extraction

Hepatic stellate cell samples are extracted "in the dark" using red light by Bligh and Dyer extraction [20].

1. Defrost the samples on ice (*see* **Note 7**).
2. Transfer 800 μL cell suspension in phosphate-buffered saline (PBS) to an amber pointed glass tube with Teflon-lined cap.
3. Add internal standards and butylated hydroxytoluene (*see* **Note 8** and **9**).
4. Add 3 mL of chloroform/methanol (1:2) (v/v) (*see* **Notes 8** and **9**).
5. Shake or vortex vigorously for 30 s.

6. Incubate for 20 min on room temperature in a rotational shaker (*see* **Note 10**).
7. Centrifuge for 5 min at 2000 RCF.
8. Transfer supernatant to clean amber pointed glass tube with Teflon lined cap (*see* **Note 11**).
9. Add 2 mL Milli-Q water to the supernatant.
10. Add 2 mL $CHCl_3$ to the supernatant.
11. Shake or vortex vigorously for 30 s.
12. Centrifuge for 5 min at 2000 RCF.
13. Transfer lower (organic) phase to a clean amber pointed glass tube with Teflon-lined cap (*see* **Note 12**).
14. To the remaining upper (water) phase, add 2 mL $CHCl_3$.
15. Shake or vortex vigorously for 30 s.
16. Centrifuge for 5 min at 2000 RCF.
17. Combine the lower (organic) phase with the previous lower phase.
18. Discard the remaining upper (water) phase.
19. Evaporate combined organic phase under a stream of nitrogen gas (*see* **Note 13**).

3.4 LC-MS Analysis

1. Set up the LC-MS system with APCI source, cooled autosampler, column oven, and reversed phase C8 column fitted with Ultra Guard column (*see* **Note 14**).
2. Set column oven temperature to 30 °C.
3. Purge LC solvent lines A and B for 5 min with solvent A and B, respectively.
4. Equilibrate the system for 10 min at 10% solvent B at a flow of 300 μL/min.
5. Set up the following method in your LCMS software: Flow of the LC, 300 μL/min; injection =10 μL sample on column.
6. The following gradient is applied with a total run time of 40 min:

Time (min)	Solvent A (%)	Solvent B (%)
0	90	10
15	90	10
35	0	100
35.1	90	10
40	90	10

7. Operate the mass spectrometer in atmospheric pressure chemical ionization (APCI) mode with spray voltage set to 3000 V (*see* **Notes 15** and **16**).
8. Capillary temperature is set to 325 °C; Probe Heater Temperature is set to 350 °C. (*see* **Note 17**).
9. Auxiliary gas flow rate is set to 20 AU and sheath gas flow rate to 60 AU.
10. Set the MS mode to full scan mode starting from 250 to 650 m/z with a scan resolution of 120 K.
11. Prepare samples by dissolving in methanol/acetonitrile/chloroform/water (46:20:17:17, v/v/v/v), and transfer to amber autosampler vials (*see* **Notes 18** and **19**).
12. Start the run with a blank (methanol/acetonitrile/chloroform/water (46:20:17:17, v/v/v/v) to equilibrate the system, followed by the samples in randomized order. End with a blank.
13. Shut down system according to factory manual.

3.5 Data Analysis

1. Extract and analyze data with processing software, for example, TraceFinder (Thermo Fisher Scientific).
2. The m/z values of relevant retinyl ester species by MS analysis are shown in Table 1.
3. LC-MS analysis of retinoids with APCI source results in source fragmentation of retinoids, generating a characteristic retinol fragment [M-H2O]+ (m/z 269.2264). The extracted peak areas of these retinol fragments are used for quantification of the retinyl ester species (Figs. 1 and 2).
4. The different retinyl esters are separated by retention time and are identified based on the [M+H]+ ions (Fig. 1).
5. LC-MS analysis of hepatic stellate cells identifies the presence of retinyl palmitate, retinyl oleate, and retinyl stearate (Fig. 3).

Table 1
m/z values of specific retinoids

Analyte	Molecular formula	Exact mass	[M+H]+	[M-$H_2$0]+
Retinyl palmitate	$C_{36}H_{60}O_2$	524.45933	525.4666	507.4560
Retinyl oleate	$C_{38}H_{62}O_2$	550.47498	551.4823	533.4717
Retinyl stearate	$C_{38}H_{64}O_2$	552.49063	553.4979	535.4873
Retinol fragment	$C_{20}H_{30}O$	286.22967	287.2369	269.2264
Retinol acetate	$C_{22}H_{32}O_2$	328.24023	329.2475	311.2369

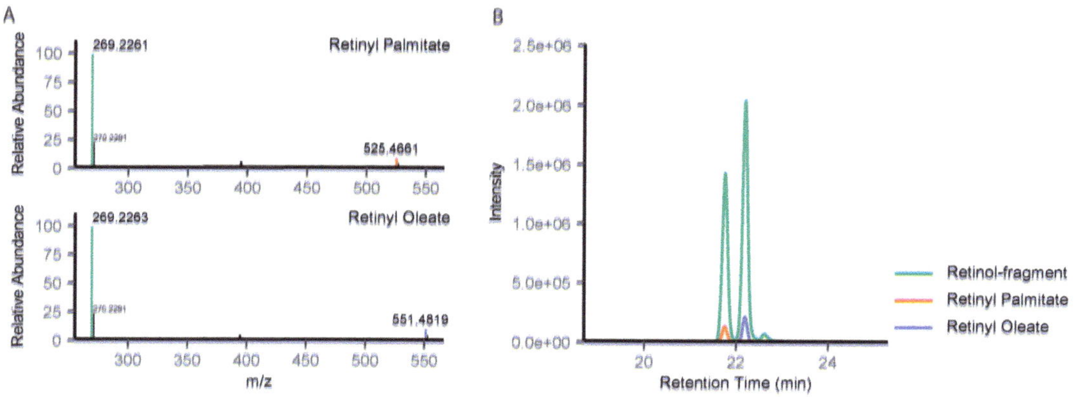

Fig. 1 Lipidomic spectra of retinyl ester standards. (**a**) Mass spectrum of retinyl palmitate (*top*) and retinyl oleate (*bottom*). (**b**) Extracted ion chromatograms of 269.2264, 525.4661, and 551.4823 (within 5 ppm), respectively (*see* also **Note 20**)

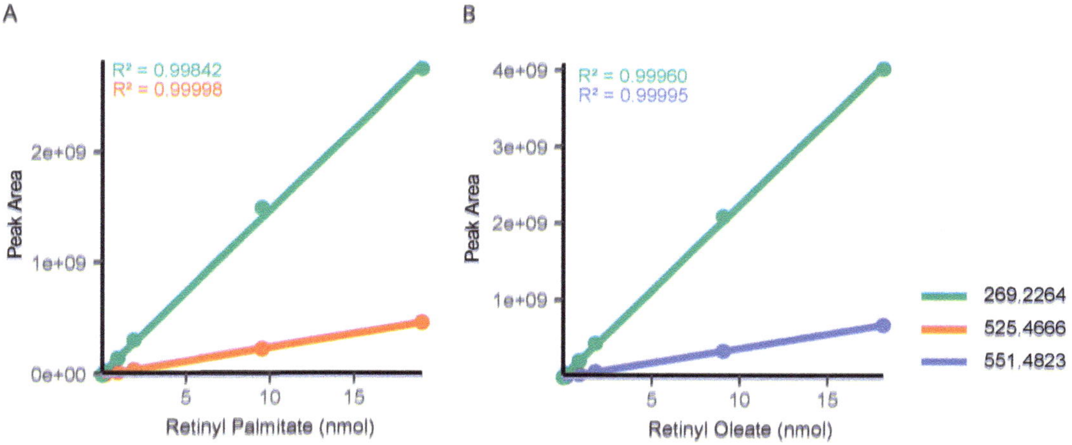

Fig. 2 Calibration curves of retinyl palmitate and retinyl oleate (for details, *see* **Notes 21** and **22**)

The extracted peak areas of the retinol fragments are directly proportional to the relative abundance of the individual retinyl ester species in biological samples (Fig. 2). For quantification, *see* **Notes 23–25**.

4 Notes

1. Exposure of retinoids to full-spectrum (white) light (regular room lights) should be avoided, even for brief periods of time. Noticeable degradation takes already place in ~10 min!

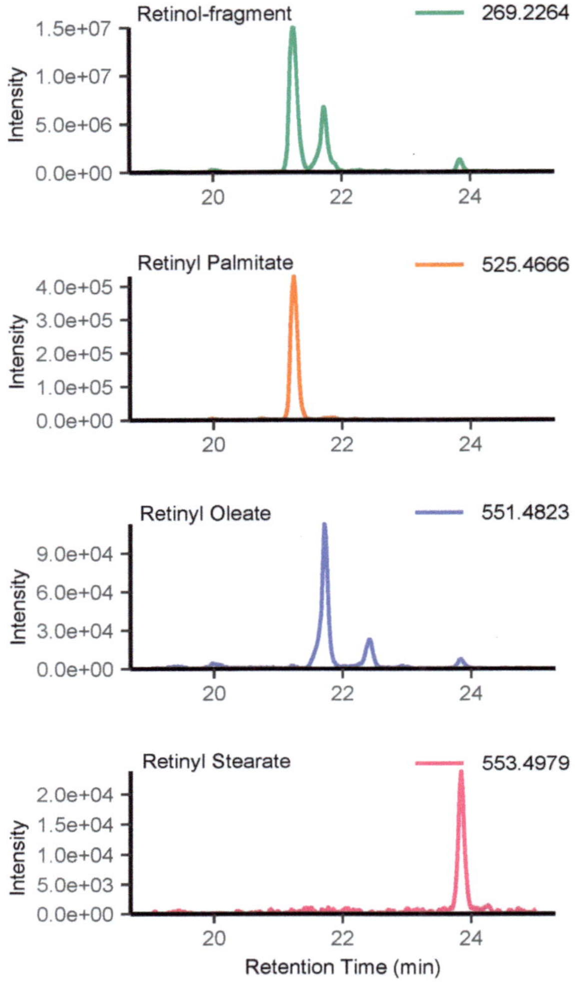

Fig. 3 LC-MS analysis of retinyl ester species in hepatic stellate cells. Example of extracted ion chromatograms of the retinol fragment (m/z 269.2264) in hepatic stellate cells (*top*) that were used for quantification of different retinyl ester species. Identification of retinyl ester species is based on the parent ions. Full scan monitoring allows monitoring of additional retinol species, i.e., retinol stearate (m/z 553.4979, *bottom*)

2. If you do not have a room with overhead yellow lights, a desk lamp outfitted with a yellow light bulb can be used in a darkened room.
3. The use of amber glassware minimizes exposure to full-spectrum light.
4. To prevent adhesion of retinoids to glassware or plastics, it is recommended to use silanized glassware.

5. Formic acid is caustic and irritant. Wear personal protective equipment (gloves and eye protection), and handle carefully.
6. Handling and preparation of organic solvents should be performed in a fume cabinet wearing chemical resistant gloves. Chloroform is a possible carcinogen in humans, so take preventive measures.
7. Frozen samples should be completely thawed (on ice) for efficient retinoid extraction.
8. Add 1 nmoL retinyl acetate and 5 nmoL of a non-biological cholesterol-d7 to monitor extraction efficiency and to determine photodegradation (*see* **Notes 23–25**). The amounts of these standards can be adjusted in subsequent experiments to be preferably in the same range as the amounts of retinyl ester and cholesterol expected in the samples.
9. Add 10 nmoL butylated hydroxytoluene to prevent oxidation and photoisomerization of retinyl esters to be quantified.
10. Make sure the caps tightly seal the tubes to prevent leakage of fluids.
11. Carefully decant supernatant into clean tube, or transfer with use of a Pasteur pipette.
12. Be careful not to disturb the protein interface between the organic (lower) phase and water (upper) phase. To transfer the lower phase more accurately, a Plastipak (3 mL) syringe adapted on a glass Pasteur pipette can be used. Apply a slight positive pressure when inserting the pipette through the upper phase into the lower phase. Replace the Pasteur pipette for every new sample to prevent cross contamination.
13. To speed up the process, you can also make use of a water bath (37 °C).
14. To prolong the lifetime of the analytical column, a guard column is advised.
15. (H)ESI (heated electro-spray ionization) is more susceptible to ion suppression effects than APCI and therefore less desirable for quantification.
16. Linear range of APCI (<10 pmoL–20 nmoL) is greater than (H)ESI (Fig. 4).
17. The listed source conditions are guidelines. Optimize conditions for your instrument/conditions. Tuning of instrumental conditions is essential to obtain sufficient sensitivity for the analysis of biological samples.
18. For better peak separation, methanol/acetonitrile/chloroform/Milli-Q (46:20:17:17) is used instead of the commonly used chloroform/methanol (1:1) to dissolve LC-MS standards/samples.

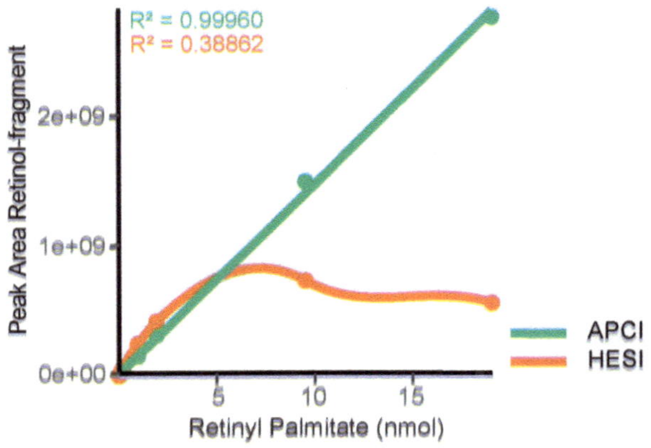

Fig. 4 Linear range of APCI vs HESI

19. For low volumes (<100 μL), use glass inserts in autosampler vials.
20. Due to in-source fragmentation, the fragment ion m/z 269.2264 is much more abundant than the precursor ions m/z 525.4661 (retinol palmitate) and m/z 551.4819 (retinol oleate). Therefore, the fragment ion m/z 269.2264 is more suitable for quantification, and the precursor ions m/z 525.4661 and m/z 551.4819 are used for identification (Fig. 1).
21. Confirm retention times and peak identity with authentic standards frequently.
22. Limitation of maximum amount is the solvation factor of retinoids in organic solvents.
23. For absolute quantitation, a standard curve is made with a stock solution containing retinyl palmitate or retinyl oleate solution (Fig. 2).
24. The extraction recovery can be determined by addition of the internal standard retinyl acetate to the samples prior to extraction. Retinyl acetate is a non-biological retinyl ester that also generates the characteristic retinol fragment [M-H_2O]+ (m/z 269.2264).
25. Potential problems with photodegradation of retinyl esters can be identified by addition of the internal standards retinyl acetate and cholesterol-d7 retinyl acetate to the samples prior to extraction. Retinyl acetate is sensitive to photodegradation, whereas cholesterol-d7 is not sensitive to photodegradation. The ratio of the retinyl acetate-derived retinol fragment [M-H_2O]+ (m/z 269.2264) and a characteristic cholesterol-d7 fragment (protonated cholestadiene) of *m/z* 376.3955 must remain constant in all samples.

References

1. Sommer A (2008) Vitamin a deficiency and clinical disease: an historical overview. J Nutr 138:1835–1839
2. Blaner WS, Li Y, Brun P-J, Yuen JJ, Lee S-A, Clugston RD (2016) Vitamin A absorption, storage and mobilization. In: Asson-Batres MA, Rochette-Egly C (eds) The biochemistry of retinoid signaling II: the physiology of vitamin A – uptake, transport, metabolism and signaling, Subcellular biochemistry. Springer, Dordrecht, pp 95–125. https://doi.org/10.1007/978-94-024-0945-1_4
3. D'Ambrosio DN, Clugston RD, Blaner WS (2011) Vitamin A metabolism: an update. Nutrients 3:63–103
4. Haaker MW, Vaandrager AB, Helms JB (2020) Retinoids in health and disease: a role for hepatic stellate cells in affecting retinoid levels. Biochim Biophys Acta Mol Cell Biol Lipids 1865:158674
5. Gudas LJ (2022) Retinoid metabolism: new insights. J Mol Endocrinol 69:T37–T49
6. O'Byrne SM, Blaner WS (2013) Retinol and retinyl esters: biochemistry and physiology. J Lipid Res 54:1731–1743
7. Goodman DW, Huang HS, Shiratori T (1965) Tissue distribution and metabolism of newly absorbed vitamin a in the rat. J Lipid Res 6: 390–396
8. Liu L, Gudas LJ (2005) Disruption of the lecithin:retinol acyltransferase gene makes mice more susceptible to vitamin A deficiency. J Biol Chem 280:40226–40234
9. O'Byrne SM, Wongsiriroj N, Libien J, Vogel S, Goldberg IJ, Baehr W, Palczewski K, Blaner WS (2005) Retinoid absorption and storage is impaired in mice lacking lecithin:retinol acyltransferase (LRAT). J Biol Chem 280:35647–35657
10. Horchani H, Bussières S, Cantin L, Lhor M, Laliberté-Gemme J-S, Breton R, Salesse C (2014) Enzymatic activity of lecithin:retinol acyltransferase: a thermostable and highly active enzyme with a likely mode of interfacial activation. Biochim Biophys Acta 1844:1128–1136
11. Yen C-LE, Monetti M, Burri BJ, Farese RV (2005) The triacylglycerol synthesis enzyme DGAT1 also catalyzes the synthesis of diacylglycerols, waxes, and retinyl esters. J Lipid Res 46:1502–1511
12. Molenaar MR, Yadav KK, Toulmay A, Wassenaar TA, Mari MC, Caillon L, Chorlay A, Lukmantara IE, Haaker MW, Wubbolts RW, Houweling M, Vaandrager AB, Prieur X, Reggiori F, Choudhary V, Yang H, Schneiter R, Thiam AR, Prinz WA, Helms JB (2021) Retinyl esters form lipid droplets independently of triacylglycerol and seipin. J Cell Biol 220(19):e202011071
13. Ajat M, Molenaar M, Brouwers JFHM, Vaandrager AB, Houweling M, Helms JB (2017) Hepatic stellate cells retain the capacity to synthesize retinyl esters and to store neutral lipids in small lipid droplets in the absence of LRAT. Biochim Biophys Acta 1862:176–187
14. Tolleson WH, Cherng S-H, Xia Q, Boudreau M, Yin JJ, Wamer WG, Howard PC, Yu H, Fu PP (2005) Photodecomposition and Phototoxicity of natural Retinoids. Int J Environ Res Public Health 2:147–155
15. Blaner WS, Obunike JC, Kurlandsky SB, al-Haideri M, Piantedosi R, Deckelbaum RJ, Goldberg IJ (1994) Lipoprotein lipase hydrolysis of retinyl ester. Possible implications for retinoid uptake by cells. J Biol Chem 269: 16559–16565
16. Guo X, Gudas LJ (1998) Metabolism of all-trans-retinol in normal human cell strains and squamous cell carcinoma (SCC) lines from the oral cavity and skin: reduced esterification of retinol in SCC lines. Cancer Res 58:166–176
17. Kim Y-K, Quadro L (2010) Reverse-phase high-performance liquid chromatography (HPLC) analysis of retinol and Retinyl esters in mouse serum and tissues. Methods Mol Biol 652:263–275
18. Mederacke I, Dapito DH, Affò S, Uchinami H, Schwabe RF (2015) High-yield and high-purity isolation of hepatic stellate cells from normal and fibrotic mouse livers. Nat Protoc 10:305–315
19. Tuohetahuntila M, Molenaar MR, Spee B, Brouwers JF, Houweling M, Vaandrager AB, Helms JB (2016) ATGL and DGAT1 are involved in the turnover of newly synthesized triacylglycerols in hepatic stellate cells. J Lipid Res 57:1162–1174
20. Bligh EG, Dyer WJ (1959) A rapid method of total lipid extraction and purification. Can J Biochem Physiol 37:911–917

Chapter 6

Studying Hepatic Stellate Cell Senescence

Sandra A. Serna-Salas, Abel A. Soto-Gámez, Zongmei Wu, Myrthe Klaver, and Han Moshage

Abstract

Hepatic stellate cells (HSCs) are the key effector cells in liver fibrosis. They are the main producers of excessive amounts of extracellular matrix components during fibrogenesis and therefore a potential target for the treatment of liver fibrosis. Induction of senescence in HSCs may be a promising strategy to slow down, stop, or even reverse fibrogenesis. Senescence is a complex and heterogeneous process linked to fibrosis and cancer, but the exact mechanism and relevant markers can be cell-type dependent. Therefore, many markers of senescence have been proposed, and many methods to detect senescence have been developed. In this chapter, we review relevant methods and biomarkers to detect cellular senescence in hepatic stellate cells.

Key words Senescence, Fibrosis, Biomarkers

Abbreviations

36B4	Ribosomal phosphoprotein P0
AMV	Avian myeloblastosis virus
ARF	ADP-ribosylation factor
Bcl2	B cell lymphoma 2
BrdU	5′-bromo-2′-deoxyuridine
CDK	Cyclin-dependent kinase
CDKN1a/P21^{CIP1}	Cyclin-dependent kinase inhibitor 1a
Cip	CDK interacting protein
CT	Threshold cycle
CXCL1	Chemokine ligand 1
CXCL10	Chemokine ligand 10
DAPI	4′,6-diamidino-2-phenylindole
DcR2	Decoy receptor 2
DMF	Dimethylformamide
DMSO	Dimethyl sulfoxide
DNA	Desoxyribonucleic acid
ELISA	Enzyme-linked immunosorbent assays

Ralf Weiskirchen and Scott L. Friedman (eds.), *Hepatic Stellate Cells: Methods and Protocols*, Methods in Molecular Biology, vol. 2669, https://doi.org/10.1007/978-1-0716-3207-9_6,

EVs	Extracellular vesicles
GAPDH	Glyceraldehyde-3-phosphate dehydrogenase
GFAP	Glial fibrillar acid protein
H3K9	Histone H3 lysine 9
HRP	Horseradish peroxidase
HSCs	Hepatic stellate cells
IGF	Insulin growth factor
IL6	Interleukin 6
IL8	Interleukin 8
INK4	Cyclin-dependent kinase inhibitors 4
Ki67	Antigen K167
Kip	Kinase inhibitory protein
LMNB1	Lamin B1
LRAT	Lecithin retinol acyltransferase
Mg	Milligrams
M-MLV	Moloney murine leukemia virus
MMP3	Metalloproteinase 3
MMP9	Metalloproteinase 9
MMPs	Metalloproteinases
P53	Tumor suppressor protein
PBS	Phosphate-buffered saline
PDGF	Platelet-derived growth factor
PFA	Paraformaldehyde
PI	Propidium iodide
qPCR	Quantitative polymerase chain reaction
Rb	Retinoblastoma
RIPA	Ratio immunoprecipitation assay
RNA	Ribonucleic acid
Rpm	Revolutions per minute
RTCA	Real-time cell analysis
RT-qPCR	Real-time quantitative polymerase chain reaction
SAHF	Senescence-associated heterochromatin foci
SASP	Senescence-associated secretory phenotype
SA-β-Gal	Senescence-associated b galactosidase
SDS	Sodium dodecyl sulfate
TEMED	Tetramethylethylenediamine
TMB	3,3′,5,5′-Tetramethylbenzidine
V	Volts
VEGF	Vascular endothelial growth factor
WB	Western blot
αSMA	Alpha-smooth muscle actin
γH2AX	Gamma-histone family member X

1 Introduction

Hepatic stellate cells (HSCs) are non-parenchymal liver cells, representing approximately 10% of the total cell population in the healthy liver. In normal conditions, HSCs have a quiescent phenotype characterized by the expression of lecithin retinol acyltransferase (LRAT) [1, 2], desmin, and glial fibrillar acidic protein (GFAP). The quiescent HSC has several functions such as storage of vitamin A and maintenance of equilibrium between the production and degradation of different extracellular matrix components [3, 4]. Under pathophysiological conditions, HSCs become activated and acquire a myofibroblast-like phenotype, characterized by increased cell proliferation, contractility, and mobility and high expression of alpha-smooth muscle actin (αSMA) and reduced expression of GFAP. In addition, they demonstrate increased expression of extracellular matrix components such as collagen type 1 at both mRNA and protein levels, thus becoming the key effector cell during development of liver fibrosis [5, 6]. Due to the involvement of HSCs in liver fibrogenesis, they are considered a target for the treatment of hepatic fibrosis or cirrhosis.

Cellular senescence is a permanent state of cell cycle arrest that limits cell proliferation and occurs in response to several stressors, among them DNA and mitochondrial damage [7–9]. Since senescent HSCs have been observed in recent studies, it is timely to clarify the methods to characterize senescence in this cell type [1, 2, 10].

Although senescence in HSCs may be a target for antifibrotic therapies, the exact potential of induction of senescence remains incompletely understood due to the lack of specific markers to discriminate between senescent and non-senescent cells [1, 11, 12]. In general, multiple markers are determined to identify senescent cells, including senescent HSCs [1].

As already noted, cell cycle arrest is one of the main characteristics observed in cellular senescence. Two cell cycle inhibitors are often expressed in senescence: cyclin-dependent kinase inhibitor 1a (CDKN1a) or $p21^{Cip1}$, which prevents cell cycle progression from phase G1 to G2, and cyclin-dependent kinase inhibitor 2a (CDKN2a) or $p16^{INK4a}$, which is related with cell cycle arrest from phase G1 to S via inhibition of cyclin-dependent kinases 4 and 6 (CDK4/CDK6). In addition to $p21^{Cip1}$, the tumor suppressor p53 is also highly expressed in senescent cells. The *p21*^{Cip1} gene is a transcriptional target of p53 and is required for p53-dependent G1 and G2 cell cycle arrest, depending on the exact context [13–15].

Activation of the lysosomal enzyme senescence-associated β-galactosidase (SA-β-Gal) is also a recognized biomarker for senescence, and detection of its activity is widely used as a marker of senescence [16].

In addition to the cell cycle markers already described as markers for senescence, expression of proliferation markers like Ki67 are decreased in senescent cells [8]. Nuclear alterations have also been observed in senescent HSCs, due to changes in heterochromatin organization and the structural proteins involved in heterochromatin organization. These alterations are known as senescence-associated heterochromatin foci or SAHFs. These foci are rich in heterochromatin-forming proteins: heterochromatin protein 1 (HP1), H3K9me3, and the histone H2A variant macroH2A. Moreover, senescent cells display phosphorylated H2A histone family member X (γ-H2AX) foci, a marker of DNA damage. These cells have a flattened morphology, enlarged nuclear size, and decreased lamin B1 content [9]. Although senescent HSCs display DNA damage, they remain active and demonstrate an overactive secretory phenotype known as the senescence-associated secretory phenotype (SASP) [8, 17, 19], which comprises growth factors, cytokines, chemokines, metalloproteinases, extracellular matrix components, and extracellular vesicles (EVs). An additional feature of senescent cells is their resistance to apoptosis, a phenomenon mediated by increased expression of antiapoptotic proteins such as Bcl-2 family members [20].

In this chapter, we will describe techniques and methods that can be used to identify senescence in hepatic stellate cells according to the different biological and morphological features described above.

2 Materials

2.1 Materials for Analysis of Primary Senescence Markers

1. Microcentrifuge.
2. Heating block.
3. Nanodrop machine.
4. Sterile tips.
5. Sterile microcentrifuge tubes 1.5 mL.
6. Pipettes range 10–1000 μL.
7. Blue plastic pestle.
8. TRI Reagent/TRIzol (Sigma T9424, Thermo Fisher Scientific #15596-018).
9. Chloroform 100%, RNAse-free.
10. Isopropanol 100%, RNAse-free.
11. Ethanol 75%, RNAse-free.
12. RNAse-free water.
13. Thermoblock.
14. Spectrophotometer.
15. Electrophoresis machine.
16. Pipettes range 10–1000 μL.
17. Blue plastic pestle.

18. Liquid nitrogen container.
19. Scraper (Greiner, 541070).
20. Eppendorf tubes (1.5 mL).
21. Liquid nitrogen.
22. PBS.
23. Lysis buffer.
24. Isobutanol.
25. Acrylamide gels.
26. Nitrocellulose membrane.
27. Milk powder.
28. Tween 20.
29. Bovine serum albumin (BSA).

2.2 Materials for Senescence-Associated β-Galactosidase (SA-β-Gal) Detection

1. For the detection of the SA-β-Gal, we recommend the usage of the senescence β-galactosidase staining kit #9860 (Cell Signalling Technology), which contains the following reagents:
    1. 10× Fixation solution.
    2. Dilute the 10× fixation solution to a 1× solution with distilled water.
    3. 10× Staining solution.
    4. Dilute the 10× staining solution to a 1× solution with distilled water.
    5. 100× Solution A.
    6. 100× Solution B.
    7. X-Gal.
2. In addition, the following materials and equipment are further necessary:
    1. Digital phase-contrast microscopy.
    2. 37 °C incubator (No CO_2).
    3. PBS.
    4. Dimethylformamide (DMF).
    5. Dimethyl sulfoxide (DMSO).

2.3 Materials for Cell Proliferation Assays

2.3.1 Material for xCELLigence Real-Time Cell Analysis

1. Flow cabinet.
2. Incubator (5% CO_2, 37 °C).
3. xCELLigence apparatus.
4. RTCA system with computer and provided software (ACEA Biosciences, Inc, San Diego, CA).
5. E-plates (ACEA Biosciences, Inc, San Diego, CA, USA).

6. Cells.
7. Culture medium.
8. Pipettes and tips.

2.3.2 Materials for BrdU Cell Proliferation ELISA

1. Incubator (37 °C).
2. Fluorescence microplate reader (e.g., Synergy H4: Gen5).
3. 96-well black or white, flat bottom culture plate.
4. Reagents kit contents (#11 669 915 001, Sigma) containing:

   Bottle 1 (red flip-up cap): BrdU labeling reagent: 1 mL, sterile; 1000× conc. 10 mmol/L 5-bromo 2′-deoxyuridine in PBS, pH 7.4.

   Bottle 2 (red): FixDenat: 2× 100 mL; ready to use.

   Bottle 3 (blue): Anti-BrdU-POD: lyophilized, stabilized; monoclonal antibody from mouse hybridoma cells (clone BMG 6H8, Fab fragments) conjugated with peroxidase (POD).

   Bottle 4 (blue): Antibody dilution solution 100 mL ready to use.

   Bottle 5 (green): Washing buffer 100 mL; 10× conc PBS.

   Bottle 6 (black): Substrate component A 100 mL; buffered solution, containing luminol and 4-iodophenol (*see* **Note 12**).

   Bottle 7 (yellow): substrate component B 1 mL; buffered solution, containing a stabilized form of H_2O_2.

   The kit is stable until expiry date at 2–8 °C.

2.4 Materials for Lamin B1 Expression Analysis

2.4.1 Materials for Analysis of Lamin B1 Loss by RT-qPCR

1. Nanodrop.
2. Sterile tips.
3. Sterile microcentrifuge tubes 1.5mL.
4. Pipettes range 10–1000 μL.
5. Ethanol 75%, RNAse-free.
6. RIPA Buffer (R0278, Sigma-Aldrich).
7. RNAse-free water.
8. RNEasy Mini Kit (QIAGEN).
9. Microcentrifuge.
10. Heating block.
11. LightCycler® 480 System (Roche) or similar.

2.4.2 Materials for Lamin B1 Detection by Immunofluorescence

1. Sterile tips.
2. Sterile microcentrifuge tubes 1.5 mL.
3. Pipettes range 10–1000 μL.

4. Microscope glass slides and coverslips.
5. Ethanol 75%, RNAse-free.
6. Anti-lamin B1 antibody (e.g., rabbit IgG, ab16048, Abcam, 1: 1000).
7. Normal rabbit IgG (isotype control, NI01, Sigma-Aldrich).
8. Fluorescently labelled anti-rabbit antibody (e.g., Alexa Fluor® 488 Donkey anti-Rabbit IgG Antibody, #R37118, Invitrogen).
9. DAPI.
10. Fluorescent microscope.

2.5 Materials for Testing of Senescence-Associated Heterochromatic Foci (SAHF) Formation

1. Sterile tips.
2. Sterile microcentrifuge tubes 1.5 mL.
3. Pipettes range 10–1000 μL.
4. Microscope glass slides.
5. 0.1% Triton X-100 solution.
6. DAPI (4′,6-diamidino-2-phenylindole).
7. RNase A (Roche, #10154105103, 100 μg/mL).
8. Fluorescent microscope.

2.6 Materials for Senescence-Associated Secretory Phenotype (SASP) Testing

2.6.1 Materials for SASP Testing by ELISA

1. Sterile tips.
2. Sterile microcentrifuge tubes 1.5 mL.
3. Pipettes range 10–1000 μL.
4. Adhesive plate covers.
5. Human IL-6 ELISA kit (Invitrogen, cat. No. KHC0061).
6. 96-well plates.
7. Standard dilution buffer.
8. Biotinylated detection antibody.
9. Streptavidin-HRP.
10. HRP diluent.
11. Wash buffer (eBioscience, Invitrogen).
12. Chromogen (Tetramethylbenzidine, TMB).

2.6.2 Materials for SASP Expression Analysis by Antibody Arrays

1. Spin columns.
2. Labelling reagent.
3. Stop solution.
4. Blocking buffer.
5. Wash buffer 1.
6. Wash buffer 2.

7. Streptavidin-conjugated fluor.
8. Adhesive plastic strips.
9. Labelling buffer.
10. Lysis buffer.
11. 30 mL centrifuge tube.
12. Distilled water.
13. Sterile tips.
14. Sterile microcentrifuge tubes 1.5 mL.
15. Pipettes range 10–1000 μL.
16. Aluminum foil.
17. Orbital shaker or oscillating rocker.
18. Gene microarray scanner or similar laser fluorescence scanner.

2.7 Materials for Identification of Morphological Changes

1. Sterile tips.
2. Round coverslips.
3. Slides.
4. Nail polish.
5. Mounting medium.
6. DAPI solution .
7. Pipettes range 10–1000 μL.
8. Formaldehyde 4%.
9. PBS 1×.
10. 0.1% Triton X-100 solution.
11. Confocal microscope.

3 Detection of Primary Senescence Markers in Hepatic Stellate Cells

3.1 Cell Cycle Arrest Markers in Senescent Hepatic Stellate Cells

The cell cycle is regulated by cyclins, cyclin-dependent kinases (CDKs), and CDK inhibitors. In conditions of differentiation and/or development, an increased expression of CDK inhibitors is observed. Furthermore, increased expression of CDK inhibitors is also observed in response to various insults, such as DNA damage and oxidative stress. There are two types of CDK inhibitors, the Inhibitor of Kinase 4/Alternative Reading Frame (INK4/ARF), which includes $p16^{INK4A}$ [21, 22], binding CDK4 and CDK6, and CDK interacting protein/kinase-inhibitory protein (Cip/Kip) corresponding to $p21^{CIP1}$, which binds the preassembled cyclin E-CDK2, cyclin A-CDK2, and cyclin D-CDK4/6 complexes [23].

Together, these pathways stop cell cycle progression from G1 phase to S phase. In HSCs, cell cycle arrest is characterized by the increased expression of p16 ^{INK4A}, p21^{Cip1}, and p53 at both protein and mRNA levels [11, 24]. These biomarkers are normally assessed by (quantitative) polymerase chain reaction (qPCR) and Western blot (WB). Here, we provide a description of the qPCR and immunoblotting methods for the detection and quantification of p16 ^{INK4A}, p21^{Cip1}, and p53 in liver tissue and hepatic stellate cells.

3.2 Detection of p16^{INK4A}, p21^{Cip1}, and p53 in Liver Tissue and Hepatic Stellate Cells

p16^{INK4A}, p21^{Cip1}, and p53 by RT-qPCR [25].

3.2.1 Preparation of Liver Tissue

1. Use approximately 30 mg tissue, and place it in a 1.5 mL tube. Add 200 μL of TRI Reagent/TRIzol, and homogenize using a blue plastic pestle; then add additional 800 μL of TRI Reagent/TRIzol, mix, and incubate for 5 min.
2. Add 200 μL of chloroform per 1 mL TRI Reagent/TRIzol, and vortex for 30 s; then incubate for 5 min at room temperature, and finally centrifuge 15 min at 12,000 rpm, 4 °C. Transfer the aqueous phase to a new and sterile 1.5 mL tube, and precipitate the RNA by adding 500 μL of isopropanol, incubate for 8 min, and centrifuge again for 10 min at 12,000 rpm, 4 °C.
3. Pellet will be visible at this point. Remove the supernatant from the previous step, and add 1 mL of cold ethanol at 75%; vortex and centrifuge for 5 min at 7500 rpm, 4 °C. Remove all ethanol, and let the pellet dry for 5–10 min. Add the appropriate amount of RNAse-free water to the RNA pellet, and incubate for 5–10 min at 65 °C. RNA samples should be stored at temperatures lower than −80 °C. To continue with the RT-qPCR, follow **step** 7 in this section.

3.2.2 Preparation of Adherent Hepatic Stellate Cells

1. HSCs seeded on plastic plates should be washed twice with cold PBS. Add TRIzol according to Table 1; scrape the cells, and transfer them to a new and sterile tube (1.5 mL). Add 100 μL of chloroform, vortex for 10–15 s, and leave the samples for 5 min at room temperature.

Table 1
mRNA isolation using TRIzol

Plate/tissue	Tri reagent	Chloroform	Isopropanol	Ethanol
Tissue 30 mg	1 mL	200 μL	500 μL	1 mL
6-well plate/per well	500 μL	100 μL	250 μL	500 μL
25 cm flasks	1.5 mL	300 μL	750 μL	1.5 mL

2. To continue, centrifuge samples for 15 min at 12000 rpm, 4 °C. Transfer aqueous phase into a new 1.5 mL tube. Precipitate the RNA by adding 250 μL of isopropanol, vortex, incubate for 10 min, and finally centrifuge the samples for 10 min at 12,000 rpm, 4 °C.
3. Remove supernatant, add 500 μL of 75% cold ethanol, and centrifuge 5 min at 7500 rpm, 4 °C. After centrifugation, remove ethanol and dry pellet for 10 min or till complete evaporation of ethanol. Dissolve RNA in RNAse-free water (amount depends on the pellet size, normally 20 μL is enough). Incubate 5 min at 65 °C. RNA samples should be stored at temperatures lower than −80 °C.
4. After RNA extraction, RNA must be quantified. Concentration and purity are determined using NanoDrop. Use 1 μL to determine the concentration. Once concentration is known, continue with the reverse transcription using a commercial reverse transcriptase (e.g., K11621, K1651, 28025013 Thermo Fisher) following the manufacturer's protocols to synthesize the first strand of cDNA using the RNA previously isolated as a template (*see* **Notes 1–3**). To continue with the gene expression analysis, RT-qPCR can be done using TaqMan probes or SYBR green, according to the suppliers' instructions (e.g., Applied Biosystems, Sigma Aldrich, or SensiMix SYBR kits from Bioline). The comparative analysis of gene expression will be done by using at least two stable reference genes (e.g., GAPDH or β-actin). The stability of the reference genes being used for normalization should be tested, and the efficiency of the primers used should be higher than 95%. qPCR technical replicates, $n \geq 2$ per biological sample, should be performed. In Table 2, the primers sequences and details for senescence markers related to cell cycle arrest in hepatic stellate cells are listed (*see* **Notes 4** and **5**).

3.2.3 Detection of $p16^{INK4A}$, $p21^{Cip1}$, and p53 by Western Blot [27]

Proteins can be isolated from different types of samples, e.g., tissue or cells. Here we will describe the isolation of protein from both tissue and adherent cells (*see* **Note 6**).

Preparation of Tissue

1. Use approximately 25–30 mg tissue, and place it in 1.5 mL tubes.
2. Add 500 μL of lysis buffer, and homogenize using a blue plastic pestle.
3. Add 500 μL more of lysis buffer.
4. Snap-freeze samples in liquid nitrogen, and defrost samples at 37 °C in a thermoblock. Repeat freeze-thawing cycle 4 times. Clarify lysates by centrifugation for 10 min at 12,000 rpm,

Table 2
Primer sequences of senescence markers

Gene	Forward 5′–3	Reverse 5′–3	Reference
Cdkn1a (rat)	TTGTCGCTGTCTTGCACTCTG	CGCTTGGAGTGATAGAAATCTGTTA	[2]
Tp53 (rat)	CCATGAGCGTTGCTCTGATG	CAGATACTCAGCATACGGATTTCCT	[2]
Cdkn1a (human)	TCACTGTCTTGTACCCTTGTGC	GGCGTTTGGAGTGGTAGAAA	[26]
P16 (human)	GAGCAGCATGGAGCCTTC	CGTAACTATTCGGTGCGTTG	[26]
LMNB1 (human)	GTGCTGCGAGCAGGAGAC	CCATTAAGATCAGATTCCTTCTTAGC	[26]
IL6 (rat)	CCGGAGAGGAGACTTCACAGA	AGAATTGCCATTGCACAACTCTT	[2]
IL6 (human)	CAGGAGCCCAGCTATGAACT	GAAGGCAGCAGGCAACAC	[26]
36b4 (rat)	GCTTCATTGTGGGAGCAGACA	CATGGTGTTCTTGCCCATCAG	[2]
GAPDH (human)	ACCCAGAAGACTGTGGATGG	TCTAGACGGCAGGTCAGGTC	[26]

4 °C. Transfer supernatant to a new tube, and store at −20 °C. Before electrophoresis, protein content must be quantified using a spectrophotometer.

Preparation of Cells

1. Wash the cells twice with cold PBS, remove PBS, and add lysis buffer with the protease inhibitor cocktail freshly prepared (e.g., RIPA buffer), 500 μL for 6-well plates or 200 μL for 12-well plates. Scrape the cells, and transfer to Eppendorf tubes of 1.5 mL.
2. Snap-freeze samples in liquid nitrogen, and defrost samples at 37 °C in a thermoblock. Repeat freeze-thawing cycle 4 times. Clarify lysates by centrifugation for 10 min at 12,000 rpm, 4 °C. Transfer supernatant to a new tube, and store at −20 °C. Before electrophoresis, protein content must be quantified using a spectrophotometer.

Sample Preparation

1. Determine the volume needed to have 10–20 μg of protein per sample.
2. Add this volume to a new tube and the equivalent volume of deionized water plus 10 μL loading buffer to have a final volume of 30 or 50 μL.
3. Heat the samples in a thermoblock at 95 °C for 5 min to allow denaturation of the proteins. Samples can be kept on ice or stored at −20 °C. Samples are now ready to continue with the electrophoresis.

Gel Preparation and Electrophoresis

1. Mount the gel electrophoresis system. Based on the molecular weight of the protein of interest, choose the acrylamide concentration, and use the corresponding reagent concentrations to prepare the separation gel (Table 3). Add the separation gel to the system, and leave approximately 1 cm free of gel at the top. Add 300 μL of isobutanol on top of the gel to level the surface. Let the gel polymerize for 30–60 min, and remove the isobutanol. Prepare the stacking gel according to Table 3, add it on top of the separation gel, and place the comb. Let the stacking gel polymerize for approximately 15 min (*see* **Note** 7).
2. Place the gels in the electrophoresis machine, add running buffer, and ensure that gels are immersed in the running buffer. Gently remove the comb, and carefully add prepared samples; place the lid on the electrophoresis machine, and run the proteins at 60–100 V for the stacking gel and 150 V for the separation gel for approximately 1 h or until the dye front runs off the bottom of the gel (*see* **Note 8**).

Table 3
Gel preparation

Separation gel							
Linear range of separation (kDa)	**10–43**	**12–60**	**20–80**	**36–94**	**57–212**		
Acrylamide concentration	**15%**	**12%**	**10%**	**7.5%**	**5%**	**Stacking gel**	
Reagents	**Volume**					**Reagents**	**Volume**
Deionized water	2.5 mL	3.5 mL	4.2 mL	5 mL	5.9 mL	Deionized water	3 mL
1.5 m Tris HCl, pH 8.8	2.5 mL	2.5 mL	2.5 mL	2.5 mL	2.5 mL	0.5 M Tris HCl, pH 6.8	1.25 mL
10% SDS	100 µL	100 µL	100 µL	100 µL	100 µL	10% SDS	50 µL
30% acrylamide	5 mL	4 mL	3.3 mL	2.5 mL	1.6 mL	30% acrylamide	666 µL
10% APS (add at the end)	50 µL	50 µL	50 µL	50 µL	50 µL	10% APS (add at the end)	50 µL
TEMED (add at the end)	5 µL	5 µL	5 µL	5 µL	5 µL	TEMED (add at the end)	10 µL

Electrotransfer

These instructions refer to a semi-dry blotting system (Pharmacia LKB, Multiphor II) and power supply (Pharmacia LKB, EPS 3500 XL).

1. Prepare the membrane according to gel size. Therefore, cut 15 filter sheets that fit with the membrane size. Make the plates (anode and cathode) wet with deionized water. Preincubate the filters according to Table 4.
2. Place the membrane in deionized water and the gel in buffer 3.
3. Make the sandwich on the lower side of the plate (anode), adding first six pre-soaked filters in buffer 1, three filters pre-soaked in buffer 2, and then the nitrocellulose membrane followed by the gel and finally the last six filters pre-soaked in buffer 3 on top. Place the other plate (cathode), and connect the wires.
4. Close the machine, and run it at 32 mA per gel for at least 1 h. Once the transfer is complete, disconnect the transfer unit from the power supply, disassemble the sandwich, and take the nitrocellulose membranes.
5. At this step, molecular weight markers must be visible. To visualize the transferred proteins, a Ponceau S staining can be performed by immersing the membranes in Ponceau S solution for 2–5 min.
6. To remove the red stain from the Ponceau S, wash the membranes with PBS/Tween 20 until the red stain has disappeared. Finally, store membranes in PBS at 4 °C or continue with the immunoblotting.

Immunoblotting

1. Prepare the blocking buffer (5% BSA or 2% milk powder +0.5% BSA in PBS-1×/Tween 0.1%) solution to block non-specific binding sites. Incubate the nitrocellulose membrane with blocking buffer for 60 min at room temperature. In the

Table 4
Filter preparation for transfer

	Solution	Required number of filters
Buffer 1	Methanol 20% Trizma base 300 mmol/L	6
Buffer 2	Trizma base 25 mmol/L Methanol 20%	3
Buffer 3	Trizma base 25 mmol/L Methanol 20% Amino-n-caproic acid 40 mmol/L	6

Table 5
Antibodies for immunoblotting

Primary antibody	Cross-reactivity	Source organism	Company	References
p21	Mouse, rat, human	Rabbit	Abcam	Ab109199
p16	Mouse, rat, human	Mouse	Invitrogen	1E12E10
GAPDH	Human, mouse, rat, porcine	Mouse	Calbiochem	CB1001
β-Actin	Human, mouse, rat, hamster, monkey	Rabbit	Cell signaling technology	4970

meantime, prepare the primary antibody, diluted in 0.1% BSA/PBS/0.1% Tween 20 (antibody dilution will depend on the antibody). After blocking, discard the buffer, and incubate with primary antibody overnight (antibodies are listed in Table 5).

2. Recover the antibodies used in a new tube, and store at −20 °C. Primary antibodies can be re-used up to 10–15 times.
3. Wash the membrane three times in PBS-1×/Tween 20 0.1% buffer (10 min, room temperature). Membranes are incubated for 45–60 min at room temperature in an orbital shaker with freshly prepared secondary antibody conjugated with horse radish peroxidase (HRP) (1:2000 dilution) in PBS/Tween 20 buffer. Discard the secondary antibodies.
4. Nitrocellulose membranes are again washed three times in PBS/Tween 20 buffer (10 min, room temperature) and once with only PBS for 5 min.
5. Membranes are then overlaid with detection reagent. Chemiluminescence is detected by cumulative acquisition using appropriate tools, e.g., ImageQuant LAS 4000. Images taken may be analyzed by densitometry using ImageJ software (NIH, Bethesda, MD, USA; freely available at http://rsb.info.nih.gov/ij/) (*see* **Note 9**).

4 Senescence-Associated β-Galactosidase (SA-β-Gal) Detection

During cellular senescence, increased activity of lysosomal β-galactosidase has been observed, associated with the expansion of the lysosomal compartment and increased content of the senescence-related isoform GLB1 [16]. Therefore, senescence-associated β-galactosidase (SA-β-Gal) activity is considered a biomarker of senescent cells and is typically measured by in situ staining using a chromogenic substrate such as X-gal. Identification of

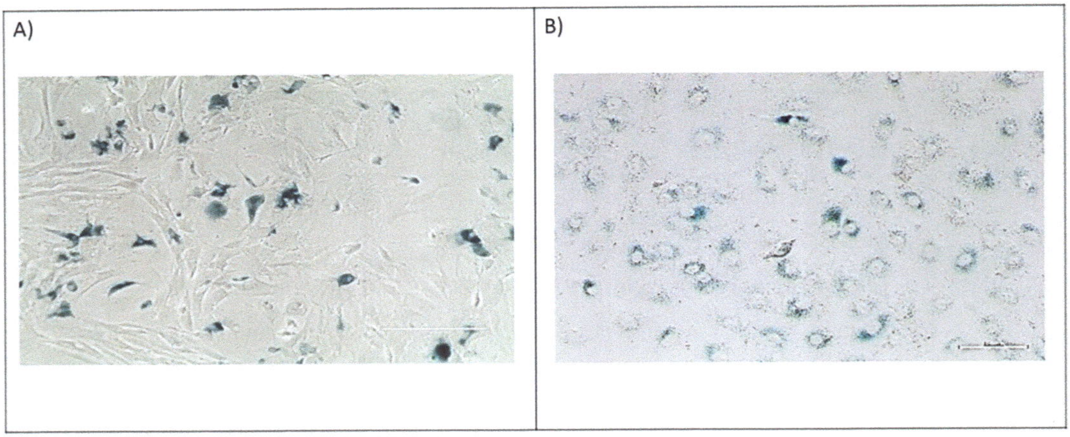

Fig. 1 Example of SA-β-Gal staining. (**a**) SA-β-Gal activity detected in hepatic stellate cells [2]. (**b**) SA-β-Gal activity present in human lung adenocarcinoma cell line A549

SA-β-Gal in tissue can be performed using histochemical techniques. Here, we provide a description of the protocol used to detect SA-β-Gal activity in hepatic stellate cells (Fig. 1).

Always use polypropylene plastic or glass to prepare and store X-Gal solution. Do not use polystyrene.

Dissolve 20 mg of X-gal in 1 mL DMF to prepare a 20 mg/mL stock solution. Excess X-gal solution can be stored at −20 °C in a light-resistant container for up to 1 month.

Work solution per mL: mix the following reagents to prepare 1 mL of work solution: β-galactosidase staining solution: 930 μL 1× staining solution, 10 μL 100× solution A, 10 μL 100× solution B, 50 μL 20 mg/mL X-gal stock solution. Adjust to pH 6.0 after preparation.

4.1 Protocol According to Kit #9860 Cell Signaling Technology

1. Wash the cells with 1× PBS, and add 200 μL (12-well plates) or 400 μL (6-well plates) of 1× fixation solution to each well according to plate used or the appropriate amount that covers the cell surface. Allow cells to fix for 10–15 min at room temperature.
2. Wash the cells twice with 1× PBS, and add 1 mL of the β-galactosidase staining solution to each well or the appropriate amount that covers the cell surface. Seal the plates with parafilm to prevent evaporation, and incubate at 37 °C overnight in a dry incubator (no CO_2) (see Solution Preparation in Methods and Equipment 5) (*see* **Note 10**).
3. While the β-galactosidase staining solution is still on the plate, evaluate the cells under a microscope (200× total magnification) for the development of blue color. For long-term storage of the plates, remove the β-galactosidase staining solution, and overlay the cells with 70% glycerol. Store at 4 °C.

In Fig. 1, an example is presented of SA-β-Gal staining, showing a comparison between two cell types, primary hepatic stellate cells from Wistar rats and human lung adenocarcinoma cell line A549 following the protocol described above.

5 Cell Proliferation Assays

5.1 xCELLigence

Cell proliferation can be measured using the xCELLigence real-time cell analysis (RTCA) system [28, 29]. xCELLigence real-time cell Analysis is provided by ACEA Biosciences, Inc, San Diego, CA, USA. The E-Plates are based on impedance sensing using gold microelectrodes covering 70–80% of 96-well-bottom area. Impedance is measured by applying a low electric current through the cell culture medium. The impedance signal is affected by cell surface area since cells act as electrical insulators. The impedance is monitored in real time for prolonged periods to determine cell index (https://www.ols-bio.de/products/live-cell-analysis-system-xcelligence). The determination of HSC proliferation should consider the starting time of the seeding and the time needed for HSCs to attach, as well as the start of the experiment. Normalization of the cell growth must be adjusted relative to the starting time of the experiments.

Setting up the E-plate (*see* **Note 11**):

1. Before adding the cells to the wells of the plate, set up the correct conditions in the software according to the experimental design.
2. Place the xCELLigence E-plate in the incubator for a minimum of 30 min in the presence of warm medium (100 μL). Click on button to start the experiment. The xCELLigence will now perform a background measurement.
3. In the meantime, quantify cells to have a total amount of 10,000–20,000 cells according to the cell type. Primary HSCs require more cells to start with than cell lines. Prepare the right dilution to add 100 μL to each well plate.
4. Once cells are added to the wells, continue with the next step.
5. After an experiment is started, it is possible to pause by clicking the button. The user can perform actions during this pause such as adding compounds, checking connections, etc. without data being collected. Click on the button to start the experiment again. At the end of the experiment, collect your data (e.g., USB stick).

5.2 Quantification of Proliferation Using BrdU Cell Proliferation ELISA

Bromodeoxyuridine (BrdU) is a synthetic nucleoside analogue with a chemical structure similar to thymidine. As such, BrdU is commonly used to determine the proliferation rate of cultured mammalian cells [30]. Proliferation is measured by determining the incorporation of BrdU into genomic DNA of proliferating cells via an ELISA assay and is based on the principle that cycling cells incorporate BrdU rather than thymidine in the labeling phase. The cells are fixed after removing the reagents, and the DNA is denaturated in one step using FixDenat. The anti-BrdU-POD antibody is added after removing FixDenat. The subsequent substrate reaction results in coloration, which is measured using a (luminescence) ELISA reader.

1. Prepare the working solutions (*see* **Note 12**) for BrdU cell proliferation ELISA as given in Table 6.
2. Culture cells in a black or white flat bottom culture plate at 37 °C until the desired confluency (usually 50%; if necessary, check this by culturing cells in a transparent 96-well culture plate at the same time). The outer rim of the plate should not contain cells, only culture medium (blank).

Table 6
Preparation of the working solutions for cell proliferation assay

BrdU labelling solution	Dilute BrdU labeling solution reagent (bottle 1; red flip-up cap) 1:100 with sterile culture medium (final concentration: 100 μmol/L BrdU). For one 96-well microplate, 1 mL diluted BrdU labeling solution is required if the cells were cultured in 100 μL/well. Stored protected from light stable for several weeks at 2–8 °C. For long-term storage, in aliquots at −15 to −25 °C
Anti-BrdU-POD stock solution	Dissolve anti-BrdU-POD (bottle 3, blue cap) in 1.1 mL deionized water for 10 min and mix thoroughly. Storage: at 2–8 °C for several months; for long-term storage, store the solution in aliquots at −15 to −25°
Anti-BrdU-POD working solution	Dilute anti-BrdU-POD stock solution 1:100 with antibody dilution solution (bottle 4, blue). Note: For one 96-well microplate, dilute 100 μL anti-BrdU POD stock solution in 10 mL antibody dilution solution. Prepare shortly before use. Do not store
Washing solution	Dilute washing buffer concentrate (bottle 5, green cap) 1:10 with deionized water. Note: for one 96-well microplate, dilute 10 mL washing buffer concentrate with 90 mL deionized water. Can be stored at 2–8 °C for several weeks
Substrate solution	Mix substrate component B (bottle 7, yellow cap) with substrate component A (bottle 6, black cap). Stir the mixture for at least 15 min at room temperature to equilibrate the components. Note: For one 96-well microplate, dilute 100 μL substrate component B in 10 mL substrate component A. can be stored at 2–8 °C for 1 week. The substrate solution should be warmed up to 15–25 °C before use, to avoid temperature effects during the measurement

Table 7
Pipette scheme for BrdU cell proliferation ELISA

Well contents	Blank	Background
Culture medium	100 μL	–
Cells	–	100 μL
BrdU	10 μL	–
Anti-BrdU-POD	100 μL	100 μL

3. Incubate cells in the presence of various concentrations of the tested substance at a final volume of 100 μL/well (quadruples) for the desired amount of time at 37 °C using the pipette scheme depicted in Table 7.
4. Add 10 μL BrdU labelling solution per well (solution 1; final concentration: 10 μmol/L BrdU), and continue incubation of the cells for at least 3 h at 37 °C.
5. Remove labelling medium by tapping off or aspiration (*see* **Note 13**). Fixate the cells by adding 200 μL FixDenat (bottle 2, red cap) per well, and incubate for 30 min at room temperature.
6. Incubation with anti-BrdU-POD: Remove the FixDenat solution, and add 100 μL/well anti-BrdU-POD working solution. Incubate cells with this solution for 90 min at room temperature. Then remove the antibody, and rinse wells three times with 200–300 μL/well washing solution (cat nr:11 669 915 001). Leave washing solution (#11 669 915 001) in the wells for 0.5–5 min to reduce background.
7. Substrate reaction: Remove washing solution (#11 669 915 001) by tapping, and add 100 μL/well substrate solution (*see* **Note 14**). ELISA can be read after this step in an appropriate spectrofluorometer.

5.3 Detection of Nuclear Alterations

In addition to cell cycle arrest and increased lysosomal activity, senescent cells develop characteristic morphological changes including enlarged and often irregularly shaped nuclei with significant chromatin reorganization. The alterations are directly correlated with the emergence of the senescent phenotype; the development of an enlarged nucleus can be partly due to cell cycle arrest in G2 phase with more nuclear content and/or the loss of the nuclear envelope, which determines nuclear morphology such as loss of lamin B1 [31]. Additionally, stable cell cycle arrest observed in senescent cells can result in the appearance of specialized domains of transcriptionally silent heterochromatin, typically referred to as senescence-associated heterochromatic foci (SAHF), which are thought to repress the expression of proliferation-promoting genes, thereby reinforcing cell cycle arrest. Both the loss of nuclear lamina components and the appearance of SAHFs have been described in HSCs and may serve as useful tools to identify senescent HSCs in samples derived from in vitro cell cultures or in vivo and clinical tissue samples.

5.4 Lamin B1 Loss

The inner surface of the nuclear envelope is delineated by a lamina that contributes to the size, shape, and stability of the nucleus. The lamina itself is a dynamic structure, as it must be disassembled and reassembled between cell cycles. Alterations in the cell cycle as observed in cellular senescence may therefore result in profound changes of the nuclear lamina. The major components of the lamina are intermediate filaments known as nuclear lamins, and specifically the progressive loss of type B lamins is associated with cellular senescence [31, 32]. Importantly, loss of lamin B1 (LMNB1) occurs at both the protein and mRNA levels, enabling detection in both fresh or frozen material, as well as fixed samples, by using diverse techniques as explained below.

5.4.1 Lamin B1 Loss by RT-qPCR

1. For transcriptional analyses using real-time quantitative PCR (RT-qPCR), both flash frozen and fresh material derived from cell culture or tissue samples may be used. Prior to extraction and for long-term storage of RNA, samples should be kept at temperatures lower than −80 °C.
2. To perform RNA extraction, the use of established protocols (e.g., TRIzol reagent-based, 2.1.1) is an option; however, the use of commercially available kits for RNA extraction (e.g., RNeasy Mini/Micro kits, from QIAGEN) is also highly recommended. These kits provide specific lysis buffers (RLT Lysis buffer), which may be used for RNA extraction as described by the manufacturers.
3. Following RNA extraction, the first strand of cDNA is synthesized using a reverse transcriptase. Contaminating genomic DNA may be eliminated using gDNA removers contained in commercially available kits (Reverse Transcription Kit, Promega).
4. Real-time PCR may be performed using commercially available TaqMan (e.g., Applied Biosystems) or SensiMix SYBR kits (e.g., Bioline) according to the manufacturers' protocols. For comparative analysis of gene expression, the use of at least two stable reference genes upon senescence induction is desired (e.g., GAPDH, 36b4, and β-actin) [41]. The stability of the reference genes being used for normalization should be tested, and the efficiency of the primers used should be higher than 95%. Lastly, qPCR technical replicates ($n \geq 2$) per biological sample should be performed.

5.4.2 Lamin B1 by Western Blotting

1. Whole-cell protein extracts are prepared from human liver tissues or hepatic stellate cells using an adequate buffer (e.g., RIPA buffer). It is possible to fractionate extracts into cytoplasmic and nuclear proteins using commercially available protein extraction kits (e.g., Bioepitope Nuclear and Cytoplasmic Extraction Kits, Bioworld Technology Co., Ltd, previously used in hepatic stellate cells).
2. In the event of fractionation, the nuclear fraction may be analyzed for lamin B1 by Western blot, with visualization and quantification according to established methods. In case of no fractionation, an invariant control, e.g., β-actin, which is typically used in HSCs, may be used for equal loading of total protein.
3. The intensity of the lamin B1 bands in senescent hepatic stellate cells may be lower compared to that of control or quiescent hepatic stellate cells (*see* **Note 9**).

5.4.3 Lamin B1 Detection by Immunofluorescence

1. If the analysis of individual cells is desired as opposed to (bulk) populations in Western blotting, lamin B1 immunofluorescence may be particularly effective in in vitro cultures. As a starting point, cells are seeded on coverslips in, e.g., 12-well plates, and allowed to adhere for at least 4 h (*see* **Note 15**).
2. The following day, the media is removed, and the cells are fixed in 2% PFA/0.2% glutaraldehyde for 3 min at room temperature. This milder fixation will enable co-staining for senescence-associated β-galactosidase (optional).
3. If SA-β-Gal staining is not required, cells may be permeabilized with 0.5% Triton X-100 (v/v) in PBS for 5 min at room temperature and washed twice with PBS at the end of the incubation period.
4. Following permeabilization, blockage of non-specific antigen sites with 15% serum of the secondary antibody species is recommended (e.g., goat serum, when using a fluorescently labelled goat anti-mouse antibody). A minimal incubation period of 30 min at room temperature is recommended (*see* **Notes 16 and 17**).
5. The coverslips are incubated with an anti-lamin B1 primary antibody for 1 h at room temperature and washed twice following the incubation period (e.g., anti-lamin B1 mouse monoclonal antibody). As a negative control, incubation with a relevant isotype control in additional coverslips is recommended (e.g., mouse IgG antibody).
6. The coverslips are now incubated with an appropriate secondary antibody (e.g., fluorescently labelled goat anti-mouse antibody) for 1 h in the dark, after which coverslips are washed twice and mounted with an appropriate volume of DAPI

mounting media solution to completely cover the surface (e.g., 20 μL).

7. Seal the edges with a sealant such as varnish, and keep the slides in the dark at 4 °C until observation (*see* **Note 18**).
8. For microscopic analysis, lamin B1 should be visible at the nuclear envelope with varying intensities. Counting at least 200 nuclei per sample is recommended. Blinded approaches with at least three data sets are also encouraged. Additionally, the morphology of the nuclei may be noted, distinguishing between a normal oval-shaped nuclei phenotype and abnormal irregular-shaped or blebbing nuclei, which typically appear more frequently in senescent samples (*see* **Note 19**).

5.5 Testing for Senescence-Associated Heterochromatic Foci (SAHF) Formation

Irregularly shaped nuclei are typically associated with cellular senescence and in addition to an enlarged size, nuclear blebbing, and loss of lamin B1, distinct heterochromatic structures accumulate in senescent cells. These structures have been designated as senescence-associated heterochromatic foci (SAHF) [33]. SAHF formation coincides with the recruitment of heterochromatin proteins and the retinoblastoma (Rb) tumor suppressor to E2F-responsive promoters, which are associated with the stable repression of E2F target genes typically involved in proliferation. Importantly, SAHF formation depends on the integrity of the Rb pathway, is influenced by p16 and p21, and does not occur in reversibly arrested cells [33]. The staining of heterochromatin proteins (e.g., HMGA2, H3K9me3) and/or cell cycle arrest markers (p16, p21) can therefore be complementary and is mostly used in in vitro cultures [29]. SAHF can be easily measured by immunofluorescence.

1. Cells are fixed in 4% formaldehyde made fresh from paraformaldehyde followed by permeabilization in 0.2% Triton X-100/0.5% normal animal serum/PBS. The species of the serum used will depend on the secondary antibody being used for detection, e.g., goat serum, for a secondary goat antibody.
2. After three washing steps, cells may be subsequently incubated for 1 h at room temperature with the desired primary antibody, followed by additional washes and incubation with any secondary antibodies.
3. DNA is visualized by using 4′,6-diamidino-2-phenylindole (DAPI) (1 μg/mL) or propidium iodide (PI) (10 μg/mL). PI should be used with RNase A (Roche, 100 μg/mL) at 37 °C for 30 min to avoid strong background fluorescence localized to the cytoplasm.
4. Fluorescent imaging can be now performed. SAHF-like chromatin spots should be evident in senescent cells, with a higher

degree of "spottiness" more evident in senescent samples compared to proliferating/control samples. The percentage of nuclei with SAHFs between conditions derived from examining more than 100 nuclei per sample is recommended. Blinded approaches with at least three data sets are encouraged. As with lamin B1 staining, the morphology of the nuclei may be noted, distinguishing between a normal oval-shaped nuclear phenotype and abnormal, irregular-shaped, or blebbing nuclei. The use of confocal images using a confocal laser-scanning microscope may also be used for increased clarity.

6 Detection of Secondary Senescence Markers

6.1 Senescence-Associated Secretory Phenotype (SASP) Markers

Unlike quiescent or terminally differentiated cells, growth arrest of senescent cells is accompanied by an altered secretome known as the senescence-associated secretory phenotype (SASP) capable of influencing the microenvironment. The SASP is composed of multiple proteins that vary largely by cell and tissue type (reviewed in [26]) but typically includes inflammatory chemokines and cytokines (IL-6, IL-8), growth factors (IGF-1, PDGF, and VEGF), and matrix remodeling proteins (MMPs). Many of these proteins (e.g., IL8, MMP1, MMP3) are also expressed in HSCs and show a clear upregulation upon senescence induction. Therefore, several of these SASP components may be measured as a secondary readout for senescent stellate cells using a variety of techniques (RT-qPCR, ELISA, antibody arrays, mass spectrometry, etc.).

6.1.1 SASP Expression by Real-Time Quantitative PCR (RT-qPCR)

1. The measurement of specific SASP markers that have been confirmed in HSCs by RT-qPCR is recommended. These include IL-6 [34], IL-8 [31], MMPl [31], MMP3 [31], CXCL1/2 [34], and CXCL10 [29, 31, 35, 36]. Furthermore, the upregulation of anti-apoptotic or immunomodulatory decoy receptors, e.g., DcR2 [12, 37, 38], is a commonly observed phenomenon in senescent cells (reviewed in [12]) and has also been described in senescent HSCs [34, 39], as well as in senescent hepatocytes and senescent cholangiocytes [40].
2. For transcriptional analyses using RT-qPCR, both flash frozen or fresh material derived from cell culture or tissue samples may be used. Before extraction and for long-term storage of RNA, all samples should be kept at temperatures lower than −80 °C. The steps of RNA extraction, cDNA synthesis, and RT-qPCR analyses may be performed as described in *see* Subheading 3.2. Similar to the previous RT-qPCR analyses, the use of at least two stable reference genes upon senescence induction is desired (e.g., GAPDH, and β-actin) [41].

6.1.2 SASP Expression by ELISA

The measurement of SASP factors at the protein level is often desirable as it may be more relevant to disease models than relative transcript levels measured by RT-qPCR. This may be achieved by enzyme-linked immunosorbent assays (ELISAs), which rely on matched antibody pairs to capture the analyte of interest and to detect the amount of captured analyte by comparing the signal ratio to a standard curve made of the spiked protein of interest at a known concentration. ELISAs are highly sensitive methods that can measure the protein of interest in a variety of specimens, for example, conditioned media from hepatocytes or HSCs, whole cell lysates derived from in vitro cultures to increase sensitivity with respect to conditioned media, as well as whole liver lysates derived from tissue samples. Recently, the intrahepatic levels of IL-6 and CXCL2 from whole liver lysates have been measured using a commercial IL-6 ELISA kit (eBioscience, Invitrogen) and a CXCL2 ELISA kit (R&D Systems) to demonstrate senescent cells as the major source of these SASP proteins after partial hepatectomy and to monitor the decrease of these proteins after the removal of senescent cells using the senolytic drug ABT-263 or ganciclovir in a transgenic mouse model [34].

The use of ELISA kits that contain pre-matched antibody pairs for capture and detection of the protein of interest is recommended. The use of high-binding multi-well plastic plates (96 well) is recommended. The plate is first coated with the capture body at the indicated concentration. This first coating step is followed by a washing step and a blocking step (e.g., 4% BSA/PBS) that will saturate the surface of the wells and minimize non-specific binding to any uncoated surface from the capture antibody. Following the blocking step, the samples to be measured are added (in technical replicates) to allow direct capture from the coated primary antibody. After typically >1 h incubation, the samples are removed, and the wells are washed to remove any leftover sample. The detection antibody is next added and incubated for an adequate time to enable binding to the captured analyte. ELISA detection antibodies are typically enzyme-labelled (e.g., horseradish peroxidase), which allows conversion of a colorimetric substrate (e.g., 3,3′, 5,5′-Tetramethylbenzidine, TMB) to produce a measurable signal. The intensity of the signal is directly proportional to the concentration of the measured compound in the original sample (*see* **Note 20**).

6.1.3 SASP Expression by Antibody Arrays

Several important SASP components can be identified by RT-qPCR or at the protein level using ELISAs. However, the SASP includes a wide variety of secreted proteins of which the expression changes over time and which may depend on the cell type and the senescence inducer [42]. The large complexity of the SASP complicates the quantification of multiple proteins using ELISA-based assays.

Therefore, the use of antibody arrays may represent a more inexpensive and less labor-intensive alternative to the use of several ELISAS for the relative quantitation of multiple SASP analytes. For antibody arrays, conditioned media derived from in vitro cultures may be employed, as well as tissue samples. For conditioned media, the number of cells should be counted for correct normalization; similarly, the amount of total protein per sample for tissue samples is required. We recommend the usage of pre-formatted antibody arrays with genes of interest kit systems (e.g., RayBio® L-Series: human L10 Array on glass slides, detects 500 human proteins) that are suitable for serum, plasma, lysates, and cell culture supernatants (*see* **Note 21**).

1. Commercially available antibody arrays (e.g., RayBiotech) may be used as instructed by the manufacturer. For conditioned media, we recommend concentrating the samples three- to fourfold using a semi-permeable membrane with a 2 kDa cutoff.
2. To minimize non-specific binding, the antibody array membranes are preincubated with blocking solution prior to incubation with the samples. Overnight incubation with the samples at 4 °C is recommended for increased sensitivity.
3. Following overnight incubation, the membranes are washed multiple times (5×) and are then typically incubated with a biotin-conjugated antibody cocktail containing a variety of antibodies (2 h at room temperature).
4. Following incubation with the biotin-conjugated antibodies, the membranes are washed again multiple times (5×) and a detection solution containing radio-labeled ^{35}S-streptavidin (for higher sensitivity) in blocking solution. Incubation is carried out (2 h at room temperature), followed by an additional five washes. Radioactivity is next detected using a phosphor imager. It is important to limit exposure to avoid any spot saturation.
5. Scanned images are then analyzed and corrected by subtracting the local background. The signal intensity is then calculated as the average of the replicate spots corresponding to the protein. Data for replicate spots with high variability should be discarded. The average intensity of the blank and negative spots on the array is subtracted from the protein intensity.
6. To ensure linearity, the protein intensity should be normalized for exposure time, antibody concentration, and streptavidin specific activity by scaling against the average value of the positive control spots on the array (which contain a constant amount of biotin). The data processing will result in numerical values that can be compared between arrays [42] (*see* **Notes 22–24**).

6.2 Identification of Morphological Changes

An important characteristic of senescent cells observed in vitro is the abnormal large and flat morphology. These morphological changes also define senescent cells [9]. The observed changes are related with an expanded nucleus. Moreover, increased granularity is also detected as a consequence of increased number and size of lysosomes. Furthermore, multinucleation can also occur in senescent cells. These morphological characteristics are often assessed using image techniques such as light or confocal microscopy confocal. Staining with DAPI can also be used to determine abnormalities in the nuclei. Morphological changes are only identified in adherent cells due to technical restrictions.

The descriptions of the techniques normally used to analyze the morphology of senescent HSCs can be found below. The method for identification of morphological changes that is described in the following relies on protocol that was previously published [43].

1. HSCs seeded on coverslips in, e.g., 12-well plates. After the cells have attached, the wells are washed twice with cold PBS.
2. Then fix the cells with 300 μL of 4% formaldehyde or the proper amount of volume ensuring that cells are completely covered with formaldehyde, and incubate for 15 min.
3. Remove the 4% formaldehyde, and wash again with PBS 3 times (*see* **Notes 15–17**).
4. Make the cells permeable by treating them with 300 μL of 0.1% Triton X-100 solution for 10 min at room temperature, and repeat the washing with PBS for 5 min.
5. Remove the PBS, add 200 μL of DAPI solution, and incubate for 15 min in the dark. All the further steps must be done in the dark. Rewash the cells with PBS twice to minimize the background signal.
6. To the glass slide, add 20 μL of mounting medium on top, place the coverslip with the cells upside down on the slide, and varnish the sides of the object glass to avoid drying of the stained cells.
7. Finally, cells can be visualized under confocal microscope by setting the excitation filter at 370 nm and emission filter at 526 nm (*see* **Notes 18** and **19**).

7 Notes

1. The reverse transcriptase (RT) is critical to the success of RT-qPCR, in addition to the selection of the DNA polymerase. To reverse transcribe mRNA, avian myeloblastosis virus (AMV) or Moloney murine leukemia virus (M-MLV) is normally used.

2. To perform RT-qPCR, total mRNA must be transcribed to cDNA, followed by reverse transcription according to the manufacturer's protocols, taking into account the recommended conditions of the reverse-transcriptase enzyme.
3. Due to the sensitivity of the mRNA to degradation, all procedures must be performed in sterile conditions and low temperatures (e.g., 4 °C), and the use of ice is required.
4. For comparative analysis of gene expression, the use of at least two stable reference genes upon senescence induction is desired (e.g., GAPDH, 36b4 and β-actin) [32]. The stability of the reference genes being used for normalization should be tested, and the efficiency of the primers used should be higher than 95%. Lastly, qPCR technical replicates ($n \geq 2$) per biological sample should be performed.
5. To analyze the data obtained by real-time RT-PCR, the comparative threshold method is normally used following the formula presented below:

$$\begin{aligned}\Delta\Delta CT = &(CTsample \ - CT\ housekeeping)\\ &- (CTreference - CTreference\ housekeeping)\end{aligned}$$

6. Be aware that growth conditions (e.g., medium composition, supplements) can influence the response to senescence stimuli. Therefore, we suggest adding the appropriate controls to the experimental protocol.
7. Some reagents used to prepare the gels are toxic. We recommend using gloves and the appropriate equipment to protect yourself while you are preparing the material.
8. To minimize proteolytic degradation, samples should always be kept on ice (until the addition of loading buffer and denaturation or storage). Lysates can be stored at −80 °C for prolonged periods (several months) without any significant degradation of their protein content.
9. The use of at least two stable reference proteins upon senescence induction is desired (e.g., GAPDH and β-actin) [32].
10. Be aware of the incubation conditions; the presence of CO_2 can cause changes to the pH, which may affect staining results.
11. Culture media must be warm before start of the experiment. If it is not warm, let the plate with the medium stand for 30 min in the incubator before you start the experiment.
12. As all commonly used biocides negatively influence the substrate reaction, substrate component A is provided as a sterile filtered solution without addition of biocides. Keep the solution at 2–8 °C, and avoid contamination when handling.

13. The assay can be interrupted after the labelling process. The dry cells can be stored up to 1 week at 2–8 °C.
14. The substrate reaction solution should be prepared in time and should be stirred for 15 min before use. Incubate at room temperature for at least 3 min on a shaker. To avoid "fading out" of substrate, samples should be quantitated within 10 min after adding the reagent.
15. During cell seeding step, do not forget to seed cells on the object glass, which must be placed into the plate before seeding. Also ensure that your plates with the object glass are sterile. Clean first with 75% ethanol.
16. Cell fixation is performed using cold 4% formaldehyde (PFA). Remove PFA gently to avoid detaching the cells from the plate.
17. All remaining steps, including washing: add (washing) solutions gently to avoid damaging or detaching the cells.
18. Immunofluorescence can be performed at room temperature.
19. To avoid photobleaching, keep the slides in the dark at 4 °C until observation, and avoid long exposure times with strong laser intensity when imaging.
20. Images can be analyzed in a confocal microscope.
21. Not all reagents are described in the commercial kit (RayBio® L-Series).
22. The assay can only measure one analyte (e.g., IL-6) present in the sample at a concentration that falls within the linear range of the assay. Therefore, the samples to be analyzed may need to be diluted or concentrated if they fall outside of the calibrated range.
23. To ensure linearity, the protein intensity should be normalized for exposure time, antibody concentration, and streptavidin specific activity by scaling against the average value of the positive control spots on the array (which contain a constant amount of biotin).
24. The data processing will result in numerical values that can be then compared between arrays [42].

8 Conclusions

The complexity of cellular senescence coupled with the heterogeneity of cells and the multitude of triggers that induce senescence causes different responses that can lead to discrepancies in cellular senescence detection. Despite these variables, there are common biomarkers representative of senescence regardless of the type of senescence and even cell type such as hepatic stellate cells, which together give a clear indication of the presence of senescent cells in in vitro and in vivo model.

Sources of Funding Dutch Cancer Foundation (KWF) and Dutch Digestive Disease Foundation (MLDS).

References

1. Zhang M et al (2021) Hepatic stellate cell senescence in liver fibrosis: characteristics, mechanisms and perspectives. Mech Ageing Dev 199:111572. https://doi.org/10.1016/j.mad.2021.111572
2. Serna-Salas SA et al (2022) α-1 adrenergic receptor antagonist doxazosin reverses hepatic stellate cells activation via induction of senescence. Mech Ageing Dev 201:111617. https://doi.org/10.1016/j.mad.2021.111617
3. Ahmad R, Ahmad A (2012) Understanding the mechanism of hepatic fibrosis and potential therapeutic approaches. Saudi J Gastroenterol 18(3):155. https://doi.org/10.4103/1319-3767.96445
4. Shang L et al (2018) Human hepatic stellate cell isolation and characterization. J Gastroenterol 53(1):6–17. https://doi.org/10.1007/s00535-017-1404-4
5. Rippe RA, Brenner DA (2004) From quiescence to activation: gene regulation in hepatic stellate cells. Gastroenterology 127(4):1260–1262. https://doi.org/10.1053/j.gastro.2004.08.028
6. Hernandez-Gea V et al (2011) Mechanism of hepatic fibrogenesis. Annu Rev Pathol 14(3):181–194. https://doi.org/10.1016/j.bpg.2011.02.005.Mechanisms
7. Campisi J (2001) Cellular senescence as a tumor-suppressor mechanism. Trends Cell Biol 11(11):27–31. https://doi.org/10.1016/S0962-8924(01)02151-1
8. Campisi J, DAdda Di Fagagna, F. (2007) Cellular senescence: when bad things happen to good cells. Nat Rev Mol Cell Biol 8(9):729–740. https://doi.org/10.1038/nrm2233
9. Herranz N, Gil J (2018) Mechanisms and functions of cellular senescence. J Clin Investig 128(4):1238–1246. https://doi.org/10.1172/JCI95148
10. Jin H et al (2016) Activation of PPARγ/p53 signaling is required for curcumin to induce hepatic stellate cell senescence. Cell Death Dis 7(4):1–11. https://doi.org/10.1038/cddis.2016.92
11. Krizhanovsky V et al (2008) Senescence of activated stellate cells limits liver fibrosis. Cell 134(4):657–667. https://doi.org/10.1016/j.cell.2008.06.049
12. Soto-Gamez A et al (2022) Enhanced extrinsic apoptosis of therapy-induced senescent cancer cells using a death receptor 5 (DR5) selective agonist. Cancer Lett 525:67–75. https://doi.org/10.1016/j.canlet.2021.10.038
13. Coppé J et al (2010) The senescence-associated secretory: the dark side of tumor suppression. Annu Rev Pathol 5:99–118. https://doi.org/10.1146/annurev-pathol-121808-102144.The
14. Davalli P et al (2016) ROS, cell senescence, and novel molecular mechanisms in aging and age-related diseases. *Oxid Med Cell Longev* 2016:2016. https://doi.org/10.1155/2016/3565127
15. MacIel-Barón LÁ et al (2018) Cellular senescence, neurological function, and redox state. Antioxid Redox Signal 28(18):1704–1723. https://doi.org/10.1089/ars.2017.7112
16. Lee BY et al (2006) Senescence-associated β-galactosidase is lysosomal β-galactosidase. Aging Cell 5(2):187–195. https://doi.org/10.1111/j.1474-9726.2006.00199.x
17. Kojima H et al (2013) IL-6-STAT3 signaling and premature senescence. Jak-Stat 2(4):e25763. https://doi.org/10.4161/jkst.25763
18. Muñoz-Espín D, Serrano M (2014) Cellular senescence: from physiology to pathology. Nat Rev Mol Cell Biol 15(7):482–496. https://doi.org/10.1038/nrm3823
19. Byun HO et al (2015) From cell senescence to age-related diseases: differential mechanisms of action of senescence-associated secretory

phenotypes. BMB Rep 48(10):549–558. https://doi.org/10.5483/BMBRep.2015.48.10.122

20. Soto-Gamez A, Demaria M (2017) Therapeutic interventions for aging: the case of cellular senescence. Drug Discov Today. Elsevier Ltd, 786–795. https://doi.org/10.1016/j.drudis.2017.01.004
21. Bruce JL et al (2000) Requirements for cell cycle arrest by p16(INK4a). Mol Cell 6(3): 737–742. https://doi.org/10.1016/S1097-2765(00)00072-1
22. Takahashi A, Ohtani N, Hara E (2007) Irreversibility of cellular senescence: dual roles of p16INK4a/Rb-pathway in cell cycle control. Cell Div 2:1–5. https://doi.org/10.1186/1747-1028-2-10
23. Hoeferlin LA et al (2011) Activation of p21-dependent G1/G2 arrest in the absence of DNA damage as an antiapoptotic response to metabolic stress. Genes Cancer 2(9): 889–899. https://doi.org/10.1177/1947601911432495
24. Panebianco C et al (2017) Senescence in hepatic stellate cells as a mechanism of liver fibrosis reversal: a putative synergy between retinoic acid and PPAR-gamma signalings. Clin Exp Med 17(3):269–280. https://doi.org/10.1007/s10238-016-0438-x
25. Adams G (2020) A beginners guide to RT-PCR, qPCR and RT-qPCR, *y Portland Press Limited under the Creative Commons Attribution Licen* [Preprint]. https://doi.org/10.1042/BIO20200034/884464/bio20200034.pdf
26. Soto-Gamez A, Quax WJ, Demaria M (2019) Regulation of survival networks in senescent cells: from mechanisms to interventions. J Mol Biol. Academic Press, 2629–2643. https://doi.org/10.1016/j.jmb.2019.05.036
27. Kurien BT, Hal Scofield R (2015) Western blotting: an introduction. In: *Western blotting: methods and protocols*. Springer, New York, pp 17–30. https://doi.org/10.1007/978-1-4939-2694-7_5
28. Martinez-Serra J et al (2014) xCELLcelligence system for real-time label-free monitoring of growth and viability of cell lines from hematological malignancies. Onco Targets Ther 7: 985–994. https://doi.org/10.2147/OTT.S62887
29. Hamidi H, Lilja J, Ivaska J (2017) Using xCELLigence RTCA instrument to measure cell adhesion. Bio Protoc 7(24). https://doi.org/10.21769/bioprotoc.2646
30. Kilgas S, Kiltie AE, Ramadan K (2021) Immunofluorescence microscopy-based detection of ssDNA foci by BrdU in mammalian cells. STAR Protoc 2(4):100978. https://doi.org/10.1016/j.xpro.2021.100978
31. Freund A et al (2012) Lamin B1 loss is a senescence-associated biomarker. Mol Biol Cell 23(11):2066–2075. https://doi.org/10.1091/mbc.E11-10-0884
32. Hernandez-Segura A, Rubingh R, Demaria M (2019) Identification of stable senescence-associated reference genes. *Aging Cell*. Blackwell Publishing Ltd 18:e12911. https://doi.org/10.1111/acel.12911
33. Dimri GP et al (1995) A biomarker that identifies senescent human cells in culture and in aging skin in vivo. Proc Natl Acad Sci 92 (September):9363–9367
34. Cheng N, Kim KH, Lau LF (2022) Senescent hepatic stellate cells promote liver regeneration through IL-6 and ligands of CXCR2. JCI Insight 7(14). https://doi.org/10.1172/jci.insight.158207
35. Narita M, Nuñez S, Heard E et al (2003) Rb-mediated heterochromatin formation and silencing of E2F target genes during cellular senescence. Cell 13:703–716. https://doi.org/10.1016/s0092-8674(03)00401-x
36. Sagiv A et al (2016) NKG2D ligands mediate immunosurveillance of senescent cells. Aging 2(2):328–344
37. Sadaie M et al (2015) Cell-based screen for altered nuclear phenotypes reveals senescence progression in polyploid cells after Aurora kinase B inhibition. Mol Biol Cell 26(17): 2971–2985. https://doi.org/10.1091/mbc.E15-01-0003
38. Madsen SD et al (2017) Decoy TRAIL receptor CD264: a cell surface marker of cellular aging for human bone marrow-derived mesenchymal stem cells. Stem Cell Res Ther 8(1): 201. https://doi.org/10.1186/s13287-017-0649-4
39. Sagiv A et al (2013) Granule exocytosis mediates immune surveillance of senescent cells. Oncogene 32(15):1971–1977. https://doi.org/10.1038/onc.2012.206
40. Ferreira-Gonzalez S et al (2018) Paracrine cellular senescence exacerbates biliary injury and impairs regeneration. Nat Commun 9(1). https://doi.org/10.1038/s41467-018-03299-5
41. Itahana K, Campisi J, Dimri GP (2007) Methods to detect biomarkers of cellular senescence: the senescence-associated β-galactosidase assay.

Methods Mol Biol 371:21–31. https://doi.org/10.1385/1-59745-361-7:21

42. Coppé JP et al (2008) Senescence-associated secretory phenotypes reveal cell-nonautonomous functions of oncogenic RAS and the p53 tumor suppressor. PLoS Biol 6(12):2853. https://doi.org/10.1371/journal.pbio.0060301

43. Im K et al. (2019) An introduction to performing immunofluorescence staining. Methods Mol Biol. Humana Press Inc., 1897: 299–311. https://doi.org/10.1007/978-1-4939-8935-5_26

Chapter 7

Isolation of Hepatic Stellate Cells and Lymphocytes for Co-culture Systems

Hee-Hoon Kim, Kyurae Kim, Song Hwa Hong, and Won-Il Jeong

Abstract

In contrast to quiescent hepatic stellate cells (HSCs), activated HSCs play crucial roles in the development of liver fibrosis by producing a huge amount of extracellular matrix such as collagen fibers. However, recent lines of evidence have also highlighted the immunoregulatory functions of HSCs, in which they interact with diverse hepatic lymphocytes to produce cytokines and chemokines, release extracellular vesicles, or express specific ligands. Therefore, to understand the exact interactions between HSCs and lymphocyte subsets in the pathogenesis of the liver disease, it is valuable to establish experimental procedures to isolate HSC and co-culture them with lymphocytes. Here, we introduce the efficient methods to isolate and purify mouse HSCs and hepatic lymphocytes using density gradient centrifugation, microscopic observation, and flow cytometry. Moreover, we provide the direct and indirect co-culturing methods of isolated mouse HSCs and hepatic lymphocytes based upon the purpose of the study.

Key words Hepatic stellate cells, Lymphocytes, Co-culture, Density gradient centrifugation, Flow cytometry

1 Introduction

Hepatic stellate cells (HSCs), first described by von Kupffer in 1876, are located in the perisinusoidal space of Disse in the liver. Although HSCs account for ~10% of resident hepatic cells, they are the main reservoir for retinoids in the body [1]. In particular, HSCs store retinol as retinyl esters in cytoplasmic lipid droplets, which help us isolate them using density gradient centrifugation [2]. Upon chronic liver injury or prolonged in vitro culture, HSCs lose their retinol lipid droplets and transdifferentiate into myofibroblast-like cell types. These "activated" HSCs become a primary fibrogenic cell type in liver fibrosis, which is characterized by producing large amounts of extracellular matrix (ECM), including collagen fibers, and by increasing the expression of α-smooth muscle actin [3]. In addition to ECM production in the liver, HSCs become proliferative, inflammatory, and chemotactic after

Ralf Weiskirchen and Scott L. Friedman (eds.), *Hepatic Stellate Cells: Methods and Protocols*, Methods in Molecular Biology, vol. 2669, https://doi.org/10.1007/978-1-0716-3207-9_7,

activation [4]. In particular, the interactions between activated HSCs and diverse hepatic lymphocytes regulate the development of liver fibrosis by producing cytokines/chemokines and expressing specific activation ligands [5]. Human and mouse HSCs can stimulate proliferation and activation of lymphocytes by expressing various membrane proteins of antigen-presenting molecules including human leukocyte antigen (HLA)-1, HLA-II, CD1b, CD1c, CD1d, and other factors such as CD40 and CD80. Their expression levels are enhanced by culture activation or prolonged exposure to inflammatory cytokines [5, 6]. In experimental models of liver fibrosis, activated HSCs produce diverse C-X-C motif chemokine ligands (CXCLs) and retinoic acid-induced ligands, including CXCL9, CXCL10, CXCL12, and retinoic acid early inducible gene 1 (RAE-1), to induce the infiltration or migration of natural killer (NK) cells, NKT cells, and T cells to the fibrotic regions [7, 8]. Moreover, activated HSCs secrete pro- or anti-fibrotic cytokines, such as interleukin (IL)-1β, IL-23, IL-17, and IL-10, which regulate the function of recruited lymphocytes [9, 10]. Conversely, lymphocytes can also regulate the activation of HSCs. For example, NK cell-derived interferon (IFN)-γ suppresses liver fibrosis by killing activated HSCs [11], while B cells aggravate the activation of HSCs by producing IL-6 in the carbon tetrachloride-induced mouse liver fibrosis model [12]. However, the roles of T cells and NKT cells in the regulation of activated HSCs are still controversial, depending on their subtypes and experimental conditions [13, 14].

Therefore, it is important to clarify the precise interactions between HSCs and lymphocytes in liver fibrosis by establishing isolation techniques for the HSCs and hepatic lymphocytes with high purity and yield, as well as by using co-culturing methods. In this chapter, we outline the isolation method of primary HSCs and lymphocytes from mouse liver using density gradient centrifugation. In addition, we briefly discuss direct or indirect co-culturing methods with isolated mouse primary HSCs and hepatic lymphocytes. Our protocols may help researchers gain a deeper understanding of the immunoregulatory roles of HSCs.

2 Materials

2.1 Isolation of Mouse HSCs

2.1.1 Reagents

1. EGTA solution: Add 0.5 mM EGTA, 5.4 mM KCL, 0.44 mM KH_2PO_4, 140 mM NaCl, 0.34 mM Na_2HPO_4, 25 mM tricine, 100 U/mL penicillin, 100 μg/mL streptomycin, and 0.25 μg/mL amphotericin B to 500 mL of distilled water. Mix in magnetic stirrer, adjust pH to 7.2 with NaOH, and sterilize the solution by filtering with a 0.22 μm filter.

2. Hanks' balanced salt solution (HBSS) solution: Add 20 μg/mL DNase I, 100 U/mL penicillin, 100 μg/mL streptomycin, and 0.25 μg/mL amphotericin B in 500 mL HBSS.
3. Collagenase I solution: Add 37.5 mg of collagenase I to 50 mL of HBSS solution.
4. Digestion solution: Add 12 mL of HBSS solution to 3 mL of collagenase I solution.
5. OptiPrep™: Dilute 60% stock OptiPrep™ density gradient medium to 20% and 11.5% in HBSS solution. Leave at room temperature.
6. RPMI medium with 100 U/mL penicillin, 100 μg/mL streptomycin, 0.25 μg/mL amphotericin B, and 10% fetal bovine serum.
7. Isoflurane.
8. 70% ethanol.
9. 0.4% trypan blue.

2.1.2 Equipment and Consumables

1. Precision balance.
2. pH meter.
3. Magnetic stirrer.
4. 500 mL 0.22 μm filter system.
5. 15 mL and 50 mL tubes.
6. Conical tube rack.
7. 42 °C water bath.
8. Peristaltic pump set to 3.5 mL/min.
9. Pump tubing, internal diameter 2.79 mm and 16″ length.
10. Absorber plate (expanded polystyrene) and dissection pins.
11. Mouse dissection tray.
12. Mouse dissection board and pins.
13. Spray bottle.
14. Sterile scissors and forceps.
15. 22G IV catheter.
16. Kimwipes™ or equivalent.
17. Non-absorbable sterile silk suture, 18 mm 3/8.
18. Bulldog clamp.
19. 60 mm petri dish.
20. Sterile 3 mL transfer pipettes.
21. Shaking incubator set to 37 °C, 90 rpm.
22. 70 μm cell strainer.

23. Clean bench.
24. Centrifuges for 15 mL or 50 mL tubes.
25. Suction pump.
26. Hemocytometer.
27. 6-well plate.
28. Bright-field microscope.
29. CO_2 incubator, 37 °C.

2.2 Isolation of Mouse Hepatic Lymphocytes

2.2.1 Reagents

1. Sterile 1 × phosphate-buffered saline (PBS).
2. Lymphocyte separation medium (Corning or equivalent).
3. RPMI medium with 100 U/mL penicillin, 100 μg/mL streptomycin, 0.25 μg/mL amphotericin B, and 10% fetal bovine serum.
4. (Optional) Magnetic-activated cell sorting (MACS) buffer: Add 5 g bovine serum albumin, 2 mM EDTA, and 9.55 g Dulbecco's PBS (DPBS) to 1 L distilled water. Mix with a magnetic stirrer, and sterilize by filtering with a 0.22 μm pore filter.
5. (Optional) MACS kit for target cells (Miltenyi Biotec).
6. Isoflurane.
7. 70% ethanol.
8. 0.4% trypan blue.

2.2.2 Equipment and Consumables

1. Mouse dissection tray.
2. Mouse dissection board and pins.
3. Sterile scissors and forceps.
4. Spray bottle.
5. Clean bench.
6. 60 mm petri dish.
7. 70 μm cell strainer.
8. Sterile 3 mL syringe plunger.
9. Sterile 3 mL transfer pipettes.
10. 15 mL and 50 mL tubes.
11. Centrifuges for 15 mL or 50 mL tubes.
12. Suction pump.
13. Hemocytometer.
14. Bright-field microscope.
15. 500mL 0.22 μm filter system.
16. (Optional) Precision balance.

17. (Optional) Magnetic stirrer.
18. (Optional) Manual cell separators for MACS and MS columns.

2.3 Co-culturing of Mouse HSCs and Hepatic Lymphocytes

2.3.1 Reagents

1. RPMI medium with 100 U/mL penicillin, 100 μg/mL streptomycin, 0.25 μg/mL amphotericin B, and 10% fetal bovine serum.
2. Trypsin-EDTA solution (1×).

2.3.2 Equipment and Consumables

1. Clean bench.
2. Transwell, 3 μm pore size, 24 mm insert for 6-well plates (Costar).
3. Culture plates.
4. Sterile twisser.
5. Microcentrifuge tubes 1.5 mL.
6. Benchtop centrifuge.
7. Bright-field microscope.

3 Methods

3.1 Isolation of Mouse HSCs

1. Pre-warm the EGTA solution, collagenase I solution, and digestion solution in a water bath at 42 °C. Connect the tubing to the peristaltic pump and circulate the EGTA solution for 30 min in a closed circuit to pre-warm the tubing. After pre-warming, completely remove the bubbles adhering to the tubing wall by tapping or rapid circulation.
2. Prepare the surgical area with a mouse dissection tray and absorber plate.
3. Anesthetize the mouse with isoflurane (*see* **Note 1**).
4. Fix the limbs of the mouse with dissection pins, and spray 70% ethanol to sterilize the abdominal area.
5. Open the abdominal cavity, and shift the intestines to the right side to visualize the infrahepatic inferior vena cava (IVC) and hepatic portal vein (Fig. 1a).
6. Gently grasp the small intestine, and slide the suture silk into the interstitial cavity. Loosely tie the suture silk over the portal vein.
7. Carefully insert the catheter into the portal vein. Since the portal vein is very thin, care must be taken not to pierce it (*see* **Notes 2** and **3**). After insertion, push the catheter sheath forward, and pull out the needle simultaneously. As fast as you can, tightly tied the suture silk (*see* **Note 4**). If the

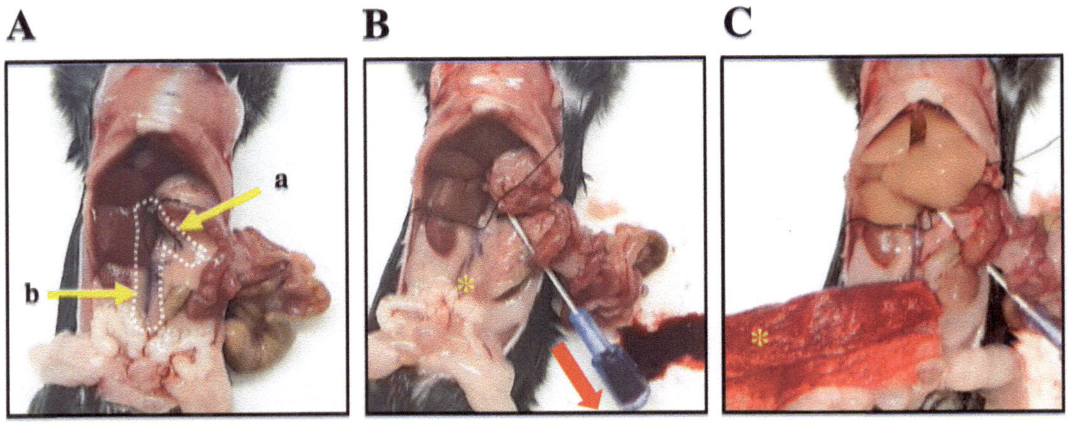

Fig. 1 Portal vein puncture and tubing connection. (**a**) The portal vein (a) and the infrahepatic inferior vena cava (b). (**b**) After checking the blood flow (arrow) which indicates successful cannulation, tightly tie suture silk, and cut the distal part of the infrahepatic inferior vena cava (asterisk). (**c**) Move the drainage to the cut area (asterisk). After 2 min of cannulation, the liver turned yellow

cannulation was successful, you can see the blood flow after the removal of the needle (Fig. 1b).

8. Connect one end of the tubing in the EGTA solution to the cannula. After confirming the swelling of the liver, immediately cut the distal end of the IVC, and place a Kimwipes™ (or equivalent) for efficient drainage. If the above processes are successful, the liver will turn white in a few seconds (Fig. 1c) (*see* **Note 5**).
9. After the circulation of approximately 40 mL of EGTA solution, stop the peristaltic pump. Transfer the tubing into the collagenase I solution, and restart the pump (*see* **Note 6**).
10. Cut the diaphragm and right ribs to expose suprahepatic IVC (Fig. 2a).
11. Clip the infrahepatic IVC with a bulldog clamp, and check the swelling of the liver. Cut the suprahepatic IVC, and move the Kimwipes™ to the right ribs (Fig. 2b) (*see* **Note 7**).
12. After circulating 50 mL of collagenase I solution, stop the pump. If the above steps are done well, the liver may be enlarged and soft (Fig. 2c). Remove the liver with sterile scissors, and transfer it into a 50 mL conical tube containing 15 mL of digestion solution. Move to a clean bench, and place the liver in a sterile 60 mm petri dish.
13. Pour 5 mL of digestion solution into a 60 mm petri dish, and briefly mince the liver with scissors. And then, gently pipet the liver with a transfer pipette 10–15 times until getting a homogeneous solution (Figs. 2d–f) (*see* **Note 8**). Move the solution into a 50 mL conical tube containing the digestion solution.

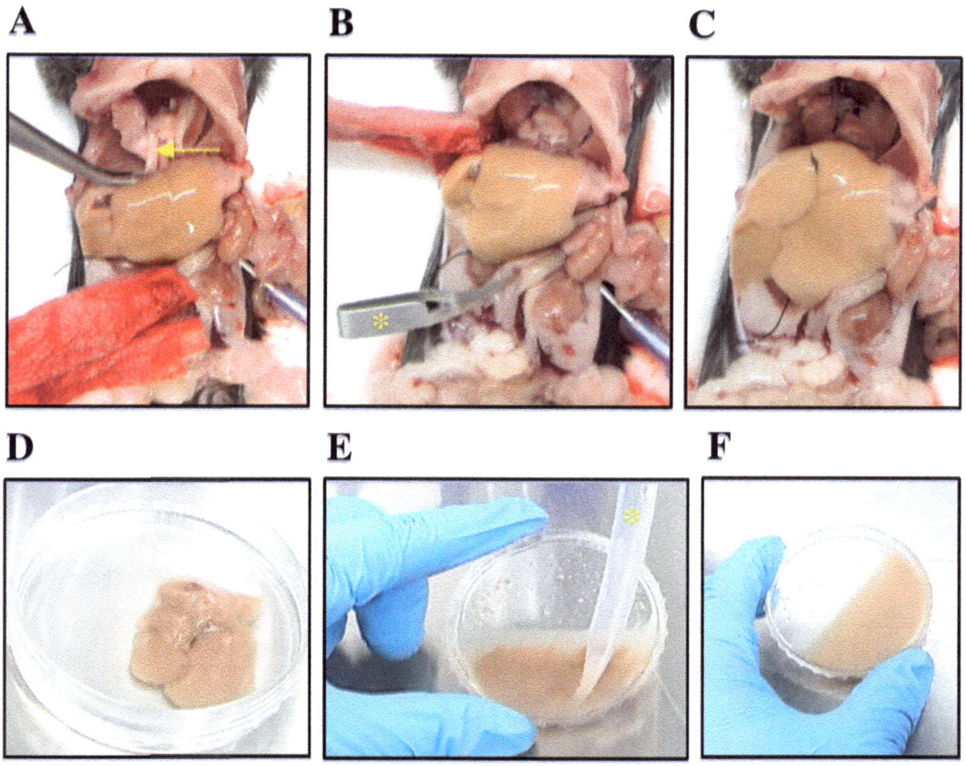

Fig. 2 Digestion of liver after collagenase perfusion. (**a**) The suprahepatic inferior vena cava (arrow). (**b**) Clip the infrahepatic IVC with a bulldog clamp (asterisk), and cut the suprahepatic inferior vena cava. Move the drainage to the upper side. (**c**) After finishing the circulation of the collagenase I solution, the liver became enlarged and limp. (**d**) Dissect the liver, and move to a 60 mm petri dish. (**e**) Gently pipet the liver with a transfer pipette (asterisk). (**f**) After 10–15 times of pipetting, the liver became shapeless

14. Further digest the liver in the shaking incubator for 20 min at 90 rpm, 37 °C, to thoroughly dissociate it at a single-cell level.
15. On a clean bench, filter the digested liver suspension through a 70 μm cell strainer on a new 50 mL tube to eliminate the undigested liver and connective tissues (Fig. 3a). Add HBSS solution to 50 mL, and centrifuge for 5 min at 50 × *g*, 4 °C (*see* **Note 9**).
16. Move the supernatant to a new 50 mL tube (*see* **Note 10**) (Fig. 3b). For the pellet, gently resuspend it to 50 mL of HBSS solution. Centrifuge both tubes for 5 min at 50 × *g*, 4 °C.
17. Move the supernatant containing the non-parenchymal cells of each tube to new 50 mL tubes (Fig. 3c). Centrifuge both tubes for 10 min at 500 × *g*, 4 °C.
18. Discard the supernatants, and gently resuspend the pellets (non-parenchymal cells) in the HBSS solution using a transfer pipette (Fig. 3d). Put the suspension from both tubes into one

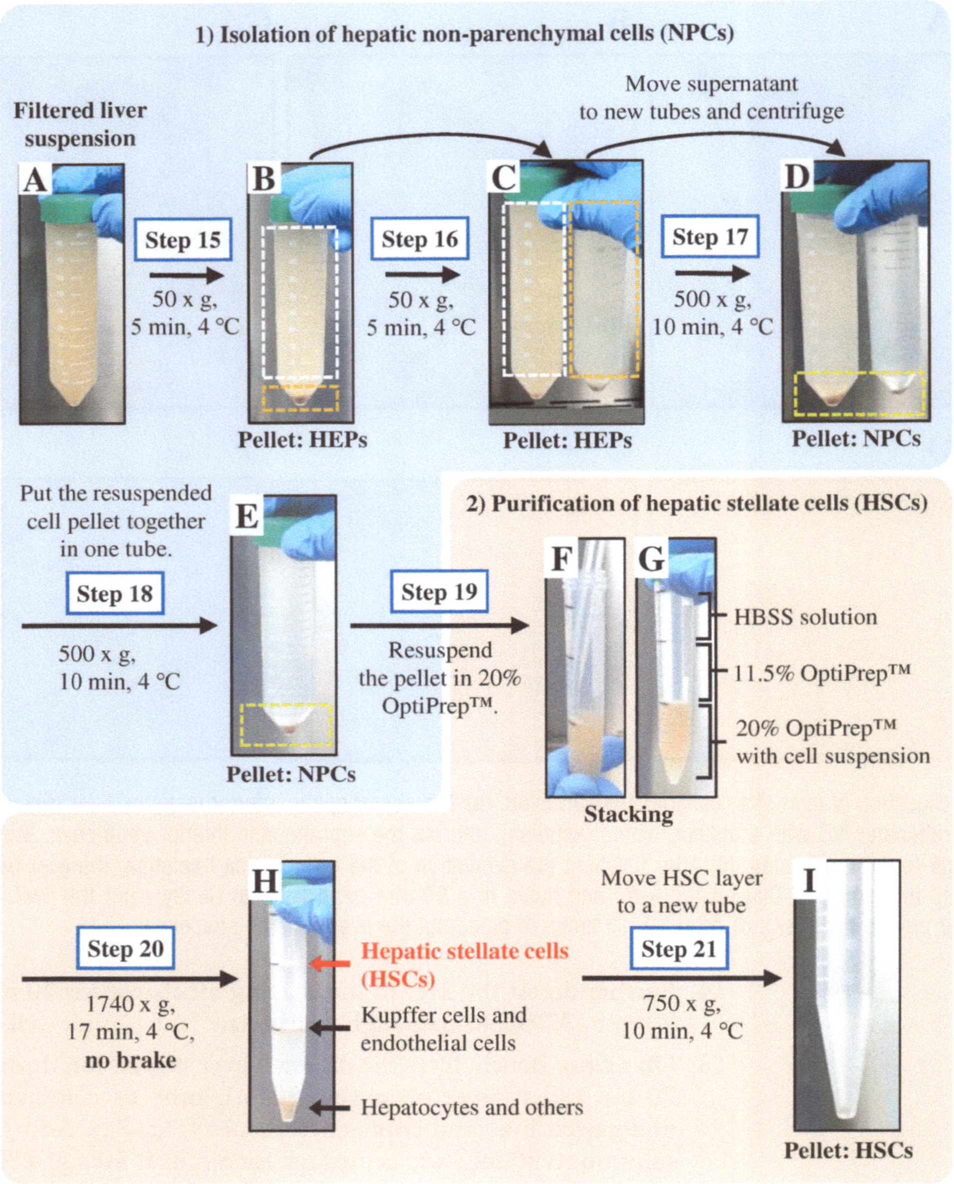

Fig. 3 Overall centrifugation procedures for the isolation of mouse hepatic stellate cells (HSCs). (**a**–**e**) After digestion of the liver by collagenase perfusion, remove the hepatocytes and cell debris by centrifugation, and isolate the hepatic non-parenchymal cells (NPCs). (**f**–**i**) Purify HSCs by centrifugation with OptiPrep™ density gradient medium

tube, and add the HBSS solution to 50 mL. Centrifuge at 500 × *g* for 10 min, 4 °C (Fig. 3e).

19. Carefully and thoroughly remove the supernatant with the aspiration, and resuspend the pellet in 5 mL of 20% OptiPrep™

using a transfer pipette (*see* **Note 11**) (Fig. 3f). Move the suspension to a new 15 mL tube, and as carefully as possible, stack the 5 mL of 11.5% OptiPrep™ over the cell suspension using a transfer pipette. And then, stack 4 mL of HBSS solution over the 11.5% OptiPrep™ in a similar way, making a total of three layers (Fig. 3g). If the layers are well stacked, you can see the clear borderline in each interface.

20. Centrifuge for 17 min at 1740 × *g*, 4 °C, without brake to prevent interlayer mixing. After centrifugation, check that each layer is not mixed.
21. Carefully take the interface between the HBSS solution and 11.5% OptiPrep™ (containing HSCs) using a transfer pipette (Fig. 3h), and move it to a new 15 mL tube. Add HBSS solution to 15 mL, and centrifuge for 10 min at 750 × *g*, 4 ° C (*see* **Note 12**).
22. Check the pellet containing HSCs (Fig. 3i). Thoroughly aspirate the supernatant to wash out the OptiPrep™ solution, and resuspend the HSCs in 1 mL of RPMI media. Keep the tube on ice until plating HSCs.
23. Mix the 10 μL of cell suspension with 10 μL of 0.4% trypan blue, and put 10 μL of the mixture in a hemocytometer. Check the viability and number of isolated HSCs in the bright-field microscope (Fig. 4a, b).
24. Plate the HSCs in RPMI media for further analysis. Recommended numbers of HSCs for various sizes of culture plates are shown in Table 1. Representative morphology of isolated HSCs is shown in Fig. 4c. In this protocol, the yield for HSCs may range from 1.0 to 1.5 million cells per liver.

3.2 Isolation of Mouse Hepatic Lymphocytes

1. Pre-warm the 5 mL of the lymphocyte separation medium and 50 mL of sterile 1 × PBS at room temperature (RT).
2. Prepare the mouse dissection board.
3. Anesthetize the mouse with isoflurane.
4. Fix the mouse with dissection pins to a board and spray 70% ethanol to sterilize the abdominal area.
5. Cut open the abdominal cavity, and carefully remove the liver.
6. Pick the liver with forceps, and wash it by shaking it in a 50 mL tube containing sterile 1 × PBS to remove the blood.
7. Weigh the liver, and put it on the 70 μm cell strainer. Then, put the strainer on a 60 mm petri dish (Fig. 5a).
8. Gently grind the liver with the bottom of the syringe plunger by adding 3–5 mL of sterile 1 × PBS (*see* **Note 13**). Periodically transfer the liver homogenates to a new 50 mL tube on ice whenever the color of the solution turns red (Fig. 5b).

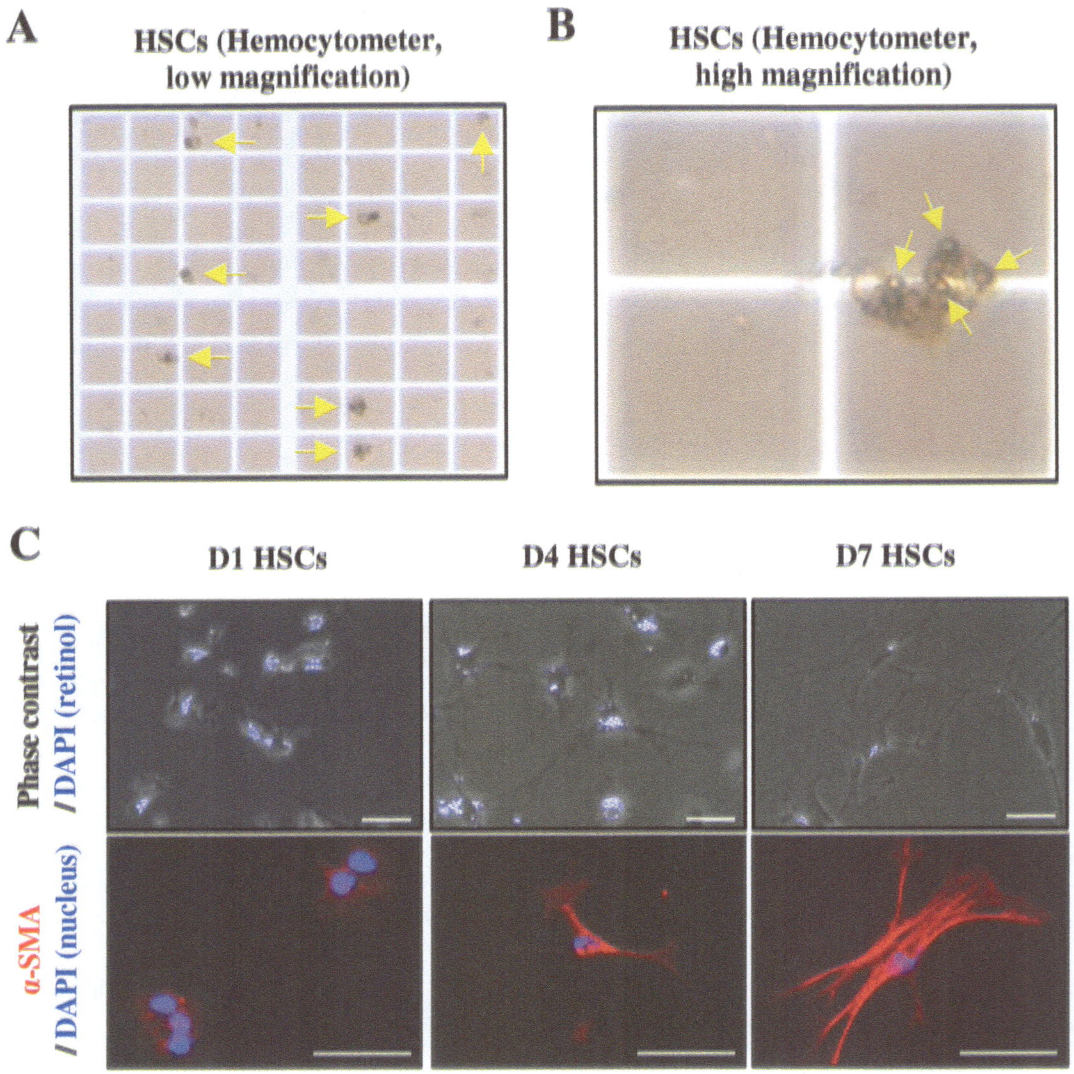

Fig. 4 Validation of isolated mouse HSCs. (**a**) Isolated HSCs (arrows) are shown on the hemocytometer under the bright-field microscope. (**b**) At high magnification, retinol droplets are identified (arrows). (**c**) Microscopic observation of isolated HSCs by the time of culture. In phase-contrast images, retinol droplets in HSCs are identified and decreased by the time of culture (upper row). Conversely, the expression of alpha-smooth muscle actin (α-SMA) in HSCs is increased by the time of culture in immunostaining. Scale bars, 50 μm

9. Add another 3–5 mL of sterile 1 × PBS to the cell strainer, and repeat **step 8** until only connective tissues remain in the cell strainer (Fig. 5c).
10. Add sterile 1 × PBS to 50 mL (Fig. 6a), and centrifuge for 5 min at 50 × *g*, 4 °C.

Table 1
Conditions for co-culturing method of isolated hepatic stellate cells (HSCs) with lymphocytes (LCs)

Co-culturing method	Purpose	Culture vessel	Medium volumes	Number of plated cells/well
1. Direct co-culture	1.1 To examine the effects of the physical interaction between the cell types	96-well plate	0.2 mL	HSCs: 5×10^4 LCs: 5×10^4
		24-well plate	0.5 mL	HSCs: 1×10^5 LCs: 1×10^5
		12-well plate	1.0 mL	HSCs: 1.5×10^5 LCs: 1.5×10^5
		6-well plate	2.0 mL	HSCs: 2×10^5 LCs: 2×10^5
	1.2 Cytotoxicity assay	12-well plate	1.0 mL	HSCs: 1.5×10^5 LCs: 1.5×10^5 to 1.5×10^6
		6-well plate	2.0 mL	HSCs: 2×10^5 LCs: 2×10^5 to 2×10^6
2. Indirect co-culture using Transwell	2.1 To examine the effects of the secreted molecules between the cell types	12-well plate, 0.4 μm pore size	(HSCs) 1.5 mL Insert (LCs) 0.5 mL	1.5×10^5 1.5×10^5 to 1.5×10^6
		6-well plate, 0.4 μm pore size	(HSCs) 2.6 mL Insert (LCs) 1.5 mL	2×10^5 2×10^5 to 2×10^6
	2.2 Migration assay	12-well plate, 3.0 μm pore size	(HSCs) 1.5 mL Insert (LCs) 0.5 mL	1.5×10^5 1.5×10^5 to 4.5×10^5
		6-well plate, 3.0 μm pore size	(HSCs) 2.6 mL Insert (LCs) 1.5 mL	2×10^5 2×10^5 to 6×10^5

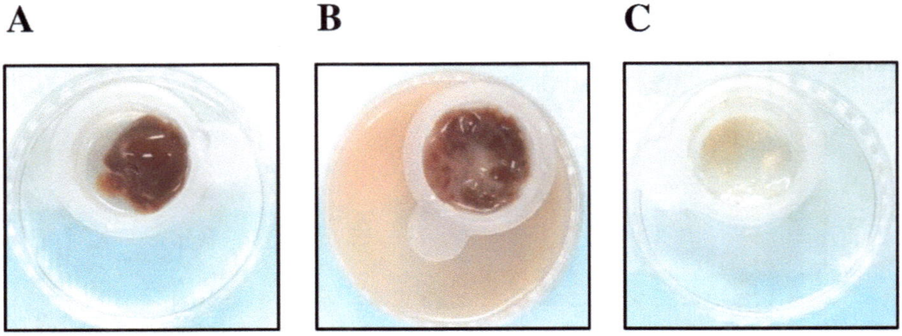

Fig. 5 Grinding of mouse liver tissue. (**a**) The liver was dissected from the mouse and put on the cell strainer. (**b**) Add 1 × PBS, and gently grind the liver with a syringe plunger on ice. Periodically move the homogenates to an ice-cold tube. (**c**) After finishing grinding, only connective tissues may remain on the cell strainer

11. Carefully move the supernatant to a new 50 mL tube (Fig. 6b). Centrifuge again for 5 min at 50 × *g*, 4 °C, to remove the cell debris and hepatocytes as much as possible.
12. Move the supernatant containing hepatic non-parenchymal cells (NPCs) to a new 50 mL tube (Fig. 6c). Centrifuge for 10 min at 500 × *g*, 4 °C.
13. Discard the supernatant thoroughly with aspiration, and resuspend the pellet in 3 mL of the pre-warmed sterile 1 × PBS in **step 1** using a transfer pipette. In a new 15 mL tube, add 3 mL of pre-warmed lymphocyte separation medium, and carefully stack over the 3 mL of cell suspension in 1 × PBS using a transfer pipette (Fig. 6d). Make sure there is no interlayer mixing. Centrifuge for 30 min at 400 × *g*, 20 °C, without brake.
14. Carefully take about 2 mL of interface, which contains lymphocytes with a transfer pipette, and move it to a new 15 mL tube (Fig. 6e). Add sterile 1 × PBS to 15 mL, and centrifuge for 10 min at 500 × *g*, 20 °C.
15. Aspirate the supernatant, and resuspend the pellet in 15 mL of sterile 1 × PBS. Centrifuge for 10 min at 250 × *g*, 20 °C. Repeat this step two times to completely wash out the lymphocyte separation medium.
16. Thoroughly aspirate the supernatant, and resuspend the cell pellet in 1 mL of RPMI media. Mix the 10 μL of cell suspension with 10 μL of 0.4% trypan blue, and put 10 μL of the mixture in a hemocytometer. Check the viability and number of isolated hepatic lymphocytes in the bright-field microscope. In this protocol, the yield may range from 0.4 to 0.6 million cells per gram of liver tissue. The flow cytometry analysis revealed significant enrichment of the lymphocytes in $CD45^+$ hepatic

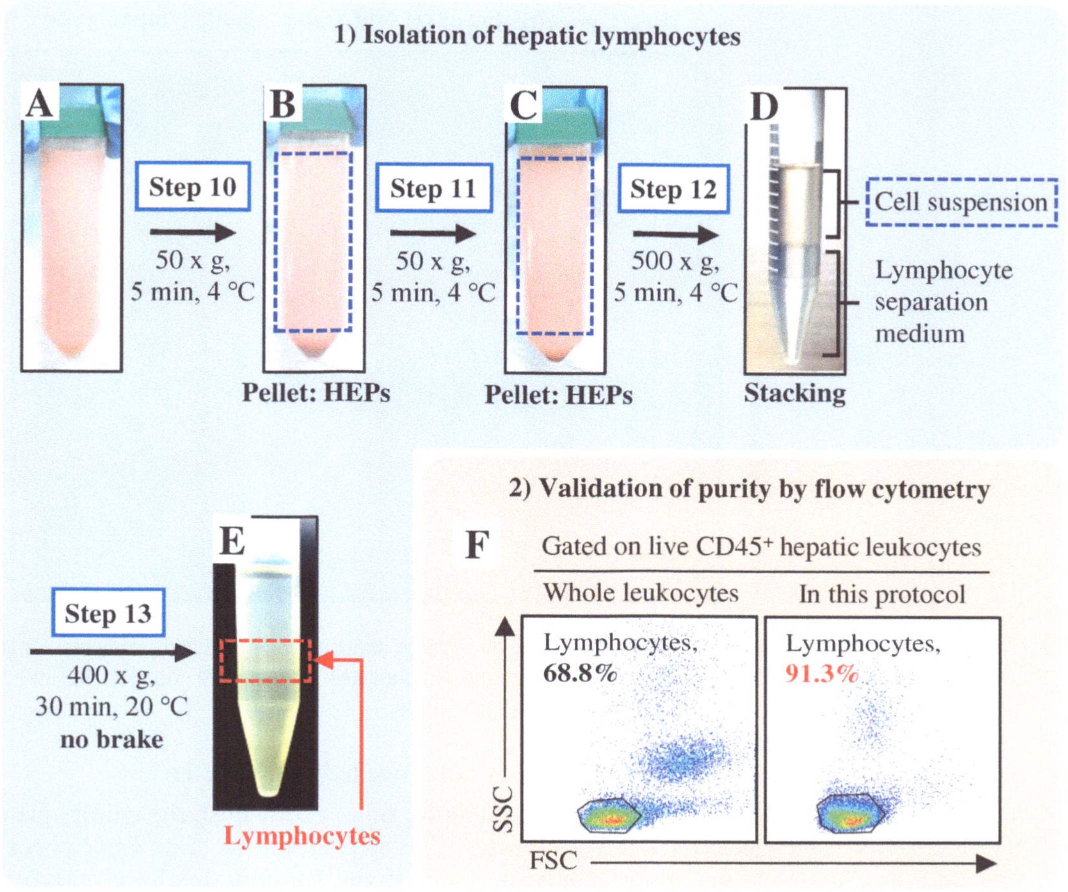

Fig. 6 Isolation and validation of mouse hepatic lymphocytes. (**a**–**e**) Mouse hepatic lymphocytes are isolated by density gradient centrifugation. (**f**) Compared to the whole hepatic leukocyte isolation method based on the Percoll® density gradient solution, hepatic lymphocytes are notably enriched in $CD45^+$ hepatic cells in this protocol

cells compared to the Percoll®-based whole hepatic leukocyte isolation method (Fig. 6f) [11].

17. (Optional) If the specific type of lymphocytes is needed, perform an additional isolation step using magnetic-activated cell sorting (MACS) with MS column (Miltenyi Biotec or equivalent).

3.3 Co-culturing of Mouse HSCs and Hepatic Lymphocytes

3.3.1 Direct Co-culture

1. Isolate and culture the HSCs on an appropriate culture disk following the steps in Subheading 3.1 (*see* **Note 14**).
2. Isolate the hepatic lymphocytes, and resuspend them in RPMI media following the steps in Subheading 3.2. Directly add appropriate numbers (volumes) of lymphocytes to the plate where HSCs are cultured (Fig. 7) (*see* **Note 15**).

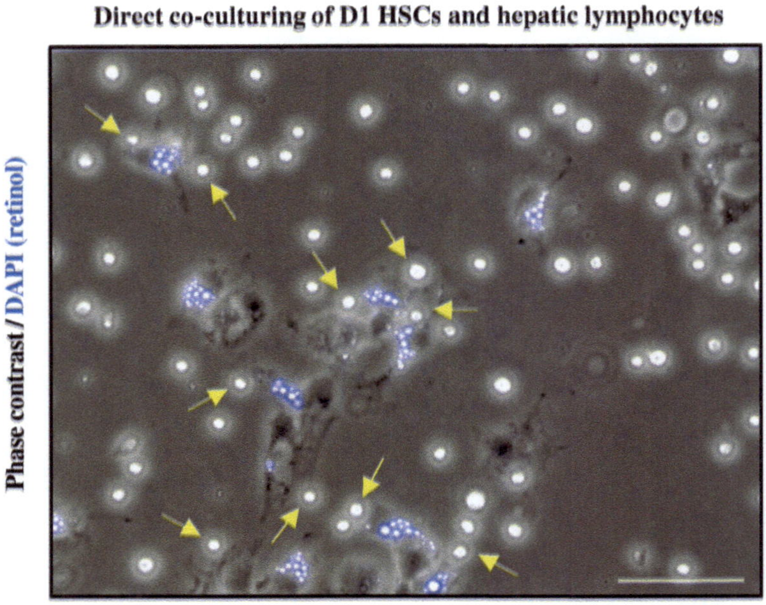

Fig. 7 Direct co-culturing of isolated mouse HSCs and lymphocytes. Close interactions between HSCs and lymphocytes (arrows) are found in the direct co-culturing method. Scale bar, 50 μm

3. Treat drugs or chemicals if needed (*see* **Note 16**).
4. At the end of the co-culture time, gently flush the culture plate to detach lymphocytes. This step should be done while checking under a microscope that the lymphocytes have detached entirely.
5. After checking full detachment under the microscope, collect the lymphocytes with medium to a sterile microcentrifuge tube and centrifuge for 3 min at 15,000 × *g*, 4 °C.
6. Take the supernatant (culture medium) and pellet (lymphocytes) separately, and analyze them immediately, or keep them at −80 °C for further analysis.
7. To harvest the remaining HSCs, add appropriate volumes of 1 × trypsin-EDTA solution, and keep in a 37 °C CO_2 incubator for 10 min.
8. Add the same volumes of RPMI media containing 10% FBS to inhibit the trypsin, and flush the plate to detach the HSCs. Since HSCs are strongly attached to the culture plate, this step should also be done under the microscope to check a full detachment of HSCs.
9. Collect the HSCs to a sterile microcentrifuge tube, and centrifuge for 3 min at 15,000 × *g*, 4 °C. Remove supernatant with

aspiration, and analyze them immediately, or keep them at − 80 °C for further analysis.

3.3.2 Indirect Co-culture Using Transwell System

1. Isolate and plate the HSCs in culture dishes following the steps in Subheading 3.1.
2. Add the appropriate volumes of RPMI medium to an empty culture plate of the same size as the plate in which HSCs were cultured.
3. Carefully add an appropriate size of Transwell insert into the plate using a sterile twisser.
4. Add the appropriate volumes of RPMI medium to the inside compartment. Recommended volumes for each compartment are shown in Table 1.
5. Incubate the plate with the Transwell insert in a humidified 37 °C CO_2 incubator for 1–12 h. This initial equilibrium step will improve cell attachment.
6. Isolate the lymphocytes following the steps in Subheading 3.2.
7. Take the plate in **step 2**, and carefully transfer the Transwell insert on to the plate incubating HSCs in **step 1** using a sterile twisser.
8. Add the appropriate number of isolated lymphocytes to the inside compartment (e.g., 96-well plate, 5×10^4 cells/well; 24-well plate, 1×10^5 cells/well; 12-well plate, 1.5×10^5 cells/well; or 6-well plate, 2×10^5 cells/well).
9. Treat drugs or chemicals if needed (*see* **Notes 17** and **18**).
10. After a period of time, carefully move the Transwell insert to a new empty culture plate.
11. Collect the lymphocytes in the Transwell insert with gentle flushing, and move to a microcentrifuge tube. After centrifugation for 3 min at 15,000 × *g*, 4 °C, aspirate the supernatant, and analyze them immediately or keep them at −80 °C for further analysis.
12. If the migration assay was performed, collect the HSCs and migrated lymphocytes separately same as in **steps 4–9** in Subheading 3.3.1.

4 Notes

1. In young mice, it is difficult to isolate the ideal number of HSCs as they may not contain enough retinol droplets in their HSCs. In addition, if mice are too old or have higher fat content in the liver, fatty hepatocytes may become contaminated with the HSC layer in density gradient centrifugation.

2. If the puncture into the portal vein fails, immediately puncture again with infrahepatic IVC, and move the drainage to the portal vein.
3. If the catheter is pushed too close to the liver, the flow of solution may be concentrated only on a certain branch of the portal vein, increasing mechanical pressure in that area. In this case, hepatic cells may be damaged and not attach efficiently to the culture plate after isolation.
4. Be careful not to dislodge the catheter sheath out of the portal vein when tying. The purpose of tying is to prevent overflow and leakage of circulating solutions.
5. If cannulation is not successful, red spots appear on the liver, indicating clotting or inefficient blood drainage. In this case, hold the infrahepatic IVC with forceps for about 10 s while gently tapping the liver with a sterile cotton swab. It may help solve the problem, but otherwise, the collagenase may not work well, and thus the yield of HSCs may decrease.
6. Before circulating the collagenase solution, air bubbles inside the tube must be removed. If air bubbles enter the tubing, carefully disconnect the tubing from the cannula, and wait for the air bubbles to come out, and then reconnect the tubing. The bubbles may block or destroy the hepatic sinusoid and decrease the yield of HSCs.
7. This step may allow the solution to reach the entire liver evenly. In addition, intermittent clamping of suprahepatic IVC with forceps may allow for efficient delivery of the collagenase solution throughout the liver, which aids in complete digestion of the liver. Once the liver is efficiently digested by the collagenase solution, it remains swollen and is dissociated when the collagenase solution is about 10–15 mL left. Continue pouring some EGTA solution on the liver during circulation of collagenase solution to prevent drying out.
8. Make sure that any piece of liver tissue must not be seen for efficient dissociation.
9. From this point, all steps have to be done on a clean bench and on ice.
10. Be careful not to mix the pellet as it contains hepatocytes.
11. Please make sure that there are no aggregates after resuspension to ensure the efficient isolation of HSCs.
12. In the interface between 11.5% OptiPrep™ and 20% OptiPrep™, there will be other hepatic non-parenchymal cells, including Kupffer cells, liver sinusoidal endothelial cells, immune cells, and red blood cells. The pellet may contain hepatocytes.

13. Care must be taken as strong crushing may decrease the cell viability through mechanical stress.
14. For day 1 HSCs, leave the cells at least 6 h to entirely attach them. And then, wash the plate with sterile 1 × PBS three times, and replenish the RPMI media.
15. If the co-culture time is too long, care must be taken in data interpretation, as the cultured HSCs are activated spontaneously over time. Therefore, it is important to set an appropriate co-culture time according to the purpose of the experiment.
16. Take pictures of the culture plate periodically to check the physical interactions between HSCs and lymphocytes.
17. Check the chemical compatibility of the used Transwell membranes from the supplier.
18. For the migration assay, periodically check the migration of lymphocytes to HSCs under the microscope. To exclude migration of lymphocytes by gravity and not HSC-derived chemokines, set the negative control to the lymphocyte-only group.

Acknowledgments

This work was supported by the National Research Foundation of Korea (NRF) grant funded by the Ministry of Science and ICT (2021R1A2C3004589, 2014M3A9D5A01073556, 2022M3A9B6017654).

References

1. Tsuchida T, Friedman SL (2017) Mechanisms of hepatic stellate cell activation. Nat Rev Gastroenterol Hepatol 14:397–411
2. Friedman SL, Roll FJ (1987) Isolation and culture of hepatic lipocytes, Kupffer cells, and sinusoidal endothelial cells by density gradient centrifugation with Stractan. Anal Biochem 161:207–218
3. Lee YS, Jeong WI (2012) Retinoic acids and hepatic stellate cells in liver disease. J Gastroenterol Hepatol 27(Suppl 2):75–79
4. Yi HS, Jeong WI (2013) Interaction of hepatic stellate cells with diverse types of immune cells: foe or friend? J Gastroenterol Hepatol 28 (Suppl 1):99–104
5. Vinas O, Bataller R, Sancho-Bru P et al (2003) Human hepatic stellate cells show features of antigen-presenting cells and stimulate lymphocyte proliferation. Hepatology 38:919–929
6. Winau F, Hegasy G, Weiskirchen R et al (2007) Ito cells are liver-resident antigen-presenting cells for activating T cell responses. Immunity 26:117–129
7. Puche JE, Saiman Y, Friedman SL (2013) Hepatic stellate cells and liver fibrosis. Compr Physiol 3:1473–1492
8. Radaeva S, Wang L, Radaev S et al (2007) Retinoic acid signaling sensitizes hepatic stellate cells to NK cell killing via upregulation of NK cell activating ligand RAE1. Am J Physiol Gastrointest Liver Physiol 293:G809–G816
9. Seo W, Eun HS, Kim SY et al (2016) Exosome-mediated activation of toll-like receptor 3 in stellate cells stimulates interleukin-17 production by gammadelta T cells in liver fibrosis. Hepatology 64:616–631
10. Byun JS, Suh YG, Yi HS et al (2013) Activation of toll-like receptor 3 attenuates alcoholic liver injury by stimulating Kupffer cells and stellate

cells to produce interleukin-10 in mice. J Hepatol 58:342–349
11. Choi WM, Ryu T, Lee JH et al (2021) Metabotropic glutamate receptor 5 in natural killer cells attenuates liver fibrosis by exerting cytotoxicity to activated stellate cells. Hepatology 74:2170–2185
12. Bhogal RK, Bona CA (2005) B cells: no longer bystanders in liver fibrosis. J Clin Invest 115: 2962–2965
13. Zhang M, Zhang S (2020) T cells in fibrosis and fibrotic diseases. Front Immunol 11:1142
14. Wang H, Park O, Gao B (2011) NKT cells in liver fibrosis: controversies or complexities. J Hepatol 55:1166. author reply 1166-1167

Chapter 8

Working with Immortalized Hepatic Stellate Cell Lines

Scott L. Friedman and Ralf Weiskirchen

Abstract

Hepatic stellate cells (HSCs) are the major cellular source of extracellular matrix production in the liver. Therefore, this cell population has received considerable attention in studies investigating fundamental features of hepatic fibrosis. However, the limited supply and ever-increasing demand for these cells, combined with the additional tightening of formal standards in animal welfare policy, make working with these primary cells increasingly difficult. Moreover, researchers working in biomedical research are challenged to implement the 3R principle of "replacement," "reduction," and "refinement" in their work. This principle, originally proposed in 1959 by William M. S. Russell and Rex L. Burch, is now widely endorsed by legislators and regulatory bodies in many countries as a roadmap to tackle the ethical dilemma associated with animal experimentation. As such, working with immortalized HSC lines is a good alternative to limit the number of animals and their suffering in biomedical research. This article summarizes issues that need to be considered when working with established HSC cell lines and provides general guidelines for the maintenance and storage of HSC lines from mouse, rat, and humans.

Key words Cell line, Cryo-conservation, Risk assessment, Cell banking, Misidentification, Marker genes, Contamination, Media, Freezing, Supplements

1 Introduction

Hepatic stellate cells (HSCs) typically contain numerous lipid droplets enriched with retinoids, a characteristic cytoplasm with a large Golgi apparatus, a well-developed endoplasmic reticulum, a few glycogen particles, a small number of mitochondria and lysosomes, and rare peroxisomes. In culture, these cells can acquire an activated phenotype and transdifferentiate into a myofibroblastic cell (MFB), which leads to the development of a dense cytoplasmic matrix containing bundles of microfilaments and microtubules, including actin and intermediate filaments. HSCs express a number of specific neurological markers such as glial fibrillary acidic protein (GFAP) and synaptophysin and further are positive for several intermediate filament proteins including desmin, nestin, and vimentin. Most

Ralf Weiskirchen and Scott L. Friedman (eds.), *Hepatic Stellate Cells: Methods and Protocols*, Methods in Molecular Biology, vol. 2669, https://doi.org/10.1007/978-1-0716-3207-9_8,

importantly, as they activate, increased expression of α-smooth muscle actin (α-SMA) and collagens are features of the MFB phenotype.

Primary HSCs have a finite life span and divide only a limited number of times before entering senescence. In addition, the isolation of these cells is experimentally challenging, and cellular features can vary when comparing cells prepared in different laboratories or from different operators. Therefore, many investigators have implemented continuous growing HSC lines from mouse, rat, and humans.

This guideline reviews the considerations of working with HSC lines, including experimental recommendations about media, culture conditions, storage, cryopreservation, banking, and aliquoting. Moreover, general guidelines are reviewed relating to safety and risk assessment when working with these modified biological materials. Information about cellular and molecular characteristics of individual HSC lines are provided and should be considered in order to prevent cell line misidentification. Finally, we detail the methods to prevent, detect, and eradicate contamination with microorganisms such as mycoplasma or other microbes.

2 General Considerations

Before working with immortalized HSC lines, we consider aspects of safety, ethical and legal issues, as well as the scientific utility of respective cell lines.

2.1 Safety Guidelines and Biosafety Risk Assessment

Most HSC lines are derived from primary HSC cultures that were either transformed with the Simian virus 40 (SV40) large T-antigen (SV40T), immortalized by introducing telomerase reverse transcriptase (TERT) activity, or by exposure to intensive radiation with ultraviolet (UV) light. Several other lines have been established from diseased livers of normal or genetically modified animals that became spontaneously immortalized in culture without further treatment. Depending on how the cell line was generated, it may need to be classified as a genetically modified organism (GMO) that carries the nucleic acid sequence encoding the transforming activity. Therefore, it is mandatory to conduct a biosafety risk assessment and to implement adequate safety measures whenever working with these lines.

2.2 Ethical and Legal Requirements

In many cases, cell lines are passed from one laboratory to another without careful consideration. However, often ignored is that cell lines can be the subject of international legally binding treaties, which define details about ownership, intellectual property rights, and patent rights. These requirements must be complied with when transferring tangible research materials between different organizations. Most often, these considerations are summarized by responsible research offices in a material-transfer agreement (MTA). Moreover, since the introduction of the Human tissue Act 2004 (HT Act) in September 2006 regulating the removal, storage, use, and disposal of biological materials obtained from humans, it may be necessary, for example, in the UK, to clarify if the line was established with informed consent of a living donor or the relatives of a deceased donor. Although the HT Act does not define cell lines resulting from expansion of primary cell cultures as relevant material, the interpretation of the HT Act may differ regionally.

2.3 Selecting the Proper Cell Line for Planned Work

From a scientific perspective, a HSC cell line should be as close as possible to its primary counterpart. However, HSC lines can be extremely variable in their biological properties. Therefore, the choice of cell line should be thoughtfully considered. Different HSC lines generated by different manipulations are available from mouse, rat, and human (Table 1).

However, from the established ~30 HSC lines, only a few have been widely used. Extensive literature is only available for a murine (GRX), two rat (HSC-T6, CFSC, and derivatives thereof), and two human cell lines (LX-2, LI90) (*see* **Note 7**).

Table 1
Hepatic stellate cell lines

Cell line	Origin	Primary source
LX-1	Human	Established from an individual clone derived after transformation of primary HSC at passage 4 with a plasmid encoding the SV40T under control of a Rous Sarcoma virus promoter [1]
LX-2	Human	Derived from spontaneous immortalization of a subset of early passaged LX-1 cells that were grown in low serum conditions (2% FBS) [1]
LI90	Human	Derived from an outgrowth of an epithelioid hemangioendothelioma found in the right liver lobe of a 55-year-old Japanese female during cholecystectomy [2]
TWNT-4	Human	Derivative of cell line LI90 produced by retroviral infection introducing human TERT (hTERT) and green fluorescent protein (GFP) cDNA flanked by a pair of loxP sites [3] (*see* **Note 1**)
TWNT-1	Human	Derivative of cell line LI90 produced by transduction with a retroviral vector expressing hTERT and GFP cDNA flanked by a pair of loxP sites [4] (*see* **Note 2**)
GREF-X	Human	Established by transfection of human myofibroblasts isolated from explants of histologically normal liver with a plasmid containing the coding sequence of polyoma virus large T antigen expressed under the control of the early promoter of cytomegalovirus (CMV) [5] (*see* **Note 3**)

(continued)

Table 1
(continued)

Cell line	Origin	Primary source
hTERT-HSC	Human	Isolated from surgical specimens of normal human liver and infected with a VSV-G pseudotyped lentiviral vector containing hTERT driven by the CMV promoter and a neomycin-selectable gene cassette [6]
HSC 180	Human	Derived from a strongly positive α-SMA positive cell isolated from cirrhotic liver [7]
NPC-hTERT	Human	Established from human non-parenchymal cells infected with a retrovirus (pFB-neo-hTERT) expressing hTERT [8]
CFSC	Rat	Spontaneous immortalized cell line derived from a cirrhotic liver of a male Wistar rat induced by interperitoneal administration of carbon tetrachloride for 5 weeks. Several derivatives of this cell line are available (CFSC-2G, CFSC-3H, CFSC-5H, and CFSC-8B) [9, 10]
NFSC	Rat	Derived by spontaneous immortalization of HSC from a normal liver of a male Wistar rat [9, 10]
HSC-T6	Rat	Established from cultured primary HSC isolated from Sprague Dawley rats transiently transfected at day 15 with a cDNA containing the SV40T driven by the *Rous sarcoma virus* promoter [11]. Originally reported to be established from a male rat, it was later shown that the cell lines has a female karyotype [12]
PAV-1	Rat	Isolated by amplification of a colony obtained by spontaneous immortalization of primary HSC isolated from an over 8-month-old male Wistar rat [13]
HSC-PQ	Rat	Derived from primary HSC isolated from Sprague Dawley rat subjected to UV light after grown to confluence [14]
BSC	Rat	Isolated from a male Wistar rat subjected to experimental biliary fibrosis rendered by obstruction of the bile duct for 18 days [15, 16]. Individual clones (e.g., BSC-C10) expressing typical HSC markers were subcultured [16]
MFBY2	Rat	Myofibroblastic-like cell line established from a 6-week-old male Wistar rat intraperitoneally injected with a mixture of carbon tetrachloride/olive oil twice a week for 1 month [17] (*see* **Note 4**)
T-HSC/Cl-6	Rat	Isolated from primary HSC isolated from Sprague Dawley rats that were transfected at day 12 after initial plating with a plasmid carrying the SV40T and a neomycin-resistance (G418) gene cassette [18]
LSC-1	Rat	Immortalized spontaneously from primary HSC isolated from normal Wistar rats [19]
RGF and RGF-N2	Rat	These two cell lines were generated from primary portal myofibroblasts isolated from adult male Sprague Dawley rat livers that underwent culture activation and subsequent immortalization by transfection with a mammalian expression vector expressing SV40T [20]

(continued)

Table 1
(continued)

Cell line	Origin	Primary source
GRX	Mouse	Established from liver of C3H/HeN mice infected by transcutaneous penetration of cercaris from the *Schistosoma mansoni* BH strain [21]
SV68c-IS	Mouse	Established from primary HSC of a 6-week-old male ICR mouse transfected with wild-type Simian virus 40 [22]
A640-IS (33 °C)/ A640-IS (39 °C)	Mouse	Originated from primary HSC isolated from a 6-week-old male ICR mouse transfected with a temperature-sensitive mutant of SV40T (tsA640) [23] (*see* **Note 5**)
M1-4HSC	Mouse	Non-tumorigenic cell line established by long-term cultivation from primary murine HSC isolated from male mice deficient for the tumor suppressor p19^{ARF} [24] (*see* **Note 6**)
A7	Mouse	Established from the transgenic mouse strain H-2K^{b}-*ts*-A58 harboring a temperature-sensitive SV40T variation directed under the control of the ubiquitous class I major histocompatibility gene H-2K^{b} promoter [25]
Col-GFP-HSC	Mouse	Primary HSCs isolated from transgenic Col-GFP mice were immortalized by infection with a lentiviral vector containing SV40T and a hygromycin resistance gene [26]
IMS/MT(−) and IMS/N	Mouse	These two cell lines were generated from primary HSCs isolated from control 129/Sv (IMS/N) and metallothionein (MT)-null mice (IMS/MT(−)) livers. The cells were immortalized by infection with SV40 [27]
JS1 (C57/Bl6 WT) JS2 (C57/Bl6 TLR4$^{-/-}$) JS3 (C57/Bl6 MyoD88$^{-/-}$)	Mouse	These three cell lines were generated from C57/Bl6 WT, TLR4$^{-/-}$, and MyoD88$^{-/-}$ mice. Therefore, primary HSCs were passaged two or three times and transfected at 50% confluency with a hygromycin B-selectable vector expressing SV40T under transcriptional control of the CMV promoter and single cell clones expanded at 33 °C [28]

2.4 Cell Line Authentication

Unfortunately, most current HSC lines have insufficient genetic characterization or short tandem repeat (STR) profiles, which are universally accepted for cell authentication. The following lines have this information available: human cell line LX-2, murine cell lines GRX, Col-GFP-HSC, and rat cell lines HSC-T6 and CFSC-2G [12, 26, 29–31]. However, when working for the first time with a new cell line, it is prudent to define the STR profile so that it can be distinguished from other cell lines used in the laboratory (*see* **Note 8**).

2.5 Testing for Contamination

Acquisition of a cell line from another laboratory presents a number of potential hazards. In particular, occult contamination with microorganisms such as mycoplasma, bacteria, and fungi are serious microbial contaminants in cell culture systems. Therefore, it is advisable to first quarantine a HSC line obtained from another researcher, commercial sources, or cell culture bank until it is tested negative for respective microbiological agents (*see* **Note 9**).

2.6 Cell Storage and Banking

Once a cell line has been procured from a reputable source, a strategy should be implemented for storage and banking to ensure a reliable and reproducible supply. Therefore, the cells are typically transferred in special freeze medium, divided into aliquots, and stored in liquid nitrogen or kept in freezers below −130 °C, as viability may progressively decrease after a few months at −80 °C. Ideally, the cells should be frozen at early passage number and divided into more than one storage vessel, preferably located in different sites and/or storage vessels (*see* **Note 10**).

2.7 Cell Culture Equipment

To avoid contamination of cultured cells with microorganisms such as bacteria, fungi, and viruses, all solutions, media, reagents, and supplies should be aseptic and sterile. Reagents and media should be stored in a domestic refrigerator at +2 °C to +8 °C. Some reagents must be kept frozen at −20 °C and/or stored in the dark (e.g., vitamin A solutions). The inventories of several cell culture reagents (trypsin, Accutase, FCS, FBS, antibiotics) are best stored at −20 °C. In addition, processing of cells should be performed in a clean cell culture hood that is properly set up and includes all necessary items (pipettor, tube racks) exclusively used in the cell culture room. Items can be disinfected by spraying with 70% ethanol and wiped clean. Cells should be cultured in a dry or humid CO_2 incubator exclusively used for cell cultures that maintains defined settings of temperature (37 °C), humidity, and CO_2 concentration (typically 5%).

3 Experimental Guidelines for Working with Hepatic Stellate Cell Lines

3.1 Culture Media for Growing Continuous HSC Lines

For the different HSC lines, a number of different culture media have been proposed. Before modifying or adapting the proposed media, it is advisable to use the media that was first defined by the laboratory that originally established the respective line (Table 2).

The exact formulation of the six basal media used for culturing of HSC lines, namely, Dulbecco's modified Eagle medium (DMEM), Dulbecco's modified Eagle medium enriched with Ham's Nutrient Mixture F-12 (DMEM/F-12), Minimal essential medium (MEM), ASF104, Waymouth MB 752/1, and William's Medium E, are listed in Table 3.

3.2 Passaging of HSC Lines

The growth rate of different HSC lines is significantly varied and can be influenced by many factors. However, the doubling time of an individual HSC line under specific conditions is well defined and an important cellular characteristic when growing exponentially. In

Table 2
Culture conditions for immortalized HSC cell lines

Cell line	Proposed culture medium[a]	References
GRX	MEM supplemented with 10% FBS, 3 μg/mL HEPES buffer, and 60 μg/mL gentamicin[b]	[21]
SV68c-IS	DMEM supplemented with 0.1% glucose, 10% heat-inactivated FCS, penicillin G (500 IU/mL), and streptomycin sulfate (100 μg/mL)	[22]
A640-IS (33 °C)/ A640-IS (39 °C)	DMEM supplemented with 0.1% glucose, 10% heat-inactivated FCS, penicillin G (500 IU/mL), and streptomycin sulfate (100 μg/mL)	[23]
M1-4HSC	DMEM supplemented with 10% FCS and antibiotics	[24]
A7	ASF104 supplemented with 2% FBS	[25, 32]
Col-GFP-HSC	DMEM supplemented with 10% FCS, 4 mM L-Glutamine, 100 IU/mL penicillin, and 100 μg/mL streptomycin	[26]
IMS/MT(−) and IMS/N	DMEM supplemented with 10% FCS and 60 mg/mL kanamycin	[27]
JS1 (C57/Bl6 WT) JS2 (C57/Bl6 $TLR4^{-/-}$) JS3 (C57/Bl6 $MyoD88^{-/-}$)	DMEM supplemented with 10% FCS	[28]
CFSC	MEM supplemented with 10% FBS	[9, 10]
NFSC	MEM supplemented with 10% FBS	[9, 10]
HSC-T6	Waymouth MB 752/1 supplemented with 10% FBS, penicillin (100 IU/mL), and streptomycin (100 mg/mL)[c]	[11]
PAV-1	DMEM supplemented with 10% FCS, 2 mmol/l glutamine, 100 U/mL penicillin, and 100 μg/mL streptomycin	[13]
HSC-PQ	DMEM supplemented with 20% FBS	[14]
BSC	Low-glucose DMEM supplemented with 10% FBS, 100 mg/mL streptomycin, 10,000 U/mL penicillin, and 25 μg/mL amphotericin B	[16]
MFBY2	DMEM supplemented with 10% FCS, 150 U/mL penicillin, and 100 mg/mL streptomycin	[17]
T-HSC/Cl-6	William's Medium E supplemented with 10% FBS, 100 IU/mL penicillin, 100 mg/mL streptomycin. The expression plasmid encoding the SV40 large T antigen carries a neomycin resistant gene. Therefore, cells are resistant to G418 (100 μg/mL)	[18]
LSC-1	DMEM supplemented with 20% FBS	[19]
RGF and RGF-N2	DMEM/F-12 supplemented with 10% FBS, 2% penicillin-streptomycin, 0.3% gentamycin, and 0.1% fungizone	[20]

(continued)

Table 2
(continued)

Cell line	Proposed culture medium[a]	References
LX-1	DMEM supplemented with 10% FBS	[1]
LX-2	DMEM supplemented with 10% FBS (or alternatively only 2%)	[1]
LI90	DMEM supplemented with 10% heat-inactivated FBS, 100 U/mL penicillin, and 100 μg/mL streptomycin	[2]
TWNT-4	ASF104 supplemented with 10% FBS, 100 U/mL penicillin G, and 100 μg/mL streptomycin sulfate	[3]
TWNT-1	ASF104 supplemented with 10% FBS, 100 U/mL penicillin G, and 100 μg/mL streptomycin sulfate	[4]
GREF-X	DMEM supplemented with 15% FCS	[5]
hTERT-HSC	DMEM supplemented with 10% FCS. The VSV-G pseudotyped vector encoding the hTERT transgene carries a neomycin resistance cassette and can be selected with 200 μg/mL G418	[6]
HSC 180	DMEM supplemented with 10% FCS, 1% antibiotic-antimycotic solution, and 1% non-essential amino acids	[7]
NPC-hTERT	Not known	[8]

Abbreviations used are as follows: *ASF104* animal cell culture medium serum-free 104, *DMEM* Dulbecco's modified Eagle medium, *DMEM/F-12* mixture of Dulbecco's modified Eagle medium and Ham's F-12 medium (1:1), *FBS* fetal bovine serum, *FCS* fetal calf serum, *MEM* minimal essential media

[a]The proposed media were taken from the original publications reporting the establishment of respective cell lines

[b]GRX cells can also be cultured in DMEM supplemented with 10% FBS, 100 IU/mL penicillin, and 100 μg/mL streptomycin [26]

[c]HSC-T6 cells can also be cultured in DMEM supplemented with 10% FBS and 1% penicillin/streptomycin (e.g., [33])

some media, pH-dependent dyes such as phenol red that exhibits a gradual color transition over a pH range of 6.2–8.2 is an effective means to detect and prevent exhaustion of media. Subculturing (i.e., passaging) of cells should be best conducted when the cells are still growing in the log phase, before reaching confluence. This prevents cell clumping and reduced growth as a consequence of contact inhibition. In most cases, it is advisable to establish a strict schedule for subculturing a specific line. Subculturing of all HSC cell lines can be achieved with Accutase or trypsin. The dissociation protocols for Accutase or trypsin that are given in the following are rather similar (*see* **Note 12**):

1. Confirm that the cells are 80–90% confluent by microscopy.
2. Pre-warm the dissociation solution (Accutase or trypsin solution) to 37 °C.
3. Remove and discard the culture media from culture disk or flask.

Table 3
Composition of the different basal media used for maintenance of hepatic stellate cell lines (*see* Note 11)

Substance class	Ingredient	DMEM (g/L)	DMEM/F-12 (g/L)	MEM (g/L)	ASF104 (g/L)[a]	Waymouth MB 752/1 (g/L)	Williams' Medium E (g/L)
Inorganic salts/ buffers	$CaCl_2$	0.2	0.1166	0.2	0.1	0.0906	0.2
	$Fe(NO_3)_3 \times 9\ H_2O$	0.0001	0.00005	–	0.0001	–	0.0000001
	$FeSO_4 \times 7\ H_2O$	–	0.000417	–	–	–	–
	$MgSO_4$	0.09767	0.04884	0.09767	0.098	0.09767	–
	MgCl	–	0.02864	–	–	0.11239	0.0977
	$MgCl \times 4\ H_2O$	–	–	–	–	–	0.0000001
	KCl	0.4	0.3118	0.4	0.4	0.15	0.4
	KH_2PO_4	0	–	–	–	0.08	–
	$NaHCO_3$	3.7	2.438	2.2[b]	1.8	2.24	2.2
	NaCl	6.4	6.9955	6.8	8	6	6.8
	NaH_2PO_4	0.109	–	0.122	0.125	–	0.122
	Na_2HPO_4	–	0.07102	–	–	0.3	–
	$NaH_2PO_4 \times H_2O$	–	0.0625	–	–	–	
	Na_2SeO_3	–	–	–	0.000004	–	–
	$ZnSO_4 \times 7\ H_2O$	–	0.000432	–	0.00001	–	0.0000002
	$CuSO_4 \times 5\ H_2O$	–	0.0000013	–	0.000001	–	0.0000001
	HEPES	–	–	–	1.2	–	–
	Sodium pyruvate	–	0.055	–	0.22	–	–
Amino acids	L-Alanine	–	0.00445	–	0.02	–	0.09
	L-Arginine × HCl	0.084	0.1475	0.126	0.2	0.075	–
	L-Arginine (free acid)	–	–	–	–	–	0.05
	L-Aspartic acid	–	0.00665	–	0.02	0.06	0.03
	L-Asparagine × H_2O	–	0.0075	–	–	–	0.02
	L-Cysteine × HCl × H_2O	–	0.01756	–	0.07	0.10029	–
	L-Cysteine (free acid)	–	–	–	–	–	0.04
	L-Cysteine	–	–	–	–	–	0.02
	L-Cystine × 2HCl	–	0.03129	0.0313	–	0.01955	–
	L-Glutamic acid	–	0.00735	–	–	0.15	0.0445
	Sodium L-Glutamate	–	–	–	0.02	–	–
	L-Glutamine	0.584[b]	0.365	0.292	–	0.35	–

(continued)

Table 3
(continued)

Substance class	Ingredient	DMEM (g/L)	DMEM/F-12 (g/L)	MEM (g/L)	ASF104 (g/L)[a]	Waymouth MB 752/1 (g/L)	Williams' Medium E (g/L)
	Glycine	0.03	0.01875	–	0.03	0.05	0.05
	L-Histidine × HCl × H_2O	0.042	0.03148	0.042	0.042	0.1641	–
	L-Histidine (free base)	–	–	–	–	–	0.015
	L-Isoleucine	0.105	0.05447	0.052	0.105	0.025	0.05
	L-Leucine	0.105	0.05905	0.052	0.105	0.05	0.075
	L-Lysine × HCl	1.46	0.09125	0.0725	0.146	0.24	0.08746
	L-Methionine	–	0.01724	0.015	0.03	0.05	0.015
	L-Phenylalanine	0.066	0.03548	0.032	0.67	0.05	0.025
	L-Proline	–	0.01725	–	0.02	0.05	0.03
	L-Serine	0.042	0.02625	–	0.08	–	0.01
	L-Threonine	0.095	0.05345	0.048	0.095	0.075	0.04
	L-Tryptophan	0.016	0.00902	0.01	0.025	0.04	0.01
	L-Tyrosine	–	–	–	0.064	0	–
	L-Tyrosine × 2 Na × $2H_2O$	0.12037	0.05579	0.0519	–	0.0577	0.05045
	L-Valine	0.094	0.05285	0.046	0.094	0.065	–
	L-Ornithine hydrochloride	–	–	–	0.1	–	–
	Glycyl L-Glutamine	–	–	–	0.5	–	–
	Alanyl L-Glutamine	–	–	–	0.5	–	–
Vitamins	Ascorbic acid × Na	–	–	–	0.005	0.0198	0.00227
	D-Biotin	–	0.0000035	–	0.0002	0.00002	0.0005
	Calciferol	–	–	–	–	–	0.0001
	Choline bitartrate	–	–	–	0.02	0.45352	–
	Choline chloride	0.004	0.00898	0.001	–	–	0.0015
	Folic acid	0.004	0.00265	0.001	0.004	0.0004	0.001
	Calcium folinate	–	0.00224	–	0.00001	–	–
	i-inositol	–	0.0126	–	0.02	–	–
	myo-inositol	0.0072	–	0.002	–	0.001	0.002
	Menadione (sodium bisulfite)	–	–	–	–	–	0.00001
	Niacinamide	0.004	0.00202	0.001	0.004	0.001	0.001

	D-Pantothenic acid (hemicalcium)	0.004	–	0.001	0.004	0.001	0.001
	Pyridoxal hydrochloride	0.004	0.002013	0.001	0.004	0.001	0.001
	Retinol acetate	–	–	–	–	–	0.0001
	Riboflavin	0.0004	0.000219	0.0001	0.0004	0.001	0.0001
	Thiamine × HCl	0.004	0.00217	0.001	0.004	0.01	0.001
	DL-α-Tocopherol Phosphate × Na	–	–	–	–	–	0.00001
	Vitamin B_{12}	–	0.00068	–	0.0001	0.0002	0.002
Others	D-Glucose	4.5	3.151	1	2	5	–
	Mannose	–	–	–	0.5	–	–
	Galactose	–	–	–	0.2	–	–
	Glutathione (reduced)	–	–	–	0.001	0.015	–
	Hypoxanthine	–	–	–	–	0.025	–
	Hypoxanthine × Na	–	0.00239	–	–	–	
	Disodium hypoxanthine	–	0.000042	–	0.002	–	–
	Linoleic acid	–	0.000105	–	–	–	–
	Lipoic acid	–	–	–	–	–	–
	Phenol Red × Na	0.0159	–	–	–	0.011	–
	Phenol Red	–	0.0081	0.011	0.005	–	–
	Putrescine × 2 HCl	–	0.000081	–	–	–	
	Pyruvic acid × Na	0.11	–	–	0	–	–
	Succinic acid	–	–	–	0.106	–	–
	Sodium succinate (hexahydrate)	–	–	–	0.027	–	–
	Disodium β-glycerophosphate	–	–	–	1.5	–	–
	Deoxyadenosine	–	–	–	0.001	–	–
	Deoxycytosine	–	–	–	0.00003	–	–
	6,8-Dihydroxypurine	–	–	–	0.0003	–	–
	Thymidine	–	0.000365	–	0.0002	–	–

(continued)

Table 3
(continued)

Substance class	Ingredient	DMEM (g/L)	DMEM/F-12 (g/L)	MEM (g/L)	ASF104 (g/L)[a]	Waymouth MB 752/1 (g/L)	Williams' Medium E (g/L)
	Uridine	–	–	–	0.005	–	–
	Putrescine dihydrochloride	–	–	–	0.0003	–	–
	Phosphoethanolamine	–	–	–	0.028	–	–
	Cholesterol	–	–	–	0.0001	–	–
	α-Cyclodextrin	–	–	–	2.2	–	–
	α-Cyclodextrin-linoleic acid	–	–	–	0.01	–	–
	Human transferrin	–	–	–	0.005	–	–
	Crystalline bovine insulin	–	–	–	0.005	–	–

[a]ASF104 medium is a serum-free medium. It is specifically intended for studies in sensitive cell lines (e.g., A7, TWNT-1, and TWNT-4) or used for special applications (e.g., studies on Vitamin A metabolism, reconstruction of liver organoid in bioreactors, or cell proliferation assays requiring serum-free conditions)

[b] should be freshly added to the medium

4. Gently rinse the cells with dissociation solution (Accutase or trypsin solution) (e.g., 2 mL for one 100 mm^2 culture dish).
5. Remove and discard the dissociation solution.
6. Add fresh dissociation solution (2 mL for one 100 mm^2 culture dish), and incubate the vessel at 37 °C for 2–3 min.
7. Confirm detachment under a microscope, and transfer the cell solution into a 50 mL falcon tube containing pre-warmed complete growth medium.
8. Pipette part of this mixture into a new culture vessel. If Accutase or trypsin should be removed, centrifuge the cell suspension (5 min at 1000 × *g*), remove the supernatant, and gently resuspend the cell pellet in desired quantity of pre-warmed culture medium. Then transfer the desired cell quantities into new vessels and gently shuffle (*see* **Note 13**).
9. Return the fresh cell vessel to the incubator.

3.3 Determination of Cell Viability Using the Trypan Blue Exclusion Test

The trypan blue exclusion test is used to determine the number of viable cells present in a cell solution. The cell viability of cells is commonly defined as the number of viable cells divided by the total number of cells. This test is based on the principle that vital cells possess intact cell membranes that exclude certain dyes, whereas damaged and nonviable cells do not. The following protocol allows quantification of cell viability by light microscopy in less than 15 min.

1. Centrifuge a defined aliquot of your cell suspension (5 min, 500 × *g*), and discard the supernatant.
2. Resuspend the cell pellet in a defined volume of phosphate-buffered saline or balanced salt solution. In this step, the cell concentration in the solution should be in the range of 1×10^6–2.5×10^6 cells/mL.
3. Mix one volume of this cell suspension with one volume trypan blue solution (0.4% w/v), and incubate this mixture at room temperature for 2 min.
4. Prepare the Neubauer chamber. Therefore, put a glass cover tightly on the central area of the Neubauer chamber.
5. Pipette an aliquot (~10 μL) of your cell suspension prepared in **step 3** close to the glass cover edge at the center of the Neubauer counting chamber.
6. Count the viable cells (unstained cells) and dead cells (stained cells) of five big squares of the hemocytometer in a light microscope.
7. Calculate the average number of viable and death cells from the five counts.

8. Calculate the total concentration of the original cell suspension prepared in **step 2** by using the following formula:

$$[\text{Cells/mL}] = [\text{Average number of living cells}] + [\text{Average number of death cells}] \times 10{,}000 \times 2$$ (*see* **Note 14**).

9. Calculate the cell viability in percentage from the calculated values of **step 7** as follows:

$$\text{Viability of cells (\%)} = \frac{[\text{Average number of living cells}]}{[\text{Average number of living cells}] + [\text{Average number of death cells}]}$$

10. Finally, calculate the original cell concentration in your original cell solution of **step 1** by multiplying with the factor resulting from **step 2**.

3.4 Freezing

Cryopreservation allows storage of cells for an indefinite period of time and allows safe and convenient shipment of cells. It further guarantees the generation of cell stocks that are protected against changes in cellular genetics, ensuring their long-term use with reproducible outcomes. Although there are general freezing media containing diverse cryoprotectants and additives commercially available, we recommend a very simple and cheap protocol for cryopreservation of HSC cell lines with high post-thaw viabilities (*see* **Note 15**).

1. Freshly prepare the required quantity of 2 × freeze medium, and store the freeze medium on ice. Two mL of freeze medium is required for a 100 mm^2 plate that has grown to 80–90% confluence. 20 mL 2 × freeze medium consists out of 11 mL fresh culture medium, 5 mL fetal calf serum (FCS), and 4 mL dimethyl sulfoxide (DMSO).
2. Gently detach the cells by using the protocol given for passing of cells (*see* Subheading 3.2), and place the cells in a sterile Falcon tube on ice.
3. Centrifuge the cells (5 min at 1000 × *g*, 4 °C), and remove the supernatant under a clean laminar flow hood.
4. Carefully resuspend the cell pellet in fresh culture medium by gently pipetting up and down in a sterile 10 mL pipette. Use 2 mL medium, for example, for cells obtained from one 100 mm^2 dish grown to 80–90% confluence.
5. Add the same amount of 2 × freeze medium, and mix the solutions to maintain a homogeneous suspension by gently pipetting the solution several times in a 10 mL pipette.

6. Aliquot 1 mL each of the cell suspension into cryogenic vials, and place them directly in a −80 °C freezer.
7. If long-term storage for more than 6 months is desired, transfer the cryogenic vials after 24 h in a liquid nitrogen tank or to a − 150 °C freezer (*see* **Note 16**).

3.5 Thawing of Frozen Cells

The following section provides a simple thawing procedure that allows a high portion of cells to recover from cryopreservation. In many protocols, the cryopreserved cells are thawed by swirling the respective cryovial in a 37 °C water bath. We strongly advise against this approach, because it is a potential source for contamination. We have adapted a general protocol for thawing of cryopreserved cell lines as follows:

1. Add 20 mL fresh culture medium to a sterile Falcon tube.
2. Remove a vial of cryopreserved cells from the nitrogen tank or −150 °C, and quickly thaw the cells by placing the vial into fingers of your hands (*see* **Note 17**).
3. Transfer the vial into a laminar flow, open the vial, and pipette the content of the vessel in the Falcon tube prepared in **step 1**.
4. Pipette 10 mL of this solution each into two 100 mm^2 petri dishes, and transfer the cell plates into the incubator.
5. Change the medium after the cells have attached to the plastic surface. This step typically takes about 1–3 h (*see* **Note 18**).

3.6 Aliquoting of HSC Lines

HSC cell lines have become attractive models for studies of HSC biology. Presently, there are only four commercially cell lines available, namely, the murine cell lines GRX and Col-GFP-HSC, rat cell line HSC-T6, and the human cell line LX-2 (*see* **Note 19**).

All other cell lines are currently not available for purchase. In some cases, HSC lines can be requested from the laboratory that originally established this line. However, there are a few legal aspects to consider.

3.6.1 Legal Aspects

1. Clarify if the necessary institutional and legal requirements are met in your laboratory to handle biological modified materials such as GMO.
2. Contact the originator of the cell line you would like to obtain and clarify whether there is a willingness to pass the cell line to your lab.
3. In most cases, it will be necessary to sign a Material Transfer Agreement (MTA) between the provider and recipient organizations in which details including ownership, intellectual property rights, and potential patent rights are legally binding determined.

4. In addition, some countries have specific regulations when importing cells or media from other countries. Clarify if there is the need to require a permit from any agency in your or the recipient country before initiating the transfer of cells.
5. Some countries have also specific regulations in using materials from humans for research purposes. Unfortunately, some authorities also misinterpret individual laws, which can lead to unnecessary discussion (*see* **Note 20**).
6. After receiving the cells, do not pass them to other laboratories without obtaining permission from the investigator who originally provided them and/or clarify the terms that enable sharing in the original MTA governing their use.

3.6.2 Guidance for Cell Line Shipping

Spontaneously established HSC lines are biological materials. Furthermore, cell lines whose DNA has been altered using genetic engineering techniques (e.g., vectors) are genetically modified organisms (GMO). In most countries, the shipment of biological materials including GMOs including packing and parceling must comply with defined requirements. Respective materials must be properly packaged to withstand all potential conditions that might occur during handling or transit. Packing materials must be leak-proof, and the materials should be best triple packaged and surrounded by absorbent material sufficient to protect the culture vessel and to absorb the total amount of liquid (media) if it should leak or break during transit. Moreover, proper labeling of packages is essential. In addition, there are two options for shipment: (i) cultured cells and (ii) frozen stocks of cells. It should be clarified in advance who will bear the costs for shipment.

In the following, guidelines and shipping instructions for shipping of HSC lines are provided.

3.6.3 Shipping Living Cells

1. Fix the date for shipment, and compile all necessary information required for sending including detailed delivery address, contact person with phone number, and email address (*see* **Note 21**).
2. Chose a suitable national or international operating courier service for your transport (*see* **Note 22**).
3. Seed the cells that should be parceled 1 or 2 days before the date of shipment. It is best to seed the cells into small flasks (e.g., T25 flasks), which can be tightly closed with screw caps and further sealed with parafilm at the top. Do not use vented caps that are equipped with gas-permeable membranes. The density of cells should be 60–70% on the day of shipment.
4. On the day of shipment, fill the flasks to capacity with pre-warmed media, close the caps tightly, and seal them with parafilm.

5. Wrap each flask in absorbent paper towels, place it into a sealed plastic bag, and place it into the transport polystyrene box.
6. To avoid moving of the flasks during shipping, fill remaining spaces in the box with bubble wrap or similar packing material.
7. Seal the polystyrene box with packing tape, and place the box inside a slightly larger cardboard box. Add a letter to the recipient in which details of the cells are provided including culture media, passage number, splitting conditions, and original publication reporting the establishment of the respective HSC line. If possible, include also a picture showing how the cells appeared microscopically before the package was sent. Seal the cardboard box with packing tape and label the package as "fragile," and attach all necessary stickers to conform to the UN packaging and labeling requirements. Ensure that the package carries also the label "do not freeze."
8. Hand over the package to the transport service, and send the recipient an email in which the tracking number and the estimated day of delivery are provided. In addition, request a confirmation when the package has arrived.

3.6.4 Shipping Cryostocks on Dry Ice

Dry ice used for coolant shipment can be hazardous if not handled correctly. Therefore, shipments with dry ice need special declaration and packing requirements for risk mitigation. In particular, the package must be capable of withstanding extremely low temperature and should allow the release of sublimated carbon dioxide gas to prevent a buildup of pressure that might rupture it (*see* **Note 23**). Cryopreserved cells can be parceled on dry ice by the following protocol:

1. Prepare the necessary paper work (*see* **steps 1** and **2** of Subheading 3.6.3).
2. Quickly remove the vials intended for parceling from the nitrogen tank or freezer, place them into a 50 mL not tightly screwed Falcon tube, and transfer them into the polystyrene box with at least 1″ (~2.5 cm) thick walls. Fill up the remaining space of the polystyrene box with dry ice (*see* **Note 24**).
3. Put the lid on the polystyrene box, but do not tape the box.
4. Place the box inside a slightly larger cardboard box, and add a letter to the recipient (*see* **step** 7 in Subheading 3.6.3). Seal the cardboard box with packing tape, label the package as fragile, and attach all necessary stickers to conform to the UN packaging and labeling requirements. It is essential to mark the package with a sticker containing the information "UN1845/dry ice," and note the quantity of dry ice (in kg) included in the package.
5. Hand over the package to the transport service, and send the recipient an email with information about the status of the delivery (*see* **step 8** in Subheading 3.6.3).

4 Choosing the Appropriate Hepatic Stellate Cell Line for Your Work

Before deciding to work with one of these cell lines, it is essential to read relevant papers, in which the establishment and characterization of the respective line are described. Based on the data provided, you can estimate how relevant this cell line might be for your work. Continuous HSC lines are available from mouse, rat (Wistar or Sprague Dawley), and humans. All these cell lines have specific morphological characteristics and grow as adherent cells (*see* **Note 25**).

Some characteristics of these cell lines that might help in choosing the appropriate cell line for your work are given in the following sections.

4.1 Mouse Origin

GRX: This cell line express collagen types I, III, and IV. In addition, this cell line is positive for fibronectin. Importantly, these cell lines express and secrete large quantities of retroviral particles. GRX cells are aneuploid with only slight deviation of normal chromosome values for mouse cells [21, 29].

SV68c-IS: This cell line is positive for expression of the large T-antigen, α-SMA, desmin, collagen type III, and GFAP [22].

A640-IS (33 °C)/A640-IS (39 °C): Under permissive temperature (33 °C), these cells express the large T-antigen, while at the non-permissive temperature (39 °C), the expression of the large T-antigen is undetectable, and cells do not grow. Expression of vimentin; collagen types I, III, and IV; fibronectin; and laminin is observed under both conditions, while the expression of α-SMA is influenced by the culture temperature. The cells are responsive toward TGF-β1 under both culture conditions [23].

M1-4HSC: This cell line expresses α-SMA, fibulin-2, vimentin, GFAP, MMP-2, collagen type I, syndecan, tenascin, rhoA, and PAI-1. The cells are responsive toward TGF-β inducing a pronounced spindle-shape morphology termed M-HT that is associated with decreased expression of desmin, but not affecting other HSC marker proteins such as GFAP or α-SMA [24]. The treatment of M1-4HSC with TGF-β (1 ng/mL) for 16 days provokes a morphological transition to a myofibroblast-like cell type, which is accompanied by enhanced cellular turnover and impaired migration (Fig. 1).

A7: This line has high retinol esterification activities, and it grows only under the permissive condition (at 33 °C) and not under the non-permissive temperature (at 37 °C) [25].

Col-GFP-HSC: This GFP- and SV40T-positive cell line expresses α-SMA, GFAP, collagen type I, fibronectin, PDGFRβ, and synaptophysin [26].

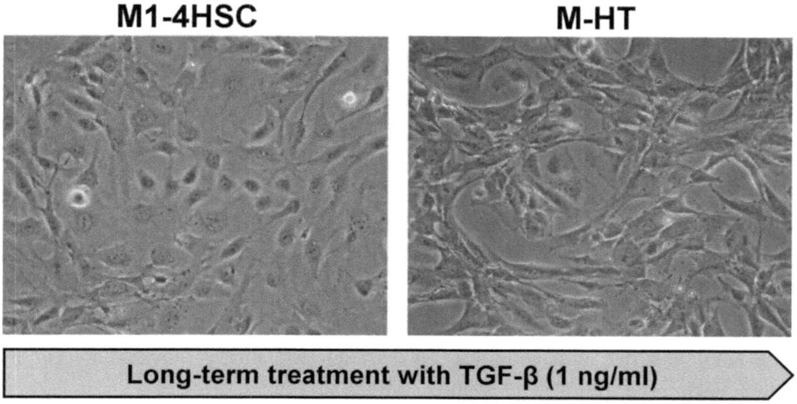

Fig. 1 Cell lines M1-4HSC and M-HT. The cell line M1-4HSC was derived by long-term passaging of hepatic stellate cells isolated from livers of adult p19^{ARF} null mice. When cultured in the presence of 1 ng/mL TGF-β, the cells transit into a myofibroblast-like phenotype termed M-HT. The image was kindly provided by Wolfgang Mikulitis (Comprehensive Cancer Center, Department of Medicine I, Institute of Cancer Research, Medical University of Vienna, Vienna, Austria)

IMS/MT(−)/IMS/N: This cell line expresses α-SMA, desmin, and SV40T and can accumulate vitamin A [27].

JS1 (C57/Bl6 WT); JS2 (C57/Bl6 TLR4$^{-/-}$); JS3 (C57/Bl6 MyoD88$^{-/-}$): HSC marker expression has not been systematically analyzed. However, these cell lines should be selectable with hygromycin at 33 °C and should be positive for SV40T [28].

4.2 Rat Origin

The following lines have been established from Wistar or Sprague Dawley rats:

CFSC and subclones (CFSC-2G, CFSC-3H, CFSC-5H, CFSC-8B): The individual CFSC clones available (i.e., CFSC-2G, CFSC-3H, CFSC-5H, and CFSC-8B) show high differences in the expression of HSC-specific markers such as collagen types I, III, and IV, fibronectin, and laminin B1. Expression of IL-6 mRNA is only detectable in clones CFSC-3H and CFSC-5H. All clones express the Gap junction protein Connexin 43 (Cx43) [9, 10].

NFSC: The cells express collagen type I, collagen type III, fibronectin, laminin B1, TGF-β, and IL-6. The expression of collagen type I mRNA can be triggered by stimulation with IL-6 [9, 10].

HSC-T6: This cell line expresses myogenic and neural crest cytoskeletal filaments including desmin, α-SMA, vimentin, and GFAP. In addition, HSC-T6 express the six retinoid nuclear receptors (RARα, RARβ, RARγ, RXRα, RXRβ, and RXRγ) and the cellular retinol-binding protein type 1 (CRBP-1),

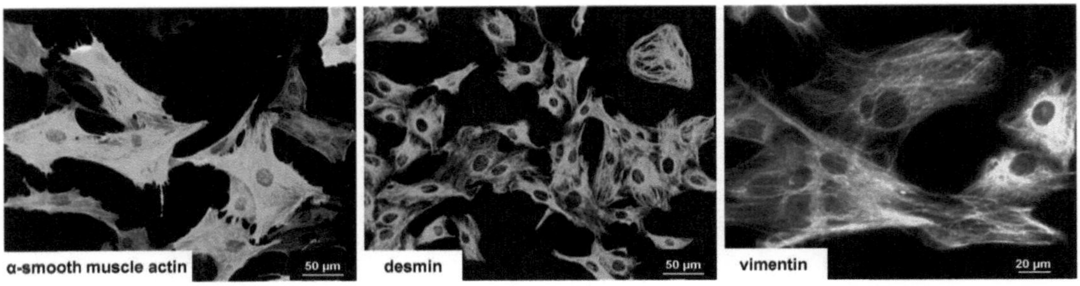

Fig. 2 PAV-1 cells express typical hepatic stellate cell markers. PAV-1 cells were immunolabelled for α-smooth muscle actin, desmin, and vimentin. This image was kindly provided by Dr. Patrick Sauvant (Bordeaux Sciences Agro, Department Feed & Food, Gradignan Cedex, France)

rendering them as a useful tool for studies of hepatic retinoid uptake, storage, and metabolism [11, 12].

PAV-1: PAV-1 cells express the typical HSC markers α-SMA, desmin, and vimentin (Fig. 2). They are also positive for fibronectin, laminin, and collagen type IV but are negative for GFAP. These cells can produce retinoic acid from retinol and store retinyl esters in cytoplasmic lipid droplet. The cell line expresses mRNAs for CRBP-1, retinol dehydrogenase II, alcohol dehydrogenase class I, RARα, and RXRα [13].

HSC-PQ: HSC-PQ express α-SMA, desmin, collagen type I, collagen type III, fibronectin, laminin, and other extracellular matrix proteins but lack expression of collagen type IV [14].

BSC and subclones (e.g., BSC-C10): BSC express GFAP, synaptophysin, vascular cell adhesion molecule, neural cell adhesion molecule, IL-6, Fibulin-2, desmin, collagen type I, α-SMA, and CD36. In addition, the cells are responsive toward TGF-β1 and are easily transfectable. The cells can be differentiated to store vitamin A [15, 16]. Importantly, BSC can undergo adipose conversation after treatment with an adipocyte differentiation cocktail composed of methylisobutylxanthine, dexamethasone, and insulin. Concomitant treatment with retinol and palmitate provokes accumulation of cellular vitamin A that can be detected by increased vitamin A-specific autofluorescence when excited with UV light (Fig. 3).

MFBY2: This cell line expresses neural cell adhesion molecule, α-SMA, TIMP-1, collagen type I, and collagen type III. The myofibroblastic cell phenotype can be reversed (e.g., the cells express GFAP, reduce expression of α-SMA, and incorporate lipids after infection with an adenovirus expressing the latency-associated peptide (LAP) that is part of the TGF-β precursor peptide) [17].

T-HSC/Cl-6: This cell line expresses α-SMA, desmin, GFAP, collagen type I, and TGF-β mRNA [18].

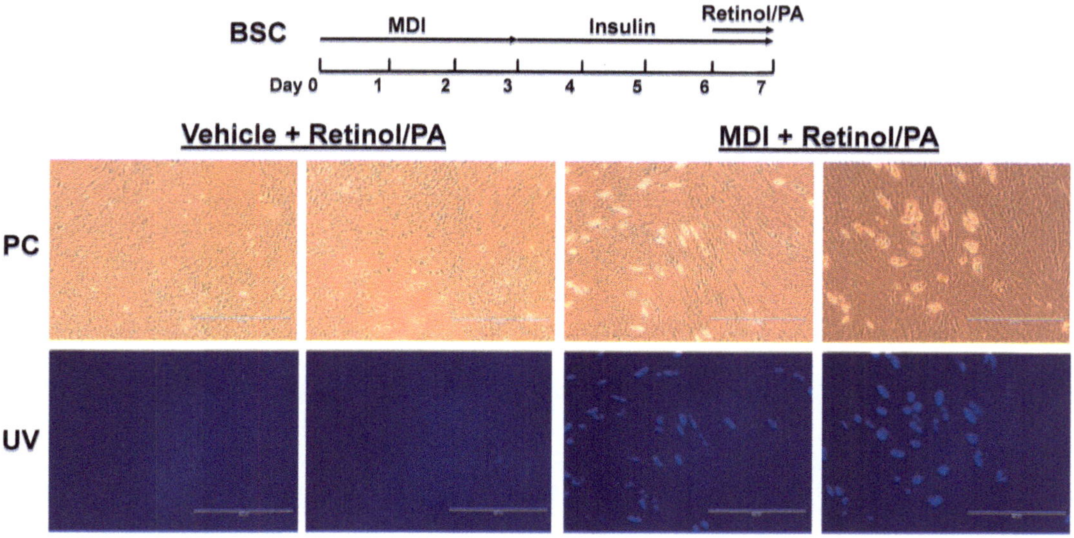

Fig. 3 BSC cells can be differentiated to store vitamin A. The BSC line was treated with the adipocyte differentiation cocktail MDI (methylisobutylxanthine, dexamethasone, insulin) and subsequently with retinol and palmitate (PA), resulting in vitamin A storage as shown by UV-excited autofluorescence. As a negative control, the cells were stimulated with the vehicle and subsequently with retinol and PA. Space bars represent 400 μm. PC, phase contrast; UV, ultraviolet light. The images were kindly provided by Hidekazu Tsukamoto and Integrative Liver Cell Core (Department of Pathology, Keck School of Medicine, University of Southern California, Los Angeles, CA, USA), supported by NIH grant 2R24AA012885

LSC-1: LSC-1 cells express desmin, GFAP, collagen type I and III, α-SMA, TGF-β1, PDGF-B, and inducible nitric oxide synthase (iNOS). In addition, LSC-1 express telomerase reverse transcriptase (TRT) and p53. This cell line is susceptible toward TGF-β1, endothelin-1, and PDGF-B. Karyotype analysis demonstrates that this cell line possesses a normal chromosome phenotype [19].

RGF, RGF-N2: These two cell lines express collagen type I, α-SMA, TIMP1, elastin, fibulin-2, collagen type XV, lysyl oxidase (lox) and lysyl oxidase-like 1–4 (loxL1–4), cytoglobin (Cygb), vimentin, GFAP, PDGFRβ, TGFβRI, and EGFR, but do not express desmin and lecithin-retinol acyltransferase (Lrat) that are characteristic for HSC [20]. This cell line is supposed to originate from portal myofibroblasts.

4.3 Human Origin

LX-1: The cells express the following proteins: SV40T, α-SMA, vimentin, GFAP, PDGFRβ, DDR2, Ob-R, MT1-MTP, TIMP1, TIMP2, and MMP-2. In addition, the cells express HSP47 and α1(I) procollagen mRNA and later can be stimulated by TGF-β1. Stimulation with PDGF-B results in increased proliferation. LX-1 can accumulate and esterify retinol [1].

LX-2: LX-2 cell express α-SMA, vimentin, GFAP, PDGFRβ, DDR2, Ob-R, MT1-MTP, TIMP1 (to a weaker content as LX-1), TIMP2, and MMP-2. The cells are highly transfectable, and proliferation can be stimulated with PDGF-B, whereas TGF-β stimulates expression of α1(I) procollagen mRNA. Although the cells were reportedly negative for SV40T (monoclonal antibody from Santa Cruz), they were later shown to be slightly positive for this protein based immunohistochemistry and Western blot analysis when a polyclonal antibody directed against amino acid 4–30 of SV40T was used, The cells can grow in low serum conditions (2% FBS). Single-locus short tandem repeat (STR) profile and some cytogenetic markers are available [1, 34]. It should be noted that the phenotypic appearance of LX-2 depends on the cell density (Fig. 4). Recently, a LX-2 derivative was established that contains an expression cassette for the clustered regularly interspaced short palindromic

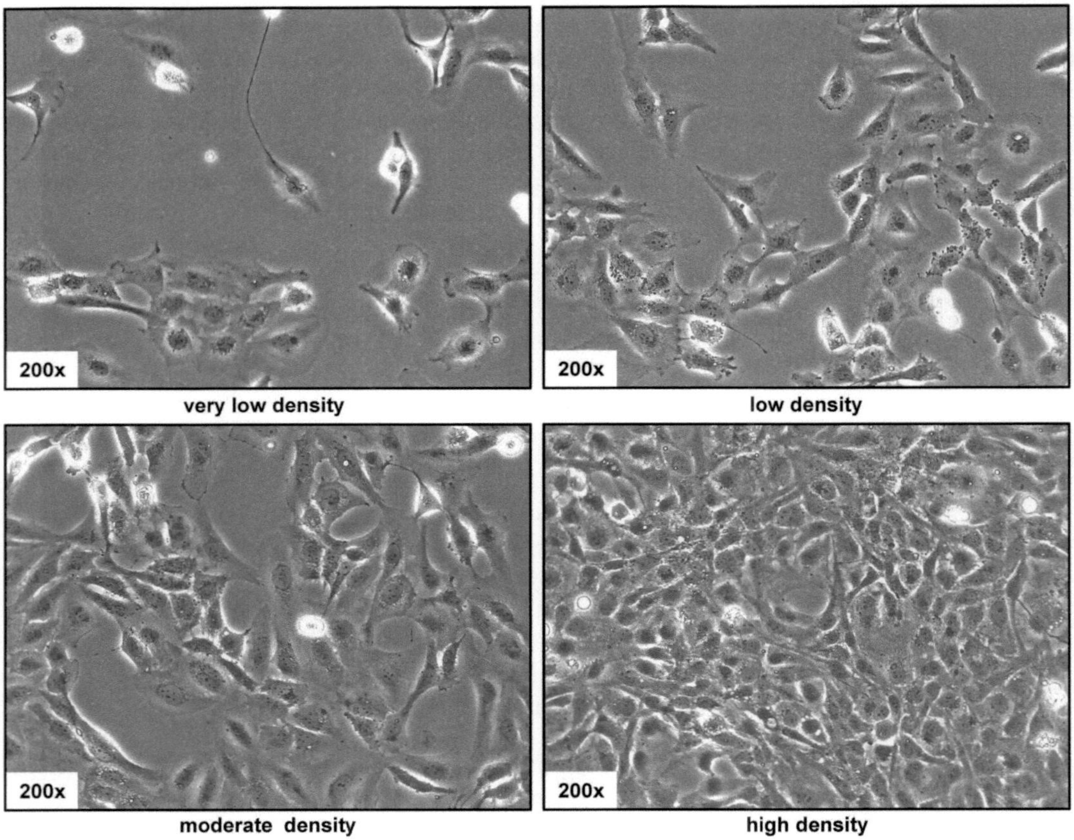

Fig. 4 Impact of cell density on morphology of cell line LX-2. The typical appearance of LX-2 cells cultured at different densities is depicted. Pictures were kindly taken by Sarah K. Schröder (Institute of Molecular Pathobiochemistry, Experimental Gene Therapy and Clinical Chemistry, RWTH Aachen University Hospital, Aachen, Germany)

repeats (*CRISPR*)-associated protein 9 (Cas9) useful for genome editing studies. This modified cell line retains key features of HSC cytokine signaling, neuronal gene expression, retinoid metabolism, and fibrogenesis, making it a highly suitable for studying aspects of gene function of human fibrosis (*see* **Note 26**).

LI90: This cell line expresses α-SMA and various types of connective tissue components (collagen types I, III, IV, V, and VI, laminin, and fibronectin). The cells are negative for desmin. In culture media containing vitamin A, the cells form cytoplasmic fat droplets. During prolonged culturing, the parenteral LI90 cell line enters replicative senescence [2, 3].

TWNT-1: This derivative of cell line LI90 constitutively expresses GFP. The cells are positive for FLT-1, KDR-flk-1, HGF, and collagen type I mRNA. The cells express hTERT, which is detectable by RT-PCR, and show high telomerase activity [4].

TWNT-4: This derivative of LI90 constitutively expresses GFP, PDGFRβ, α-SMA, and type I collagen. Expression of PDGFRβ, α-SMA, and collagen can be decreased by stimulation with IFN-γ (10 U/mL) or rapamycin (0.1 ng/mL). Moreover, apoptosis can be induced in these cells by TRAIL [3].

GREF-X: GREF-X cells are positive for α-SMA. They express collagen types I, IV, V, and VI, as well as laminin and fibronectin. The cells are responsive to TGF-β1 and are able to incorporate and esterify retinol. Although the cells were immortalized by transfection with a plasmid containing the coding sequence of polyoma virus large T antigen, the cells are negative in immunofluorescence, Southern blot, and PCR for the polyoma virus large T antigen [5]. This cell line was later shown to be of rat origin and listed by the International Cell Line Authentication Committee (ICLAC) under Registration ID: ICLAC-00123 as a misidentified cell line (*see* **Note 3**).

hTERT-HSC: This cell line expresses mRNAs for IL-6, IL-8, IL-10, PDGFRα, PDGFRβ, and GFAP and expresses α-SMA and collagen type I proteins. The cells contain functional telomerase and are capable of incorporating exogenous retinol [6]. In this cell line, the reconstitution of telomerase activity by ectopic expression of human telomerase reverse transcriptase (hTERT) is an effective mean to maintain telomerase length in human hTERT-HSC. While telomerase-negative HSC enter a non-dividing, senescence state after 9–15 passages, the introduction of hTERT into activate human HSCs results in an immortalized phenotype that exhibits morphologic and functional characteristics of activated HSC (Fig. 5).

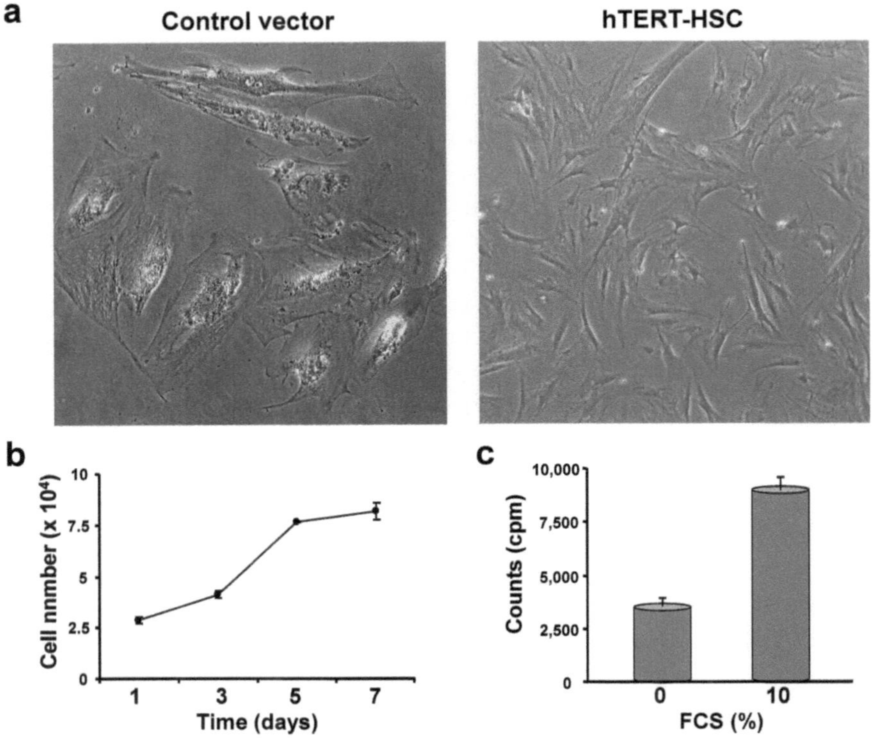

Fig. 5 Ectopic expression of functional human telomerase reverse transcriptase (hTERT-HSC) immortalizes activated HSC in culture. (**a**) Ectopic expression of functional telomerase in HSC promotes cell survival and results in immortalization. Compared to cells that were infected with a control vector (left panel), hTERT-positive cells (right panel) retain the morphologic characteristics of activated HSC. While the vector control-infected cells enter a non-dividing sate after 9–15 passages in culture, hTERT-infected cells continue to proliferate (original magnification: ×100). (**b**) To investigate the growth characteristics of hTERT-infected cells, cells were plated in 35 mm dishes, and cells were counted in duplicate from day 1 to day 7 in culture using a hemacytometer. Data represent the mean ± SD. (**c**) For [^{3}H]thymidine incorporation assay, hTERT-positive cells were serum-starved for 24 h in serum-free medium and afterward grown in medium containing 10% FCS and supplemented with 1 μCi/mL [^{3}H]thymidine for another 24 h. HSCs maintained in 0% FCS served as a control. Experiments were performed in triplicate. Data represent the mean ± SD. The image was provided by Bernd Schnabl, modified and published with permission. For more details on this cell line, refer to [6]

HSC-180: This cell line produces and secretes large quantities of functional urokinase plasminogen activator (u-PA) [7].

NPC-hTERT: This cell line expresses α-SMA, collagen type I, HGF, and hTERT [8].

4.4 Authentication Methods

Cultured cell lines are susceptible to contamination by microbes and cross-contamination with other cells. Therefore, periodic monitoring for characteristics is advised to authenticate their identity and prevent loss of time, money, and erroneous experimental data. Unfortunately, short tandem repeat profiles are only available for human cell LX-2 [34] and mouse cell line GRX [29]. In addition,

Table 4
Useful characteristics that can be used for cell authentication

Method	Suitable for
STR profile	LX-2, GRX, Col-GFP-HSC, HSC-T6, CFSC-2G
Expression of SV40T	LX-1, SV68c-IS, A640-IS (33 °C), IMS/MT(−), IMS/N, HSC-T6, Col-GFP-HSC, JS1 (C57/Bl6 WT), JS2 (C57/Bl6 TLR4$^{-/-}$), JS3 (C57/Bl6 MyoD88$^{-/-}$)
Expression of GFP	TWNT-1, TWNT-4, Col-GFP-HSC
Expression of GFAP	LX-1, LX-2, hTERT-HSC, HSC-T6, BSC, MFBY2, T-HSC/Cl-6, LSC-1, RGF, RGF-N2, M1-4HSC, Col-GFP-HSC
Expression of α-SMA	LX-1, LX-2, LI90, TWNT-4, GREF-X, hTERT-HSC, NPC-hTERT, HSC-T6, PAV-1, HSC-PQ, MFBY2, T-HSC/Cl-6, LSC-1, RGF, RGF-N2, SV68c-IS, A640-IS (33 °C), A640-IS (39 °C), M1-4HSC, Col-GFP-HSC, IMS/MT(−), IMS/N
Expression of Collagen type I	LX-1, LX-2, LI90, TWNT-4, TWNT-1, GREF-X, hTERT-HSC, NPC-hTERT, CFSC and derivatives (to different content), NFSC, PAV-1, HSC-PQ, BSC, MFBY2, T-HSC/Cl-6, LSC-1, GRX, A640-IS (33 °C)/ A640-IS (39 °C), M1–4HSC, Col-GFP-HSC
Expression of desmin	LI90, HSC-T6, PAV-1, HSC-PQ, BSC, T-HSC/Cl-6, LSC-1, SV68c-IS, M1-4HSC, IMS/MT(−), IMS/N

we have recently established STR profiles for the mouse cell line Col-GFP-HSC and rat cell lines HSC-T6 and CFSC-2G [12, 30, 31]. Apart from a specific STR profile, there are some cellular and growth characteristics that might help recognize cross-contamination or misidentification. Suitable cell characteristics for the different HSC lines that can be used to prevent misidentification are depicted in Table 4 (*see* **Note 27**).

4.5 Detection of Lipid Droplets in Hepatic Stellate Cell Lines

Some HSC lines have retained the ability to enrich retinoids (vitamin A and its metabolites) within cytoplasmic lipid droplets. In particular, the uptake of retinoids can be significantly fostered in some lines by addition of vitamin A (e.g., as a retinyl ester) to the culture medium. In unfixed cultured cells, the retinoid-containing lipid droplets can be easily detected because they exhibit a striking, blue-green autofluorescence when excited with a light of approximately 330 nm. However, this autofluorescence is rapidly fading, and fat droplet documentation is therefore somewhat limited. In contrast, there are specific stains that are ideally suited to document lipid droplet accumulation in cells.

In the following, we describe well-established procedures for lipid droplet staining, namely, the Oil Red O protocol and the BODIPY or Nile Red protocols. Based on their different chemical structures, the spectral properties of these three dyes are different (Fig. 6) (*see* **Note 28**).

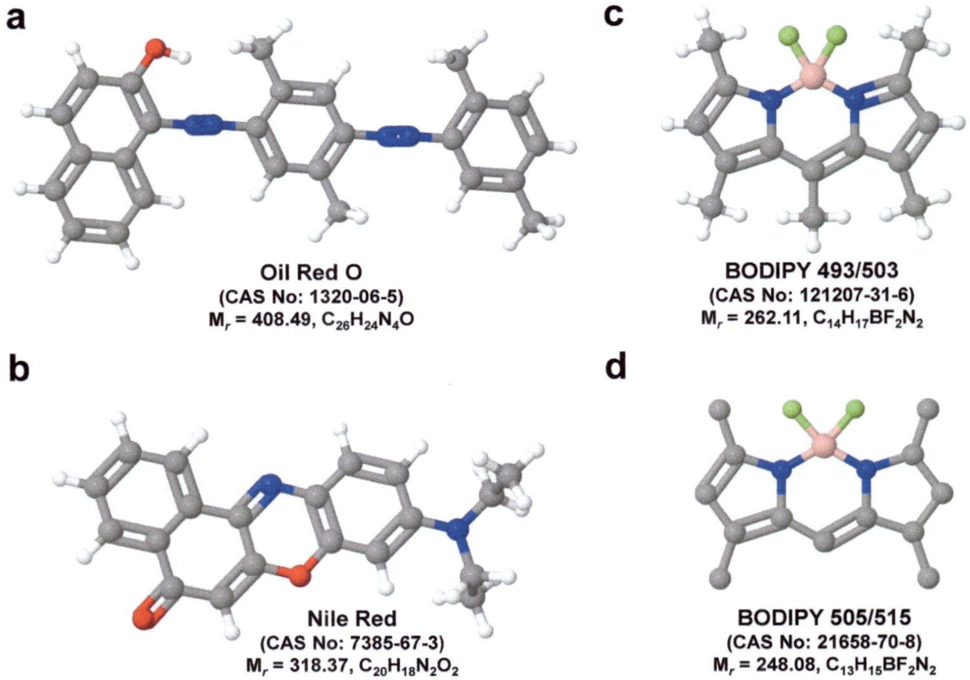

Fig. 6 Chemical structures of (**a**) Oil Red O, (**b**) Nile Red, (**c**) BODIPY 493/503, and (**d**) BODIPY 505/515. All four dyes are useful for the staining of fat and lipid droplets. Structure, CAS number, molecular weights, and chemical composition of each dye are given

4.5.1 Oil Red O Staining

Oil Red O (CAS No. 1320-06-5), also known as Solvent Red 27 or Sudan Red 5B, is a fat-soluble red diazo dye that can be generally used for selective staining and detection of neutral triglycerides and lipids on frozen sections, paraffin section, and cells. Combined with hematoxylin as a nucleic counterstain, the dye allows excellent coloration of fat droplets. In the following, a rapid Oil Red O staining procedure is described.

1. Freshly prepare a working solution of Oil Red O (2 mg/mL Oil Red O in 60% isopropanol). Therefore, dilute three parts of an Oil Red O stock solution (3 mg/mL solved in 100% isopropanol) in two parts of sterile water.
2. Remove the culture media from the cells, and gently wash the cells twice with phosphate-buffered saline.
3. Remove the washing solution, and fix the cells by adding 10% formalin solution. Incubate the cells for 30 min.
4. Remove the fixative, and gently wash the cells two times with distilled water.
5. Incubate the fixed cells in 60% isopropanol for 5 min.

6. Thereafter, remove the 60% isopropanol, and add a suitable volume of the Oil Red O working solution.
7. Incubate the cells for 10 min with gentle rotation (use a rotator at low speed).
8. Remove the staining solution, and remove excess dye by washing the cells several times with distilled water.
9. Counterstain the cells with Mayer's hematoxylin solution for 1 min (*see* **Note 29**).
10. Remove the hematoxylin solution, and wash the cells several times with distilled water.
11. Document the staining of the wet cells under microscope. Lipids will appear in red, and the nuclei will appear in blue. Avoid drying of cells.

4.5.2 BODIPY and Nile Red Stains

BODIPY and many of its derivatives are fluorescent dyes with unique properties. They can absorb UV light and emit very narrow wavelength below 600 nm. In particular, low-molecular-weight, neutral conjugates can effectively permeate to live cells. In particular, BODIPY 493/503 (CAS no: 121207-31-6) and BODIPY 505/515 (CAS no: 21658-70-8) are perfectly suited to stain neutral lipids and lipid droplets. Similarly, Nile Red (CAS no: 7385-67-3) is a phenoxazine dye which shows strong fluorescence enhancement upon transition from aqueous to hydrophobic environments and is useful to detect intracellular lipid droplets.

For staining, it is best to culture the cells to be stained on sterile glass coverslips positioned in six-well tissue plates.

1. Grow cells on a coverslips placed in a well of a 6-well plate to a density of 50–60%.
2. Freshly prepare a BODIPY (~10 μM) or Nile Red working solution (1 μM) in PBS (*see* **Note 30**).
3. Remove the media, and fix the cells by incubating in 2 mL 3% paraformaldehyde at room temperature for 30 min.
4. Aspirate the fixing solution, and wash coverslips three times each with 2 mL PBS.
5. Cover the cells with BODIPY or Nile Red staining solution, and incubate in the dark for 5–10 min at room temperature.
6. Aspirate the staining solution, and wash the cells three times with PBS.
7. Mount the coverslips onto glass slides with fluorescence mounting media.
8. BODIPY-stained fat droplets can be visualized by their green fluorescence, while the Nile Red fluorescence is detectable in both red and green channels.

5 Notes

1. This cell line is less prone to replicative senescence compared to parenteral LI90 cells.
2. Similar to TWNT-4, this cell line is less prone to replicative senescence than parenteral LI90 cells.
3. In a study of 252 consecutive human cell lines that were submitted to 6 cell repositories, this cell line was shown by DNA profiling and molecular cytogenetics to be of rat origin [35].
4. Infection with an adenovirus expressing the latency-associated peptide (LAP) from TGF-β reverts the HSC phenotype.
5. At the permissive temperature (33 °C), the SV40T is expressed, and cells grow, while at the non-permissive temperature (39 ° C), the SV40T is undetectable, and cells arrest.
6. In response to TGF-β treatment for 3 weeks, the cells acquire a MFB phenotype termed M-HT.
7. Typically, "established" HSC lines are used in investigations studying aspects of cellular toxicity, proliferation, cell adhesion, migration, extracellular matrix synthesis, and retinol metabolism. The transfection efficiency of these cells is commonly high, enabling studies investigating the function of genes or gene products. Nevertheless, it is advisable to critically review the literature before selecting one HSC line for research studies. Of note, researchers tend to use a specific HSC line as the "default" cell line to investigate specific aspects of HSC biology. It is advisable to use this same HSC line when conducting similar studies and then further validate the relevance of the biologic response by testing in primary stellate cells, where available.
8. Even after obtaining the cell line directly from the originator, this is no guarantee of the authenticity of the cell line. In the case that a cell line should be used without a defined STR profile, at least the existence of some characteristic markers of this cell line should be confirmed.
9. Mycoplasma infection can be easily detected in cellular supernatants and further classified using self-made or preformatted commercially available PCR kit systems, by indirect staining with a fluorescent dye such as Hoechst 33528 or alternatively by the agar and broth culture methods. Detailed protocols on these methods can be found elsewhere [36].
10. To overcome genetic drift in the continuous line due to a high number of cell passages, it is advisable to regularly discard cultures that have often been passaged and replenish the stock with one of the cryopreserved cell sources.

11. All basal media except ASF104 medium must be further supplemented with serum (FCS, FBS). Addition of antibiotics (penicillin, streptomycin, or others) and in some cases L-glutamine or non-essential amino acids is necessary. Some HSC lines carry vectors conferring resistance against selection antibiotics (e.g., G418, gentamycin, and kanamycin). These antibiotics can be added to the media to select for cells carrying the respective selection marker. For some HSC lines, it is proposed to use heat-inactivated serum (56 °C for 30 min) to inactivate the complement system and undetermined inhibitors of cell growth. Based on our experience, heat inactivation is not necessary for cell lines GRX, CFSC, HSC-T6, PAV-1, Col-GFP-HSC, and LX-2.
12. Detachment of cells from plasticware with Accutase is the gentlest way to dissociate the cells. However, trypsin solution (0.25–0.5% (w/v) prepared in balanced salt solution without Ca^{2+} and Mg^{2+}) is also well tolerated by most HSC lines.
13. Based on the different growth rates, we typically split the cell lines for maintenance in a ratio of 1:5–1:20. However, if an experiment is planned, it is advisable to determine the number of cells in a Neubauer chamber or an automated cell counter and seed a specific number of cells suitable for the planned experiments. Moreover, if experience with the line is limited, it is recommended to determine the total number of cells and the percent viability of detached cells by using the trypan blue exclusion procedure. A short protocol is detailed in the next paragraph.
14. One big square corresponds to a volume of 0.1 μL. The obtained value must be multiplied by a factor of 10,000 to extrapolate to 1 mL. The factor 2 corresponds to the dilution factor resulting from adding of trypan blue.
15. In the past, this protocol was successfully applied by the author to freeze GRX, Col-GFP-HSC, HSC-T6, CFSC, PAV-1, and LX-2 cells.
16. Liquid nitrogen should be handled only in well-ventilated areas, and boiling and splashing should be prevented by care handling. Moreover, liquid nitrogen should only be filled into approved containers with loose-fitting lids not filled to more than 80%. It is mandatory to protect the eyes and face during handling of liquid nitrogen by a full-face shield over safety glasses. Always wear gloves during handling and safety shoes when working with large volumes of liquid nitrogen.
17. Regularly transfer the vessel between different finger areas, and wear a specialized, insulated glove. This will prevent cold damage to fingers.

18. The final concentration of DMSO in the cell suspension prepared in **step 3** is 1% (v/v). Although cell lines can tolerate this concentration for some time without severe cytotoxicity, DMSO can be removed before plating by centrifugation of the cell solution prepared in **step 3**. Therefore, spin the Falcon tube in a centrifuge at 100 × *g* for 5 min. Then, remove the supernatant, gently resuspend the cells in only 10 mL growth medium, transfer the cells into one 100 mm^2 petri dish, and transfer the plate into the incubator.
19. The following five cell lines are commercially available: (i) GRX (Banco de Células do Estado do Rio de Janeiro, Rio de Janeiro, Brazil; BCRJ code: 0094) [37]; (ii) Col-GFP-HSC (DSMZ, German Collection of Microorganisms and Cell Cultures, Braunschweig, Germany, #ACC 941) [38]; (iii) HSC-T6 (Merck, Sigma-Aldrich, Taufkirchen, Germany, #SCC069) [39]; (iv) LX-2 (Sigma-Aldrich, #SCC064) [40]; and (v) LX-2 Cas9 (Merck, Sigma-Aldrich, #SCC613) [41]. All these cell lines are classified as genetically modified organisms (GMO) with risk group 1 (Biosafety level 1, BSL1).
20. In some countries, primary human tissues and cells used for production of cell lines for research purposes are classified as "relevant material" for which strict regulations are defined. For example, such laws are specified in the UK in the Human Tissue Act of 2004. However, once established, a cell line (even when from human origin) is classified as "non-relevant material," and the usage should not be restricted.
21. Monday is the best day for parceling; shipments over the weekend should be avoided.
22. Not all courier services transport biological materials or deliver to all countries. Fill out the required documents (e.g., waybill, customs formalities), and organize necessary packing materials (polystyrene box, bubble wrap, plastic bags, safety packs, etc.) in advance.
23. When deciding to send cryopreserved stocks on dry ice, please consider the fact that based on the quality (density) and quantity of dry ice chosen, "re-icing" during transports lasting longer than 2–4 days might be necessary. This may incur transport costs.
24. Wear insulated gloves when handling dry ice.
25. A major disadvantage of using immortalized HSC cell lines is that they are not normal, in that they divide indefinitely and have anomalies in chromosome number and structure. It is also known that these lines sometimes do not express common HSC gene expression signatures and have different cellular morphologies or properties. Therefore, several aspects of HSC biology cannot be addressed in a continuous HSC line.

Furthermore, inherent intellectual property rights and special institutional requirements must be considered when working with genetically modified material. Be sure that the cells you work with are the desired ones. Take care of proper cell authentication.

26. The CRISPR/Cas9 transgene in this cell is an RNA-guided genome editing tool, which allows specific, efficient, and versatile gene editing. The CRISPR/Cas9 system consists of two parts: the Cas9 endonuclease that can unwind the DNA duplex and cleave both strands of a target sequence and the guide RNA (gRNA) that binds to Cas9 and determines (based on its sequence) the precise cleavage location at which Cas9 will cut the DNA.

27. Typically, all HSC lines have either an activated or even a transdifferentiated myofibroblastic phenotype. The cellular characteristics of some cell lines can be specifically modulated by culturing the cells in media containing vitamin A, TGF-β1, or adipocyte differentiation cocktail. In addition, there are several lines in which the growth, differentiation, and functions can be significantly modulated by carrying either a temperature-sensitive mutant of the SV40 large T-antigen or by functional expression of active telomerase that stabilizes telomere length, thereby extending the cellular life span. These cellular characteristics are also good features that can be used for cell authentication. Some cell lines can be kept "quiescent" by growth on a soft substratum (e.g., Matrigel) or in suspension culture.

28. While Oil Red O can be detected in visual light, the BODIPY and Nile Red dyes can be detected under UV light in which lipid-bound Nile Red emits a very broad light that can be detected as red or green fluorescence, while the BODIPY dyes emit less intense fluorescence light at more defined wavelengths (BODIPY 493/503, excitation 493 nm, emission 503 nm; BODIPY 505/515, excitation 505 nm, emission 515 nm).

29. Mayer's hematoxylin solution is commercially available from different suppliers at low prices. We recommend buying this solution in a preformed format. To prepare this solution on your own, dissolve 50 g of aluminum potassium sulfate in 1 L distilled water, and add 1 g of hematoxylin. After it is completely dissolved, add 0.2 g of sodium iodate and 20 mL of acetic acid (100%). Shortly boil the solution, cool, and filter it through a 0.2 μm filter.

30. The stock solution of BODIPY493/503 (M_r = 264.12) or BODIPY505/515 (M_r = 248.08) is prepared by solubilizing 1 mg/mL in anhydrous DMSO or absolute ethanol. This

corresponds to a concentration of ~3.79 mM and ~ 4.03 mM for BODIPY 493/503 and BODIPY 505/515, respectively. The Nile Red (Mr = 318.37) stock solution is prepared by dissolving 1 mg/mL dye in anhydrous DMSO. This corresponds to a concentration of ~3.14 mM. All three stock solutions should be stored frozen (−20 °C) in smaller aliquots protected from light.

Acknowledgments

The authors are grateful to Bernd Schnabl, Wolfgang Mikulits, Sarah K. Schröder, and Hidekazu Tsukamoto for providing images for this article. Ralf Weiskirchen is supported by grants from the German Research Foundation (WE2554/13-1, WE2554/15-1, WE2554/17-1) and a grant from the Interdisciplinary Centre for Clinical Research within the faculty of Medicine at the RWTH Aachen University (grant PTD 1-5). Work from Dr. Friedman's laboratory is supported by NIH Grant 5R01DK128289-03.

References

1. Xu L, Hui AY, Albanis E, Arthur MJ, O'Byrne SM, Blaner WS, Mukherjee P, Friedman SL, Eng FJ (2005) Human hepatic stellate cell lines, LX-1 and LX-2: new tools for analysis of hepatic fibrosis. Gut 54(1):142–151. https://doi.org/10.1136/gut.2004.042127
2. Murakami K, Abe T, Miyazawa M, Yamaguchi M, Masuda T, Matsuura T, Nagamori S, Takeuchi K, Abe K, Kyogoku M (1995) Establishment of a new human cell line, LI90, exhibiting characteristics of hepatic Ito (fat-storing) cells. Lab Investig 72(6):731–739
3. Shibata N, Watanabe T, Okitsu T, Sakaguchi M, Takesue M, Kunieda T, Omoto K, Yamamoto S, Tanaka N, Kobayashi N (2003) Establishment of an immortalized human hepatic stellate cell line to develop antifibrotic therapies. Cell Transplant 12(5):499–507. https://doi.org/10.3727/000000003108747064
4. Watanabe T, Shibata N, Westerman KA, Okitsu T, Allain JE, Sakaguchi M, Totsugawa T, Maruyama M, Matsumura T, Noguchi H, Yamamoto S, Hikida M, Ohmori A, Reth M, Weber A, Tanaka N, Leboulch P, Kobayashi N (2003) Establishment of immortalized human hepatic stellate scavenger cells to develop bioartificial livers. Transplantation 75(11):1873–1880. https://doi.org/10.1097/01.TP.0000064621.50907.A6
5. Weill FX, Blazejewski S, Blanc JF, Huet S, Gauthier JM, Neaud V, Olaso E, Dubuisson L, Azais-Braesco V, Vidal-Vanaclocha F, Balabaud C, Bioulac-Sage P, Rosenbaum J (1997) Characterization of a new human liver myofibroblast cell line: transcriptional regulation of plasminogen activator inhibitor type I by transforming growth factor beta 1. Lab Investig 77(1):63–70
6. Schnabl B, Choi YH, Olsen JC, Hagedorn CH, Brenner DA (2002) Immortal activated human hepatic stellate cells generated by ectopic telomerase expression. Lab Investig 82(3):323–333. https://doi.org/10.1038/labinvest.3780426
7. González-Cuevas J, Bueno-Topete M, Armendariz-Borunda J (2006) Urokinase plasminogen activator stimulates function of active forms of stromelysin and gelatinases (MMP-2 and MMP-9) in cirrhotic tissue. J Gastroenterol Hepatol 21(10):1544–1554. https://doi.org/10.1111/j.1440-1746.2006.04398.x
8. Waki K, Anno K, Ono T, Ide T, Chayama K, Tahara H (2010) Establishment of functional telomerase immortalized human hepatocytes and a hepatic stellate cell line for telomere-targeting anticancer drug development. Cancer Sci 101(7):1678–1685. https://doi.org/10.1111/j.1349-7006.2010.01576.x

9. Greenwel P, Schwartz M, Rosas M, Peyrol S, Grimaud JA, Rojkind M (1991) Characterization of fat-storing cell lines derived from normal and CCl4-cirrhotic livers. Differences in the production of interleukin-6. Lab Investig 65(6):644–653
10. Greenwel P, Rubin J, Schwartz M, Hertzberg EL, Rojkind M (1993) Liver fat-storing cell clones obtained from a CCl4-cirrhotic rat are heterogeneous with regard to proliferation, expression of extracellular matrix components, interleukin-6, and connexin 43. Lab Investig 69(2):210–216
11. Vogel S, Piantedosi R, Frank J, Lalazar A, Rockey DC, Friedman SL, Blaner WS (2000) An immortalized rat liver stellate cell line (HSC-T6): a new cell model for the study of retinoid metabolism in vitro. J Lipid Res 41(6): 882–893. https://doi.org/10.1016/S0022-2275(20)32030-7
12. Nanda I, Steinlein C, Haaf T, Buhl EM, Grimm DG, Friedman SL, Meurer SK, Schröder SK, Weiskirchen R (2022) Genetic characterization of rat hepatic stellate cell line HSC-T6 for in vitro cell line authentication. Cell 11(11): 1783. https://doi.org/10.3390/cells11111783
13. Sauvant P, Sapin V, Abergel A, Schmidt CK, Blanchon L, Alexandre-Gouabau MC, Rosenbaum J, Bommelaer G, Rock E, Dastugue B, Nau H, Azaïs-Braesco V (2002) PAV-1, a new rat hepatic stellate cell line converts retinol into retinoic acid, a process altered by ethanol. Int J Biochem Cell Biol 34(8): 1017–1029. https://doi.org/10.1016/s1357-2725(02)00023-7
14. Pan Q, Li DG, Lu HM, Wang YQ, Zhang WZ, Xu QF (2005) A new immortalized rat cell line, hepatic stellate cell-PQ, exhibiting characteristics of hepatic stellate cell. Hepatobiliary Pancreat Dis Int 4(2):281–284
15. Xiong S, Yavrom S, Hazra S, Wu D, She H (2001) Spontaneously immortalized cell line from rat liver fibrosis with phenotypes of both hepatic stellate cells and myofibroblasts. Hepatology 34:520A
16. Sung CK, She H, Xiong S, Tsukamoto H (2004) Tumor necrosis factor-alpha inhibits peroxisome proliferator-activated receptor gamma activity at a posttranslational level in hepatic stellate cells. Am J Physiol Gastrointest Liver Physiol 286(5):G722–G729. https://doi.org/10.1152/ajpgi.00411.2003
17. Isono M, Soda M, Inoue A, Akiyoshi H, Sato K (2003) Reverse transformation of hepatic myofibroblast-like cells by TGFbeta1/LAP. Biochem Biophys Res Commun 311(4): 959–965. https://doi.org/10.1016/j.bbrc.2003.10.093
18. Kim JY, Kim KM, Nan JX, Zhao YZ, Park PH, Lee SJ, Sohn DH (2003) Induction of apoptosis by tanshinone I via cytochrome c release in activated hepatic stellate cells. Pharmacol Toxicol 92(4):195–200. https://doi.org/10.1034/j.1600-0773.2003.920410.x
19. Li T, Leng XS, Zhu JY, Wang FS (2015) Establishment and characterization of an immortalized rat hepatic stellate cell line. Int J Clin Exp Pathol 8(10):12064–12074. PMID: 26722391
20. Fausther M, Goree JR, Lavoie ÉG, Graham AL, Sévigny J, Dranoff JA (2015) Establishment and characterization of rat portal myofibroblast cell lines. PLoS One 10(3):e0121161. https://doi.org/10.1371/journal.pone.0121161
21. Borojevic R, Monteiro AN, Vinhas SA, Domont GB, Mourão PA, Emonard H, Grimaldi G Jr, Grimaud JA (1985) Establishment of a continuous cell line from fibrotic schistosomal granulomas in mice livers. Vitro Cell Dev Biol 21(7):382–390. https://doi.org/10.1007/BF02623469
22. Horie S, Kitamura Y, Kawasaki H, Terada T (2000) Inhibitory effects of antisense oligonucleotides on the expression of procollagen type III gene in mouse hepatic stellate cells transformed by simian virus 40. Pathol Int 50(12): 937–944. https://doi.org/10.1046/j.1440-1827.2000.01146.x
23. Kitamura Y, Tanigawa T, Katsumoto T, Tomita K, Wang HR, Hirai K, Ichihara K, Terada T (1997) Cell growth and differentiation of a novel mouse Ito (fat-storing) cell line transformed by a temperature-sensitive mutant of simian virus 40. Hepatology 26(2): 323–329. https://doi.org/10.1002/hep.510260211
24. Proell V, Mikula M, Fuchs E, Mikulits W (2005) The plasticity of p19 ARF null hepatic stellate cells and the dynamics of activation. Biochim Biophys Acta 1744(1):76–87. https://doi.org/10.1016/j.bbamcr.2004.12.009
25. Matsuura T, Kawada M, Sujino H, Hasumura S, Nagamori S, Shimizu H (1999) Vitamin A metabolism of immortalized hepatic stellate cells in the bioreactor. In: Wisse E, Knook DL, De Zanger R, Arthur MJP (eds) Cells of the hepatic sinusoid 7. Kupffer Cell Foundation, Leiden, pp 88–89
26. Meurer SK, Alsamman M, Sahin H, Wasmuth HE, Kisseleva T, Brenner DA, Trautwein C, Weiskirchen R, Scholten D (2013) Overexpression of endoglin modulates TGF-β1-signalling

pathways in a novel immortalized mouse hepatic stellate cell line. PLoS One 8(2): e56116. https://doi.org/10.1371/journal.pone.0056116

27. Miura N, Kanayama Y, Nagai W, Hasegawa T, Seko Y, Kaji T, Naganuma A (2006) Characterization of an immortalized hepatic stellate cell line established from metallothionein-null mice. J Toxicol Sci 31(4):391–398. https://doi.org/10.2131/jts.31.391
28. Guo J, Loke J, Zheng F, Hong F, Yea S, Fukata M, Tarocchi M, Abar OT, Huang H, Sninsky JJ, Friedman SL (2009) Functional linkage of cirrhosis-predictive single nucleotide polymorphisms of Toll-like receptor 4 to hepatic stellate cell responses. Hepatology 49(3):960–968. https://doi.org/10.1002/hep.22697
29. Schröder SK, Schüler HM, Petersen KV, Tesauro C, Knudsen BR, Pedersen FS, Krus F, Buhl EM, Roeb E, Roderfeld M, Almeida JL, Weiskirchen R (2022) Genetic and molecular characterization of the immortalized murine hepatic stellate cell line GRX. Cell 11:1504. https://doi.org/10.3390/cells11091504
30. Meurer SK, Brenner DA, Weiskirchen R (2022) Multiplex short tandem repeat profiling of immortalized hepatic stellate cell line col-GFP HSC. PLoS One 17(9):e0274219. https://doi.org/10.1371/journal.pone.0274219
31. Nanda I, Schröder SK, Steinlein C, Haaf T, Buhl EM, Grimm DG, Weiskirchen R (2022) Rat hepatic stellate cell line CFSC-2G: genetic markers and short tandem repeat profile useful for cell line authentication. Cell 11(18):2900. https://doi.org/10.3390/cells11182900
32. Saito M, Matsuura T, Masaki T, Maehashi H, Shimizu K, Hataba Y, Iwahori T, Suzuki T, Braet F (2006) Reconstruction of liver organoid using a bioreactor. World J Gastroenterol 12(12):1881–1888. https://doi.org/10.3748/wjg.v12.i12.1881
33. Zhu L, Mou Q, Wang Y, Zhu Z, Cheng M (2020) Resveratrol contributes to the inhibition of liver fibrosis by inducing autophagy via the microRNA-20a-mediated activation of the PTEN/PI3K/AKT signaling pathway. Int J Mol Med 46(6):2035–2046. https://doi.org/10.3892/ijmm.2020.4748
34. Weiskirchen R, Weimer J, Meurer SK, Kron A, Seipel B, Vater I, Arnold N, Siebert R, Xu L, Friedman SL, Bergmann C (2013) Genetic characteristics of the human hepatic stellate cell line LX-2. PLoS One 8(10):e75692. https://doi.org/10.1371/journal.pone.0075692
35. MacLeod RA, Dirks WG, Matsuo Y, Kaufmann M, Milch H, Drexler HG (1999) Widespread intraspecies cross-contamination of human tumor cell lines arising at source. Int J Cancer 83(4):555–563. https://doi.org/10.1002/(sici)1097-0215(19991112)83:4<555::aid-ijc19>3.0.co;2-2
36. Young L, Sung J, Stacey G, Masters JR (2010) Detection of mycoplasma in cell cultures. Nat Protoc 5(5):929–934. https://doi.org/10.1038/nprot.2010.43
37. BCRJ 0094 – GRX cell line. Available at: https://bcrj.org.br/celula/grx/. Last accessed 13.3.2023
38. DSMZ ACC 941 – Col-GFP-HSC cell line. Available at: https://www.dsmz.de/collection/catalogue/details/culture/ACC-941. Last accessed 13.3.2023
39. Merck SCC069 – HSC-T6 Rat Hepatic Stellate Cell line. Available at: https://www.sigmaaldrich.com/DE/de/product/mm/scc069. Last accessed 13.3.2023
40. Merck SCC064 – LX-2 Human Hepatic Stellate Cell line. Available at: https://www.merckmillipore.com/DE/de/product/LX-2-Human-Hepatic-Stellate-Cell-Line,MM_NF-SCC064. Last accessed 13.3.2023
41. Merck SCC613 – LX-2 Cas9 Human Hepatic Stellate Cell Line. Available at: https://www.emdmillipore.com/US/en/product/LX-2-Cas9-Human-Hepatic-Stellate-Cell-Line,MM_NF-SCC613?ReferrerURL=https%3A%2F%2Fwww.google.com%2F. Last accessed 13.3.2023

Chapter 9

Induction of Obstructive Cholestasis in Mice

Ralf Weiskirchen, Sabine Weiskirchen, Carmen G. Tag, and Steffen K. Meurer

Abstract

Experimental bile duct ligation (BDL) in rodents causes cholestatic liver injury characterized by structural and functional alterations that include periportal biliary fibrosis. These changes are time-dependent and based on excess accumulation of bile acids in the liver. This in turn causes damage of hepatocytes and functional loss, leading to recruitment of inflammatory cells. Liver resident pro-fibrogenic cells facilitate extracellular matrix synthesis and remodeling. The proliferation of bile duct epithelial cells provokes a ductular reaction characterized by bile duct hyperplasia. Experimental BDL surgery is technically simple and quick to perform and reliably generates progressive liver damage with a predictable kinetics. The cellular, structural, and functional alterations induced in this model are similar to that in humans suffering from diverse forms of cholestasis including primary biliary cirrhosis (PBC) or primary sclerosing cholangitis (PSC). Therefore, this extrahepatic biliary obstruction model is used in many laboratories worldwide. Nevertheless, BDL can result in significant variations and high mortality rates when surgery is carried out by untrained or inexperienced personnel. Here we present a detailed protocol to achieve a robust experimental obstructive cholestasis in mice.

Key words Inflammation, Fibrosis, Hepatic stellate cells, Portal myofibroblasts, Immune cells, Hepatocytes, Ductular reaction, Serum markers, Bile acids, Bile duct epithelial cells

1 Introduction

Historically, the ligation of the common bile duct (BDL) has been carried out as an experimental procedure for many years in species including dog, cat, rabbit, dove, guinea pig, lamb, hen, frog, rat, and mouse [1]. In normal conditions, the bile produced by hepatocytes is stored in the gallbladder and released upon demand into the duodenum. The surgical prevention of bile flow from the liver to the duodenum provokes a massive increase in bile acid concentrations in the liver and the serum, which undermine hepatocyte integrity (Fig. 1).

Progressive hepatic damage is indicated by elevated serum activities of aspartate aminotransferase (AST), alanine

Ralf Weiskirchen and Scott L. Friedman (eds.), *Hepatic Stellate Cells: Methods and Protocols*, Methods in Molecular Biology, vol. 2669, https://doi.org/10.1007/978-1-0716-3207-9_9,

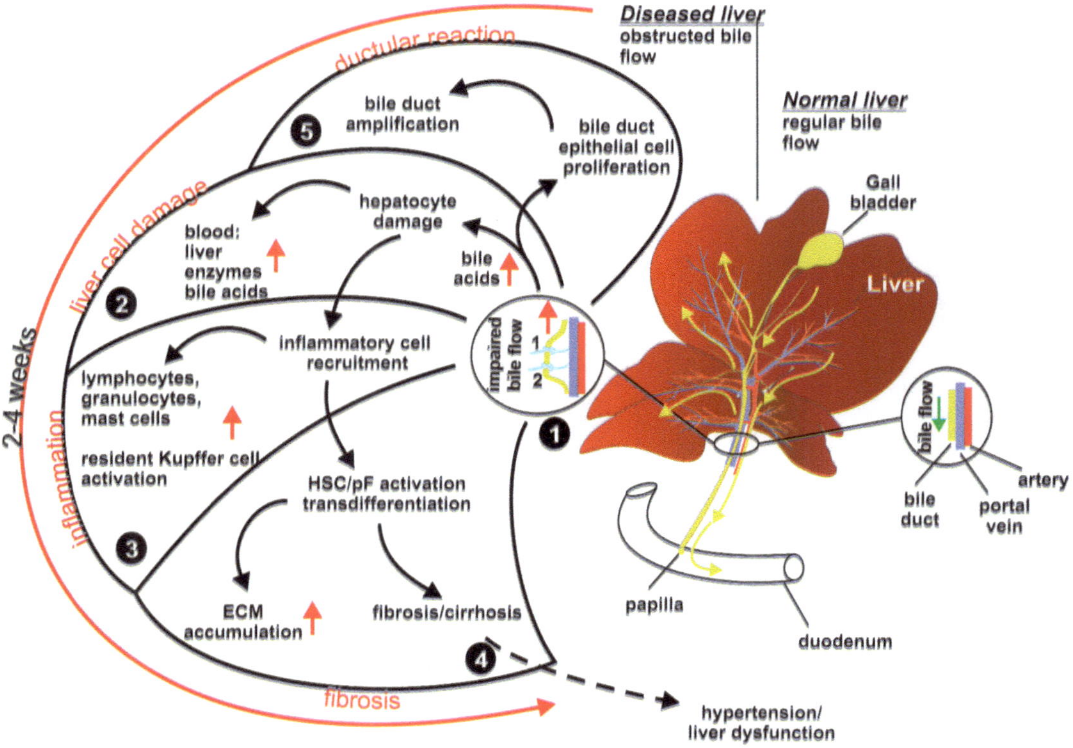

Fig. 1 The bile duct ligation (BDL) surgery and pathological consequences. (*right*) In the normal liver, the bile, produced by hepatocytes, is collected in the canaliculi and transported to and stored in the gallbladder. Upon stimulation, the bile is released in the duodenum for digestion purposes. (*left*) In the BDL surgery, the common bile duct is ligated by two sutures that are each fixed in close proximity by two surgical knots (➊). The biliary obstruction prevents proper bile flow from the liver to the intestine, causing an accumulation of bile in the liver and serum. Hepatocytes are affected (➋), and immune cells are recruited to the liver (➌). The release of pro-fibrogenic cytokines (e.g., TGF-β1) from hepatocytes, liver-resident cells (i.e., Kupffer cells), and infiltrating immune cells results in activation and transdifferentiation of hepatic stellate cells (HSC) and portal fibroblasts (pF) affecting matrix turnover and causing massive accumulation of extracellular matrix (ECM) and hypertension (➍). In addition, bile duct epithelial cells proliferate, and new intrahepatic bile ducts are formed (➎)

aminotransferase (ALT), γ-glutamyltranspeptidase (γ-GT), alkaline phosphatase (AP), and lactate dehydrogenase (LDH), released by dying hepatocytes. In turn, this leads to increased infiltration of immune cells and a massive atrophy of hepatocytes. At the same time, the number of extracellular matrix (ECM)-producing, α-smooth muscle actin (α-SMA)-positive cells is increased [2]. At later time points, the concentration of total bilirubin increases, and jaundice develops. To cope with the excess of bile acid, a ductular reaction develops, which is characterized by increased numbers of biliary epithelial cells and amplification of intrahepatic bile ducts. One of the most striking features is the activation of liver resident fibroblasts, namely, hepatic stellate cells (HSC) and portal fibroblasts (pF), which "transdifferentiate" into contractile myofibroblasts that produce large quantities of ECM proteins such as collagens and inhibit their degradation provoking an accumulation and remodeling of the hepatic matrix scaffold [3, 4].

This model accurately mimics human cholestasis in many aspects. Therefore, prolonged bile duct obstruction is globally used for studying fundamental aspects of cholestatic liver disease. In the following, we provide a detailed protocol for the BDL surgery procedure in mice. The correct application of this protocol results in the reproducible time-dependent morphological, structural, and functional alterations in the liver associated with cholestasis.

The described surgical procedure is simple and rapidly applicable. In the protocol, the animals are anesthetized, and a midsection laparotomy is performed. Through the opened abdominal cavity, the bile duct is exposed and ligated twice using surgical twine. This intervention prevents normal bile flow from the liver to the intestine.

Finally, the abdominal organs (liver and small intestine) are placed back into their physiological positions, the abdomen is closed, and the animals allowed to recover. Following the surgery, the animals typically develop a strong progressive fibrotic reaction after 2 weeks originating from the periportal fields.

2 Materials

2.1 Animals

To guarantee high reproducibility and low mortality rates, it is essential that the animals used are kept permanently under specific-pathogen-free (SPF) conditions. If animals are received from a commercial supplier or a collaborator, they should acclimatize to the local housing conditions for at least 1 week (*see* **Note 1**).

2.2 Antiseptics, Narcotics, and Analgesics

1. Polyalcohol hands antiseptic.
2. General anesthetic.
3. Narcotic analgetic (e.g., buprenorphine).
4. Eye and nasal ointment.

2.3 Instruments and Surgical Equipment

All instruments should be as sterile and clean as possible. Scissors, forceps, needle holders, and Colibri retractors can be sterilized by autoclaving, and all other instruments should be disinfected by cleaning with 70% ethanol solution or another appropriate antiseptic.

1. Heating pad including temperature controller.
2. Infrared lamp (100–150 W).

3. Vaporizer for general anesthetic or equivalent anesthesia system.
4. Small animal shaver.
5. Scissors.
6. Graefe forceps.
7. Moria iris forceps.
8. Microserrations forceps.
9. 6-0 Mersilk suture.
10. 5-0 Mersilene suture.
11. Needle holder Mathieu.
12. Colibri retractor.
13. Heat plate.

2.4 Glassware, Disposables, and Solutions

1. 25 mL beaker.
2. 5 mL syringe.
3. 1 mL syringe.
4. Hypodermic needles (1 inch 26 G).
5. Scotch tape.
6. Tissue paper.
7. Silk tape.
8. Cotton gauze swabs.
9. Fluid-impermeable drapes with excellent adsorptive capacity.
10. Soft cotton swabs.
11. Sterile isotonic 0.9% (w/v) NaCl solution.

3 Methods

3.1 Workplace and Presurgical Animal Preparation

1. Set up the workplace. To avoid any unnecessary handling of animals, all equipment and necessary reagents should be clearly arranged prior to the beginning of the experiment.
2. Place and heat the heating pad. Throughout the surgical procedure, mice should be placed on a controlled heating pad adjusted to 37 °C.
3. Mice subjected to BDL should be examined for signs of physical or behavioral abnormalities.
4. Document the weight of the mice before starting with the surgery.
5. Anesthetize the mice by using an inhalation anesthetic (*see* **Note 2**).

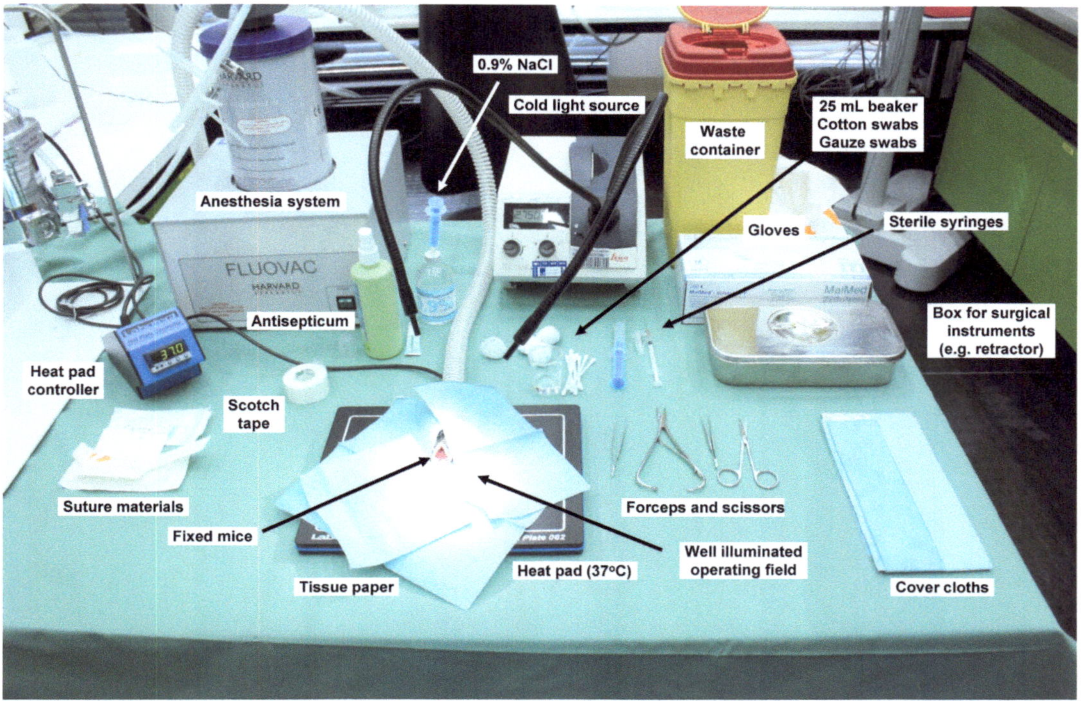

Fig. 2 Example of a properly arranged workplace for performing the BDL surgery in mice. Instruments and surgical equipment are arranged clearly, and the mouse that will be subjected to BDL is well fixed on the heating plate. The areas surrounding the operation sides are covered with tissue paper

6. Administer analgesia in proper concentration (*see* **Note 3**).
7. Shave the abdominal fur with an electric fur shaver.
8. Softly, cream the eyes of the mouse with an eye and nose ointment to protect the eyes from drying.
9. Place the mouse on the pre-heated hot plate, connect the snout to the inhalation anesthetic system, and fix the legs with silk tape stripes.
10. Sterilize the shaved abdominal skin with a gauze swab moistened with a standard, ready-to-use antiseptic.
11. Cover the adjacent body areas of the surgical site with tissue paper.
12. Before starting the surgery, establish that the workplace is properly arranged and necessary surgical instruments are properly prepared (Fig. 2).

3.2 General Overview of the Surgical Procedure

1. Familiarize yourself with all work steps in advance so that the operation does not take too long (*see* **Note 4**).
2. It should be possible to complete the subsequent experimental procedure for one mouse in 20–30 min (*see* **Note 5**).

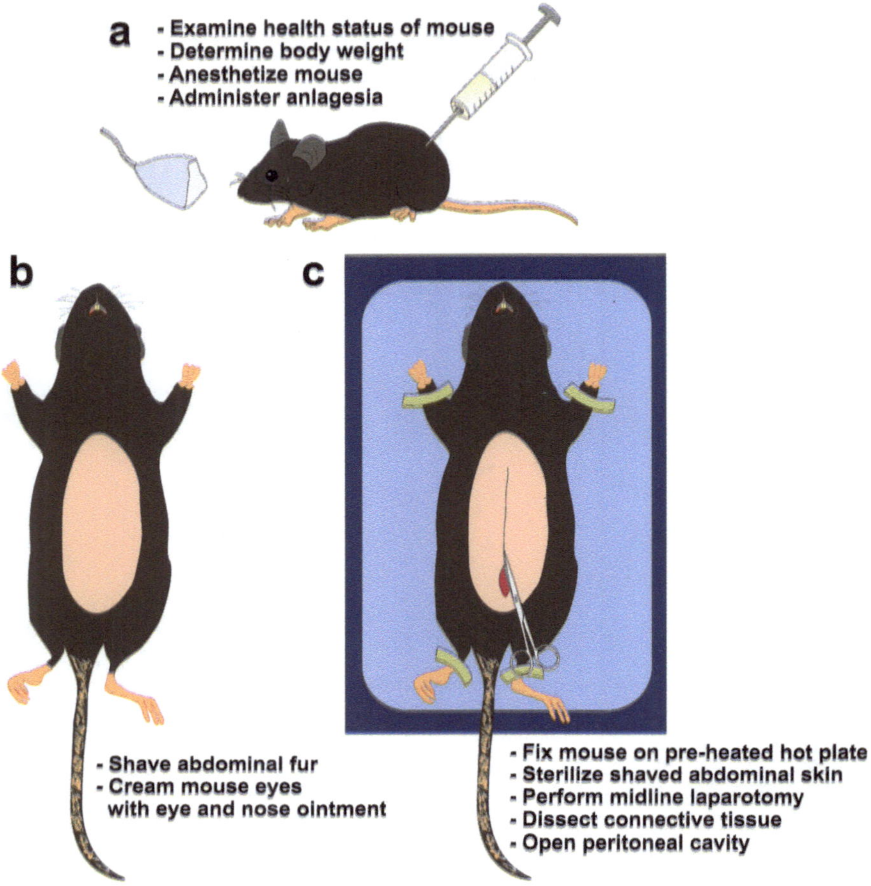

Fig. 3 Individual steps for performing a bile duct ligation in mice (part I). (**a**) Anesthetization and administration of analgesia. (**b**) Exposure of the surgical area and eye protection of the mice subjected to the surgery. (**c**) Fixing on a heat plate and opening of the peritoneal cavity by midline laparotomy

3. In the first part of the protocol, the animal is prepared for the intervention. In particular, the health status of the mouse is examined and the body weight determined (Fig. 3a). This step is necessary for calculation of the proper concentration of analgesia.
4. In the next step, the operation area is exposed by shaving the abdominal fur (Fig. 3b). In addition, the eyes of the mouse are treated with a protective eye and nose ointment.
5. Then, the mouse is fixed on a pre-heated (37 °C) heat plate, the abdominal skin sterilized, and the peritoneal cavity opened by a midline laparotomy (Fig. 3c).
6. In the second part of the protocol, the cavity is spread, the liver lifted and attached to the diaphragm so that the ventral side of the liver is exposed, and the bile duct uncovered (Fig. 4a).

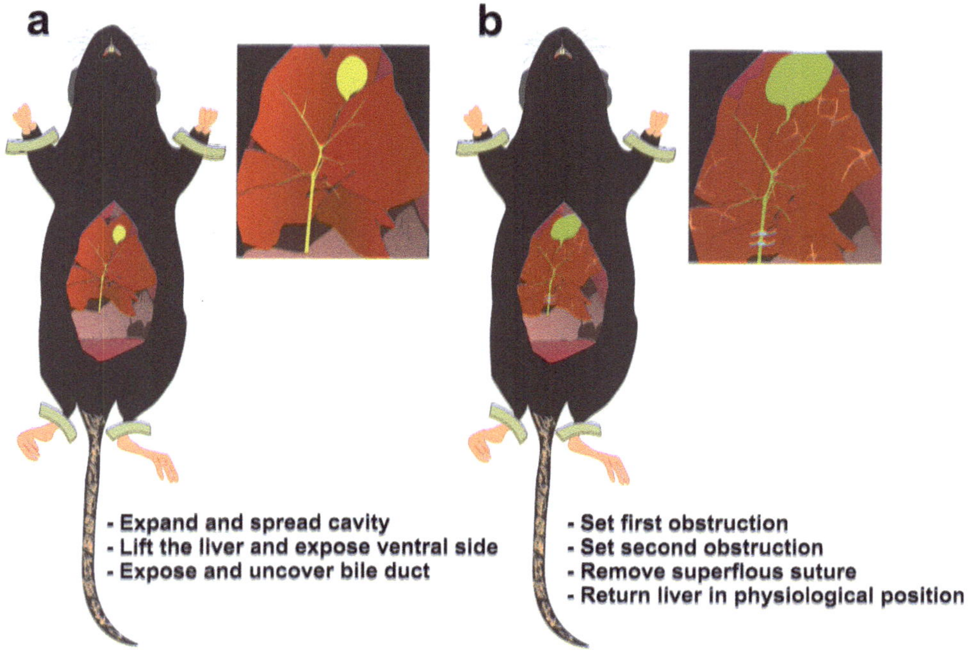

Fig. 4 Individual steps for performing a bile duct ligation in mice (part II). (**a**) Uncover the bile duct. (**b**) Place the bile duct obstruction

7. Bile duct obstruction is accomplished by setting two sutures, each with two surgical knots (Fig. 4b).
8. Thereafter, all organs are returned to and arranged in their physiological position in the body cavity.
9. The abdominal cavity is closed and sewn with surgical thread.

3.3 Surgical Procedures in Practice

1. Cut open the fur using surgical scissors.
2. Carefully dissect the connective tissue located on the top of the peritoneum by using scissors as a spreader.
3. Open the peritoneal cavity by cutting precisely along the *linea alba* to prevent damage of blood vessels and bleeding. The length of the vertical incision line along and through the *linea alba* should be about 2 cm in length.
4. Expand the opening of the cavity by inserting a holding suture in the sternum, raise the filament of the suture, and fix it on the top of the inhalation mask.
5. Spread the opening of the peritoneal cavity by inserting a Colibri retractor (Fig. 5a, b).
6. Moisten a cotton swab with 0.9% NaCl solution, and lift the liver with the moistened cotton swab so that the liver attaches to the diaphragm and the ventral side of the liver is exposed (Fig. 5c).

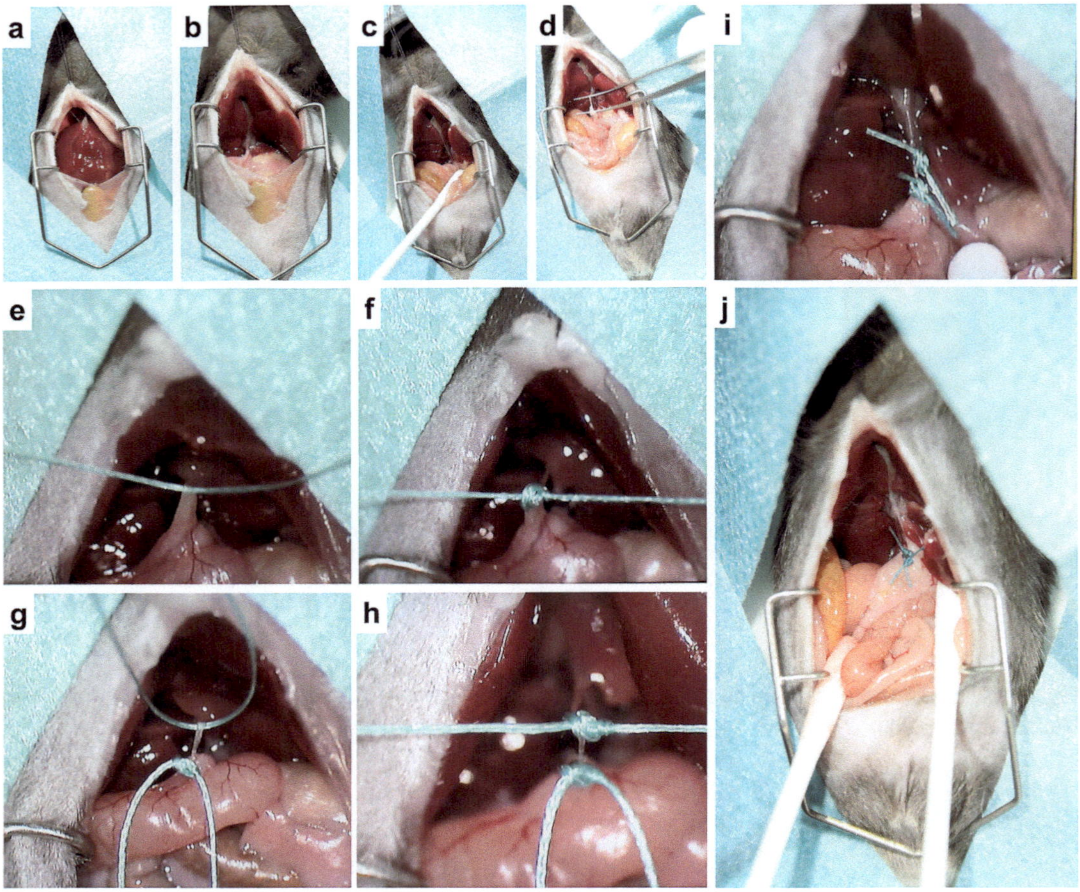

Fig. 5 The bile duct surgery in mice. (**a**, **b**) The abdominal cavity is opened and spread with a Colibri retractor and the liver lifted and attached to the diaphragm. (**c, d**) The small intestine is moved caudally, and the bile duct is detached from the portal vein and artery. (**e, f**) The first obstruction is set. (**g, h**) The second obstruction is set. (**i, j**) Superfluous suture is removed, and the internal organs are moved back to their physiological position

7. Move the gut (small intestine) caudally to expose the bile duct.
8. Detach the bile duct from the flanking portal vein and hepatic artery using microserrations forceps (Fig. 5d).
9. Obstruct the bile duct with a 5-0 suture, and fix it with two surgical knots (Fig. 5e, f).
10. Place a second cranial ligation, and fix it in the same manner as the first one (Fig. 5g, h) (*see* **Note 6**).
11. Cut off as much of the sutures as possible (Fig. 5i).
12. Carefully lower the sternum, and remove the Colibri retractor.
13. Rinse the peritoneal cavity with 0.9% NaCl solution, and move back all abdominal organs in their physiological positions with wetted swaps (Fig. 5j).

14. Close both abdominal layers (i.e., peritoneum and cutis plus facia) with separate running 6-0 Mersilk sutures.
15. Remove superfluous sutures, and sterilize the complete operation area with a gauze swab wetted with antiseptic solution.

3.4 Postoperative Treatment

1. After finalizing the surgical procedure, replace the mouse into a cage warmed up by an infrared lamp until it has fully recovered.
2. Thereafter, the mouse is placed back into a normal cage with ad libitum access to tap water and food.
3. Apply analgesic therapy on a regular basis for up to 3 days, and monitor the animals at regular intervals (*see* **Note 7**).
4. After 3–4 weeks, when jaundice and fibrosis have fully developed, sacrifice the animal, collect blood for subsequent measurements of clinical chemistry parameters, and retrieve the liver for histochemical and/or biochemical analysis. Collect also secondary affected organs for analysis if needed.

3.5 Essential Controls

Animals subjected to sham surgery are essential controls to identify potential alterations that are not induced by the BDL per se. The application of inhalation anesthetics, the intraperitoneal injection of analgesia, the opening of the body cavity, and the handling of organs outside the body are stressful for the animal and can cause extensive inflammatory responses and postoperative pain. Therefore, sham-operated mice should be included as surgical control group to identify these factors. The sham operation is done essentially as outline in Subheading 3.3 with omitting **steps 9–11**.

3.6 Readouts for Fibrosis

BDL leads to an acute obstructive jaundice and secondary biliary fibrosis that can progress to cirrhosis. The pathological consequences of the BDL treatment can be recognized visually by onset of jaundice in skin, blood, and urine. Moreover, the induced changes depicted in Fig. 1 can be traced by alterations of specific blood parameters. Liver damage correlates with alterations in alanine-aminotransferase (ALT), aspartate-aminotransferase (AST), lactate dehydrogenase (LDH), alkaline phosphatase (AP), γ-glutamyl transferase (γ-GT), total bilirubin, and total protein (or albumin). To monitor the inflammatory response, the expression of pro-inflammatory cytokines as TNF-α and IL-6 can be evaluated. In addition, increased expression of pro-fibrogenic genes such as collagen type I, fibronectin, matrix metalloproteinase-2 (MMP-2), and tissue inhibitor of matrix metalloproteinase-1 (TIMP-1) indicate ongoing hepatic fibrosis. Hepatic injury (cellular) and fibrosis (structural) can be documented in sections of formalin-fixed and paraffin-embedded liver specimens by using hematoxylin and eosin (H&E), Sirius Red, and Masson's trichrome stain. Furthermore, histopathological alterations associated with fibrosis can be documented by

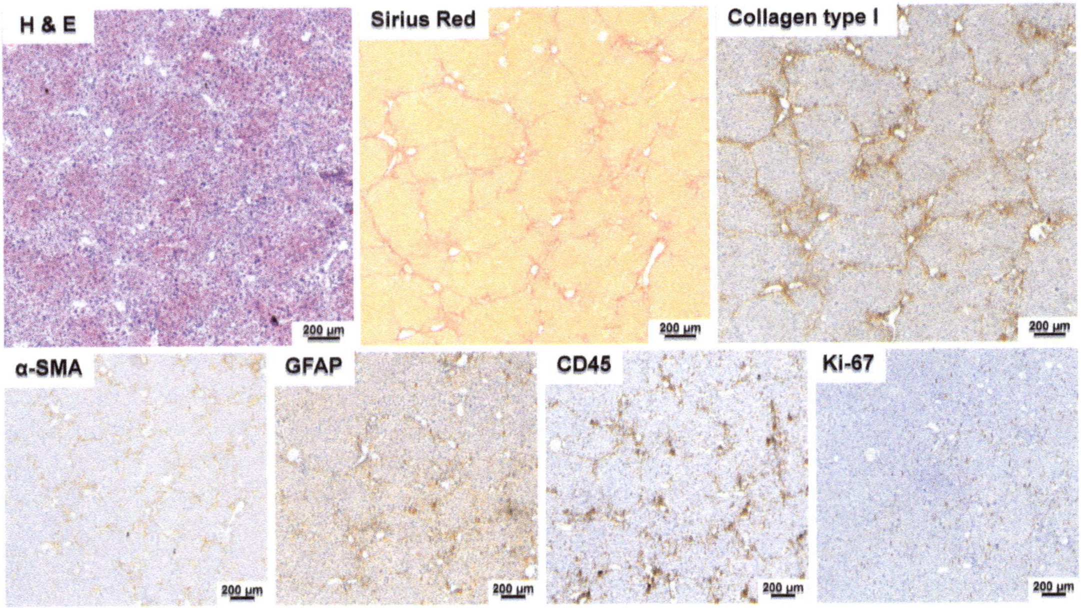

Fig. 6 Histopathological stains and immunological stains of hepatic fibrosis after BDL surgery. Serial liver sections were prepared from a liver of a mouse subjected to BDL for 3 weeks. Hematoxylin-eosin staining (H&E) was carried out to assess tissue morphology, histopathological lesions, and cellularity. Collagen deposits were visualized either by Sirius Red stain or by immunohistochemistry using an antibody specific for collagen type 1. In addition, immunohistochemistry was performed for α-SMA, GFAP, the leukocyte common antigen CD45, and the proliferation marker Ki-67. Suitable protocols for tissue staining and immunohistochemistry are given elsewhere [5–8]

immunohistochemistry using specific antibodies for collagen type I, α-SMA, GFAP, or CD45 representing a pan-leukocyte marker, as well as Ki-67, indicating increased bile duct epithelial cell proliferation (Fig. 6).

Increased expression of pro-inflammatory and pro-fibrotic markers can also be documented and quantified by reverse transcription quantitative PCR (RT-qPCR) and/or Western blot analysis (Fig. 7).

3.7 Classification of Severity of Procedure

The BDL-induced cholestatic liver disease model of up to 3 or 4 weeks with no major impairment of hepatic function is classified as a "moderate" procedure according to Article 15 of the EU Directive 2010/63 in which procedures are classified as "non-recovery," "mild," "moderate," or "severe" [9].

4 Notes

1. The protocol described can be performed with all genetic backgrounds and gender. However, the degree of fibrosis obtained

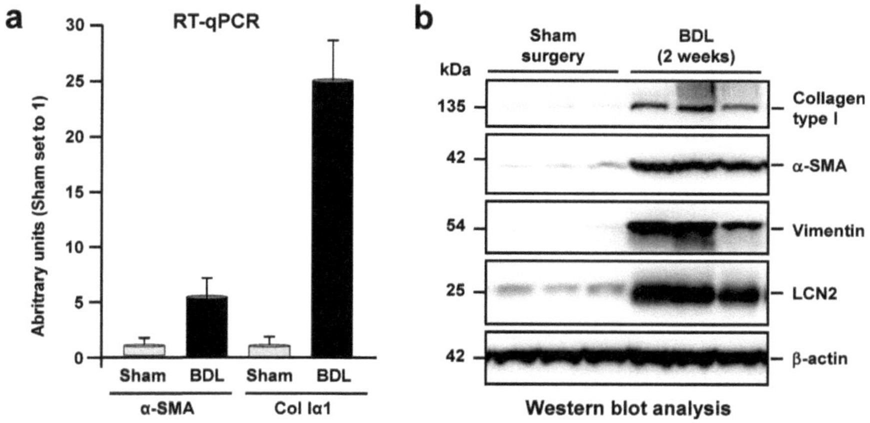

Fig. 7 RT-qPCR and Western blot analysis of markers associated with inflammation and fibrosis. (**a**) A typical RT-qPCR result of α-SMA and collagen type I (Col Iα1) expression obtained after 2 weeks BDL is depicted. The expression in sham-operated controls is set to 1 in this analysis. (**b**) Expression of Col Iα1, α-SMA, Vimentin, and Lipocalin 2 (LCN2) in livers of animals subjected to BDL or sham operation analyzed by Western blot analysis (each $n = 3$). The expression of β-actin served as loading control. Suitable primer combinations for α-SMA and Col Iα1 mRNA expression analysis and antibodies for Western blot analysis can be found elsewhere [5, 10]

by BDL strongly depends on the genetic background of the mice. Animal research must comply with legal requirements that vary greatly in different countries. In most countries, any animal research requires permission from an Institutional Animal Care and Use Committee and a project license for conducting animal experiments. Pictures depicted in this article were taken from experiments that were approved by the respective official State Animal Care and Use Committee (LANUV, Recklinghausen, Germany) and conducted in accordance with the German federal law regarding the protection of animals. Animals were housed under SPF conditions according to the guidelines of the Federation for Laboratory Animals Science Association (FELASA).

2. Based on our experience, we recommend 4% isoflurane in 100% oxygen with a flow rate of 4 L/min as inhalation anesthesia. The advantage of isoflurane is that it not only reduces pain but also acts as a relaxing agent on muscles, which is advantageous for conducting intra-abdominal operations. After placing and fixing the mouse on the pre-heated plate, the isoflurane concentrations and flow rate can be reduced to 1.5–3.0% and 1 L/min.

3. We recommend intraperitoneal injection of buprenorphine hydrochloride solution at a concentration of 0.1 mg/kg body weight. The stock solution of the analgesic is prepared in 0.9% NaCl solution.

4. To get familiar with the physiological location of the bile duct in mice, it might be helpful to study respective details on a dead animal first before doing the surgery on a living animal.
5. The anesthetic may need to be re-administered if the duration of anesthesia is too long.
6. Some other protocols recommend after this step to dissect the bile duct between the two ligatures. We recommend leaving the bile duct intact because potential bile acid leaks will induce severe acute peritonitis, ascites, systemic endotoxemia, and renal impairment.
7. Ongoing cholestasis is indicated by progressive jaundice. It is essential that the animals do not suffer unnecessarily. We recommend following the guidelines for body condition scoring (BCS) for mice, which is a noninvasive method for assessing health status during the course of the experiment and providing a general framework for the definition of humane endpoints for laboratory animals [11].

Acknowledgments

Authors' laboratory is supported by grants from the German Research Foundation (WE2554/13-1, WE2554/15-1, WE2554/17-1, and ME 3431/2-1) and a grant from the Interdisciplinary Centre for Clinical Research within the faculty of Medicine at the RWTH Aachen University (grant PTD 1-5).

References

1. Cameron GR, Oakley CL (1932) Ligation of the common bile duct. J Pathol Bacteriol 35(5):769–798. https://doi.org/10.1002/path.1700350512
2. Weiskirchen R, Weiskirchen S, Tacke F (2019) Organ and tissue fibrosis: molecular signals, cellular mechanisms and translational implications. Mol Asp Med 65:2–15. https://doi.org/10.1016/j.mam.2018.06.003
3. Acharya P, Chouhan K, Weiskirchen S, Weiskirchen R (2021) Cellular mechanisms of liver fibrosis. Front Pharmacol 12:671640. https://doi.org/10.3389/fphar.2021.671640
4. Weiskirchen R, Weiskirchen S, Tacke F (2018) Recent advances in understanding liver fibrosis: bridging basic science and individualized treatment concepts. F1000Res 7:F1000 Faculty Rev-921. https://doi.org/10.12688/f1000research.14841.1
5. Borkham-Kamphorst E, Haas U, Pinoé-Schmidt M, Abdallah AT, Weiskirchen R (2021) Chronic mineral oil administration increases hepatic inflammation in wild type mice compared to lipocalin 2 null mice. Lab Investig 101(12):1528–1539. https://doi.org/10.1038/s41374-021-00672-9
6. Salguero Palacios R, Roderfeld M, Hemmann S, Rath T, Atanasova S, Tschuschner A, Gressner OA, Weiskirchen R, Graf J, Roeb E (2008) Activation of hepatic stellate cells is associated with cytokine expression in thioacetamide-induced hepatic fibrosis in mice. Lab Investig 88(11):1192–1203. https://doi.org/10.1038/labinvest.2008.91
7. Klüber P, Meurer SK, Lambertz J, Schwarz R, Zechel-Gran S, Braunschweig T, Hurka S, Domann E, Weiskirchen R (2021) Depletion of Lipocalin 2 (LCN2) in mice leads to dysbiosis and persistent colonization with segmented filamentous bacteria. Int J Mol Sci 22(23):13156. https://doi.org/10.3390/ijms222313156

8. Iwasaki J, Afify M, Bleilevens C, Klinge U, Weiskirchen R, Steitz J, Vogt M, Yagi S, Nagai K, Uemoto S, Tolba RH (2019) The impact of a nitric oxide synthase inhibitor (L-NAME) on ischemia-reperfusion injury of cholestatic livers by Pringle maneuver and liver resection after bile duct ligation in rats. Int J Mol Sci 20(9):2114. https://doi.org/10.3390/ijms20092114
9. EUR-Lex Access to European Union law. Directive 2010/63/EU of the European Parliament and of the Council of 22 September 2010 on the protection of animals used for scientific purposes. The document can be downloaded from: https://eur-lex.europa.eu/legal-content/EN/ALL/?uri=celex%3A32010L0063. Last assessed 13 Mar 2023
10. Borkham-Kamphorst E, van de Leur E, Zimmermann HW, Karlmark KR, Tihaa L, Haas U, Tacke F, Berger T, Mak TW, Weiskirchen R (2013) Protective effects of lipocalin-2 (LCN2) in acute liver injury suggest a novel function in liver homeostasis. Biochim Biophys Acta 1832(5):660–673. https://doi.org/10.1016/j.bbadis.2013.01.014
11. Ullman-Culleré MH, Foltz CJ (1999) Body condition scoring: a rapid and accurate method for assessing health status in mice. Lab Anim Sci 49(3):319–323. PMID: 10403450

Chapter 10

Mouse Models for Hepatic Stellate Cell Activation and Liver Fibrosis Initiation

Yulia A. Nevzorova, Ralf Weiskirchen, and Christian Liedtke

Abstract

Liver fibrosis is a severe health problem worldwide with increasing incidence. However, specific drugs for treatment of hepatic fibrosis are currently not available. Accordingly, there is a strong need to conduct intensive basic research, which also includes the necessity to use animal models to evaluate new anti-fibrotic therapy concepts. Numerous mouse models of liver fibrogenesis have been described. This involves chemical, nutritional, surgical, and genetic mouse models, which involve also activation of hepatic stellate cells (HSCs). However, for many investigators, it may be challenging to identify the most suitable model for a specific question on liver fibrosis research. In this chapter, we will provide a brief overview about the most common mouse models of HSC activation and liver fibrogenesis and thereafter provide detailed step-by-step protocols of two selected mouse fibrosis models based on own experience, which in our opinion are best suited to cover many current scientific issues. On the one hand, there is the classical carbon tetrachloride (CCl_4) model; this model of toxic liver fibrogenesis is still one of the best suited and most reproducible models for basic features of hepatic fibrogenesis. On the other hand, we also introduce the novel DUAL model of alcohol plus metabolic/alcoholic fatty liver disease developed in our laboratory, which mimics all histological, metabolic, and transcriptomic gene signatures of human advanced steatohepatitis and related liver fibrosis. We describe all the information required for proper preparation and detailed implementation of both models including animal welfare aspects, thereby serving as a useful laboratory guide for mouse experimentation in liver fibrosis research.

Key words Alcohol, Liver damage, Hepatotoxin, Carbon tetrachloride, Hepatic fibrosis

1 Introduction

Liver fibrosis occurs in the setting of chronic liver injury in a multistep process involving hepatitis (often associated with alcoholic or nonalcoholic steatosis), eventually progressing to liver cirrhosis and hepatocellular carcinoma (HCC). The incidence of end stage liver disease associated with fibrosis is steadily increasing and a major cause of morbidity and mortality. Yet, no specific

Ralf Weiskirchen and Scott L. Friedman (eds.), *Hepatic Stellate Cells: Methods and Protocols*, Methods in Molecular Biology, vol. 2669, https://doi.org/10.1007/978-1-0716-3207-9_10,

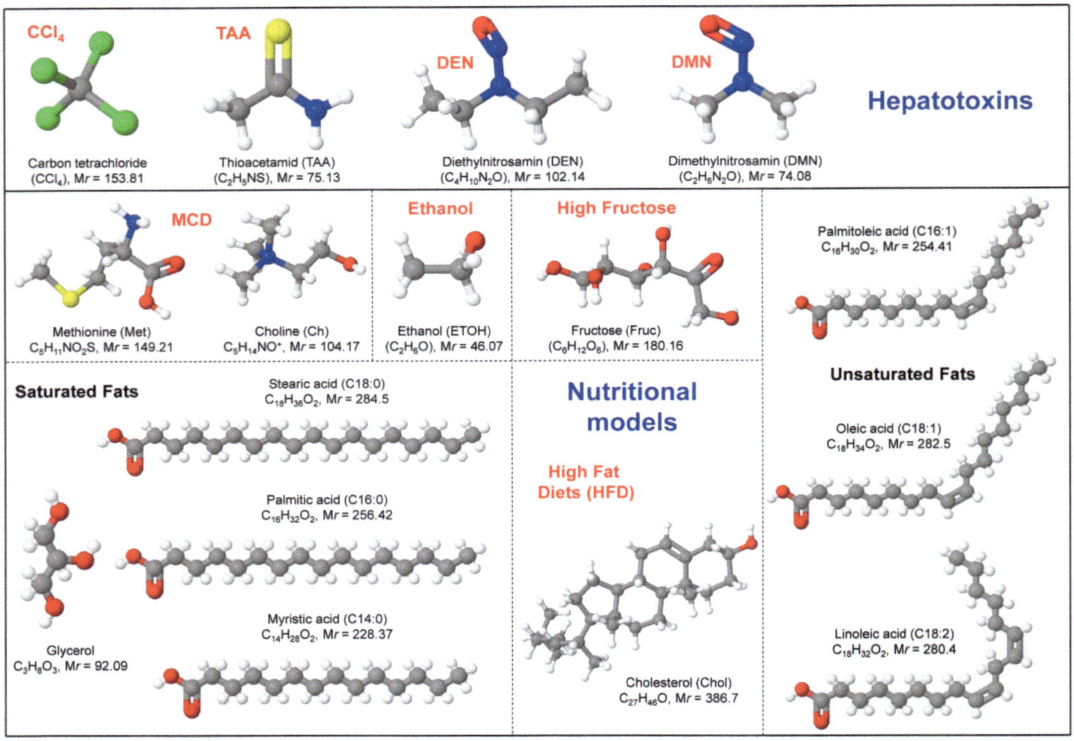

Fig. 1 Experimental liver fibrosis. In rodents, experimental liver damage and fibrosis can be induced by various hepatotoxins (e.g., carbon tetrachloride, thioacetamide, diethylnitrosamine, dimethylnitrosamine), special diets that either lack essential nutrients (e.g., methionine, choline) or are enriched in ethanol or fructose or in fats

anti-fibrotic drugs are available for treatment of hepatic fibrosis. Consequently, there is an urgent need of extensive research also involving animal models of liver fibrosis.

Animal models to mimic the different pathogeneses of human liver fibrogenesis are performed according to European/international animal welfare regulations as outlined elsewhere [1]. These models can be categorized into chemical models (alone or in combination with a second hit), nutritional models (involving specialized diets with our without alcohol), or genetic models (Fig. 1). The latter models induce chronic liver injury due to loss of a gene required for liver homeostasis. In addition, few surgical methods have been described that induce liver fibrosis in mice due to ligation of the common bile duct or through implantation of a mechanical pump which challenges the liver intrinsically with ethanol [2]. However, the latter model will not be addressed here in more detail as it is probably not widely applicable due to worldwide enhancing animal welfare rules [1].

An overview about the most common methods for activating HSCs and initiation of liver fibrosis is listed in Table 1. All of these

Table 1
Rodent models of liver fibrosis

Category of fibrosis model	Examples	References	Advantages	Limitations
Chemical/toxic models	CCl_4	[1]	Fast, well reproducible, easy to quantify	Not physiological, does not mimic a concrete human pathogenesis
	CCl_4/DEN	[1]	Fast, reflects human pathogenesis with fibrosis and hepatocellular carcinoma (HCC)	Work intensive, requires weekly injections of CCl_4 for up to several weeks
	TAA	[3]	Can be easily applied i.p. or orally in drinking water, well reproducible	Molecular mechanism of TAA-induced liver fibrosis is not completely understood, long latency of fibrosis
Nutritional models, nonalcoholic	HFD	[4]	Simply feeding of a commercially available diet, easy to apply	Mice develop steatosis, sometimes steatohepatitis, but moderate fibrosis after long latency
	MCD	[4]	Simply feeding of a commercially available diet, easy to apply; mice develop steatohepatitis and fibrosis within 4–10 weeks	Not physiological as mice develop lower plasma triglyceride and cholesterol levels, weight loss, and a metabolic profile opposite to that observed in humans
Nutritional models, alcoholic	Lieber-DeCarli diet	[5]	Liquid diet, easy to apply	Daily change of feeding tubes required, animals only develop steatosis, but not steatohepatitis or fibrosis
	DUAL diet	[6]	Easy to apply, reflects human pathogenesis with obesity, liver damage, steatohepatitis, and prominent hepatic fibrosis	Feeding may take up to 1 year for strong fibrosis; besides that, no limitation reported so far
Surgical models	Bile-duct ligation	[7, 8]	Well established, fast model; cholestatic liver fibrosis develops within 28 days	Surgical intervention necessary; training and experience required
	Tsukamoto-French model	[2]	Mice develop alcoholic liver fibrosis	Surgical intervention required, difficult to get ethical approval in some countries due to animal welfare issues

(continued)

Table 1 (continued)

Category of fibrosis model	Examples	References	Advantages	Limitations
Genetic models	$Mdr2^{-/-}$ mice	[9]	Mice develop spontaneous cholestasis, fibrosis, and HCC	High variation; difficult to quantify, long lasting model; phenotype strongly depends on genetic background, gender, animal keeping conditions, microbiome
	$NEMO^{\Delta hepa}$ mice	[9]	Spontaneous phenotype of mice reflects human pathogenesis involving steatohepatitis, fibrosis, HCC	Pathogenesis is caused by lack of NF-κB activation and spontaneous apoptosis, therefore in part unphysiological. Fibrotic liver is largely heterogeneous, therefore difficult to quantify

models have advantages and limitations, and it remains a major challenge to select the most appropriate fibrosis model for each scientific question.

The single or repeated application of the hepatotoxin carbon tetrachloride (CCl_4) in mice is still a leading model worldwide to induce liver fibrosis [1]. Although this does not mimic a specific liver fibrogenesis of human, the model is robust and leads to highly reproducible results. In our view, the best treatment route is the application of CCl_4 via intraperitoneal (i.p.) injection. It is the model of choice, if one wants simply to induce HSCs and liver fibrosis in genetically modified mice in order to evaluate whether the included manipulation results in modified fibrosis development. The model is also useful for testing of potential anti-fibrotic drugs in mice.

Most metabolic models have strong limitations, as they frequently induce steatosis without developing strong fibrosis in part due to the rapid hepatic metabolism of mice. Recently, we developed a preclinical mouse model of alcohol-associated liver disease plus metabolic-associated fatty liver disease (DUAL model) [6], which generates lobular inflammation and advanced hepatic fibrosis after 12 months of treatment with 10% (vol/vol) alcohol in sweetened drinking water in combination with a Western diet.

Genetic liver fibrosis mouse models, i.e., genetically engineered mice that spontaneously develop liver fibrosis, are increasingly available. Some models enable study of steatosis, inflammation, and fibrosis, such as liver-specific nuclear factor (NF)-κB essential modulator knockout mice (NEMO) or constitutive $Mdr2^{-/-}$ mice [9]. However, patent regulations and material transfer constraints might limit the availability of these models that reduce its use.

In this chapter, we will provide two established step-by-step protocols that induce activation of HSCs in mice with subsequent liver fibrosis. One protocol describes the classical CCl_4-induced liver fibrosis model, which has been applied in our laboratory successfully for many years. In addition, we provide details of the new DUAL model of ethanol/Western diet-induced liver fibrogenesis, which has been established very recently [6].

2 Materials

2.1 CCl_4-Mediated Induction of Liver Fibrosis

1. Laboratory balance.
2. Heating pad at 37 °C.
3. Disposable, sterile syringes.
4. 27 gauge needle.
5. CCl_4 (*see* **Note 1**).
6. Corn oil.

2.2 DUAL Diet-Mediated Induction of Liver Fibrosis

1. Laboratory Balance
2. Graduated cylinder, serological pipette, and pipette aid.
3. Sterile glass or plastic jar for preparation of solution.
4. Laboratory agitator.
5. Western diet (WD) enriched in fat (40% kcal), cholesterol (2% by weight), and fructose (22% by weight) (*see* **Note 2**).
6. A regular chow diet for rodents used as a control diet (*see* **Note 3**).
7. Ethanol (EtOH) for molecular biology, >99.8% (absolute alcohol, without additive) (*see* **Note 4**).
8. D-glucose powder (*see* **Note 5**).
9. Fresh, filtered, uncontaminated tap drinking water, according to particular requirements of the animal facility. Alternatively, autoclaved water can be used.

3 Methods

3.1 CCl_4-Mediated Induction of Liver Fibrosis

1. Weigh the mice and examine the mice for signs of distress (i.e., changes in respiration, rough hair coat, unusual behavior, hunched posture) (*see* **Notes 6** and **7**).
2. CCl_4 concentrations usually are 0.5–0.7 μL/g body weight diluted in corn oil depending on the desired degree of fibrosis and the genetic background of the mice used. Additional parameters to adjust the strength of liver fibrosis are the application intervals (i.e., two versus three CCl_4 injections per week) as well as duration of treatment (which is typically between 4 and 6 weeks; Fig. 2). According to our experience, i.p. injection of 0.5 μL CCl_4/g body weight in C57BL/6 inbred mice three times a week for 6 weeks results in pronounced liver fibrosis resembling human stage 4 according to Ishak scoring. If a milder fibrosis induction is desired, i.p. injection of 0.5 CCl_4/g body weight in C57BL/6 mice can be applied two times a week for only 4 weeks.
3. Calculate the required volume of CCl_4. For practical reasons, we recommend to always inject a constant volume of 50 μL containing a solution of CCl_4 in corn oil. Accordingly, the required volume of CCl_4 (0.5–0.7 μL/g mouse weight) is made up to 50 μL of total volume with corn oil. As an example, one would inject 10 μL CCl_4 solved in 40 μL corn oil into a mouse that weighs 20 g to obtain a final dosage of 0.5 μL CCl_4/g body weight (*see* **Note 8**). An example for calculation is given in Table 2.

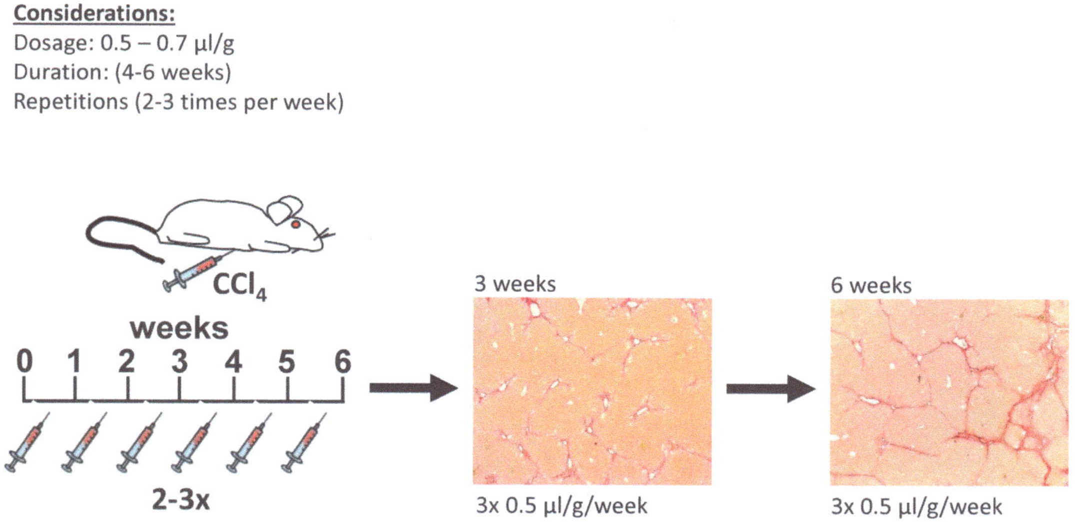

Fig. 2 Application of the carbon tetrachloride model to mice. *Left*: At the beginning of the experiment, investigators should determine the desired degree of fibrosis at the end point of treatment. Parameters to adjust are dosage, duration, and number of weekly repetitions of CCl_4 injection. *Right*: A typical CCl_4 experiment is shown using three injections per week into C57BL/6 mice with 0.5 μL CCl_4/g body weight for a total of 6 weeks. Representative Sirius Red stainings of paraffin-embedded liver sections after 3 weeks and 6 weeks of treatment are shown

4. Inject 50 μL of CCl_4/corn oil solution in mice using disposable, sterile syringes. To this end, the mouse is scuffed behind the neck region between the thumb and forefinger. After turning over the hand, the animal rests on the palm against the base of the thumb using a third finger to stabilize the pelvic region. CCl_4 is injected into the lower side of the abdomen with the mouse head down avoiding injury to bladder, diaphragm, or intestine using a 27 gauge needle.
5. Animals are reinspected 1 h after injection for abnormalities and every 24 h thereafter. In the rare case that an animal develops severe, long lasting complications, euthanasia of that animal has to be considered according to previously defined termination criteria in agreement with local animal welfare regulations (*see* **Note 9**).
6. At the end of the experimentation (i.e., after 4–6 weeks), animals are sacrificed by cervical dislocation, and the livers are removed for histological and molecular analyses thereafter (*see* **Note 9**). A typical liver histology of mice subjected to CCl_4 is given in Fig. 2.

3.2 DUAL Diet-Mediated Induction of Liver Fibrosis

Recently we developed a preclinical DUAL (ALD (alcohol-associated liver disease) plus MAFLD (metabolic-alcoholic fatty liver disease)) model in mice (Fig. 3). Our DUAL model is

Table 2
Calculation of CCl_4/corn oil ratio in injection solution for a given mouse weight and desired treatment concentration between 0.5 and 0.7 μL/g body weight

Mouse weight (g)	18			19			20			21			22			23			24			25		
Desired CCl_4 concentration (μl/g)	0.5	0.6	0.7	0.5	0.6	0.7	0.5	0.6	0.7	0.5	0.6	0.7	0.5	0.6	0.7	0.5	0.6	0.7	0.5	0.6	0.7	0.5	0.6	0.7
Injection volume CCl_4 (μl)	9	10.8	12.6	9.5	11.4	13.3	10	12	14	10.5	12.6	14.7	11	13.2	15.4	11.5	13.8	16.1	12	14.4	16.8	12.5	15	17.5
Injection volume corn oil (μl)	41	39.2	37.4	40.5	38.6	36.7	40	38	36	39.5	37.4	35.3	39	36.8	34.6	38.5	36.2	33.9	38	35.6	33.2	37.5	35	32.5
Total injection volume (μl)	50	50	50	50	50	50	50	50	50	50	50	50	50	50	50	50	50	50	50	50	50	50	50	50

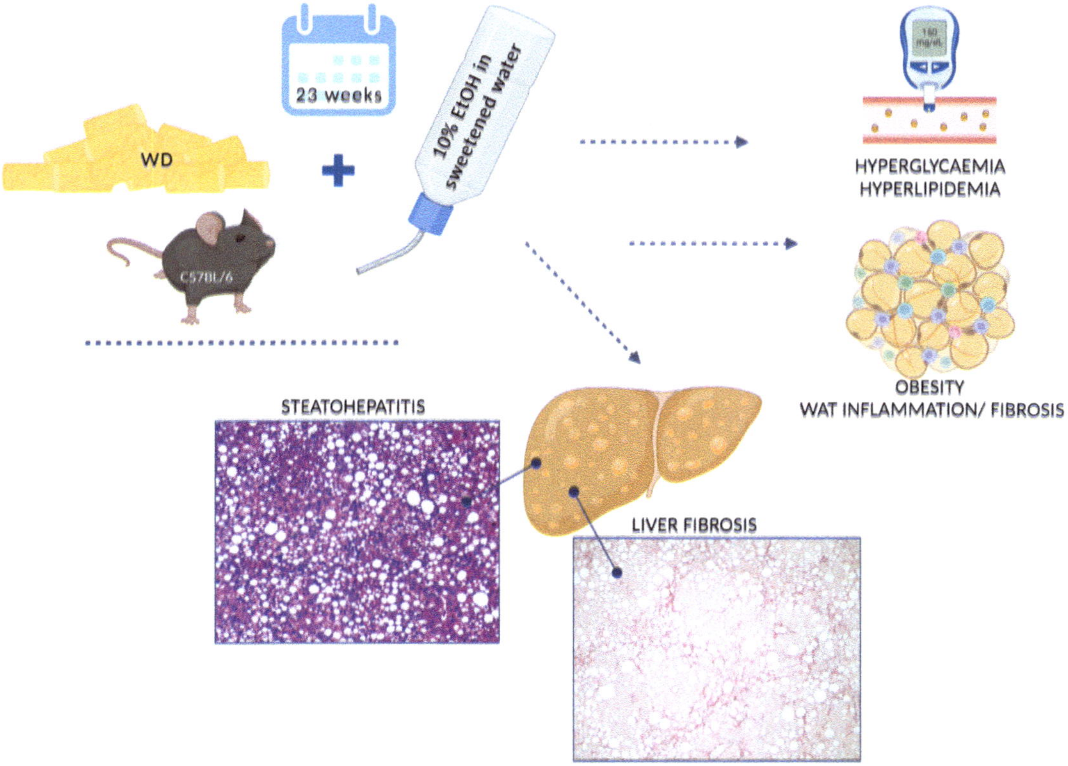

Fig. 3 DUAL liver damage model. The DUAL murine model perfectly mimics all histological, metabolic, and transcriptomic gene signatures of human advanced steatohepatitis and fibrosis and thus serves as a preclinical tool for the development of therapeutic targets (*WAT* white adipose tissue, *WD* Western diet). Created with BioRender.com

characterized by obesity, glucose intolerance, liver damage, steatohepatitis, and prominent hepatic fibrosis, as well as inflammation and fibrosis in white adipose tissue [6]. The DUAL model mimics all histological, metabolic, and transcriptomic gene signatures of human advanced steatohepatitis and therefore can serve as an excellent preclinical tool for the development of so much needed new therapeutic targets. Before the beginning of the DUAL feeding, the impact of several factors should be seriously taken into account (*see* **Note 10**).

1. House the animals in a temperature (20–23 °C)- and humidity (30–40%)-controlled room with a 12-h light/dark cycle (*see* **Note 11**).
2. Randomize 10-week-old mice to two groups (treatment vs. control group) (*see* **Note 12**).
3. For DUAL feeding, it is obligatory to introduce alcohol gradually and always start with an adaptation period (Table 3). The

Table 3
Scheme of the DUAL feeding with gradual incorporation of alcohol

Time point	Diet	Water	Ethanol
Day 1–4	Start with WD	*Ordinary water*	–
Day 5–7	WD	6.75% D-glucose in the drinking water	–
Day 8–10	WD	6.75% D-glucose in the drinking water	1% (vol/vol)
Day 11–12	WD	6.75% D-glucose in the drinking water	2% (vol/vol)
Day 13–14	WD	6.75% D-glucose in the drinking water	4% (vol/vol)
Third week	WD	6.75% D-glucose in the drinking water	5% (vol/vol)
From fourth week	WD	6.75% D-glucose in the drinking water	10% (vol/vol)

WD Western diet

Western diet (WD) is given ad libitum in mice and placed directly in the cage dieting dish.

4. Record the weight of the food before giving it to mice! Reweigh food weekly to determine the food intake. The WD should be replaced every 10 days to avoid spoilage. For C57BL/6NCrl mice, the average food intake is approximately 4 g/day.
5. To prepare sweetened water, 67.5 g D-glucose powder should be carefully mixed in 1 L of water by, first, preparing the sweetened water, and only after the complete dissolving of D-glucose powder, add EtOH. The sweetened water is given ad libitum in the drinking bottles (*see* **Note 13**).
6. The control group receives chow diet and ordinary tap water during all the period of treatment.
7. Record the chow diet food intake correspondingly in the control group.
8. Check the health of the animals every day and weigh all mice weekly.
9. At time point of choice, mice should be humanely euthanized by carbon dioxide (CO_2) or cervical dislocation according to the animal protocol and ethical regulation (*see* **Note 14**).
10. After euthanasia, carefully perform external examination of mice. In particular, pay attention to the general aspects of the mouse health (e.g., obese or malformed) and to the state of superficial tissues and organs (e.g., icterus of the skin). Measure and record the body weight of each animal. Immediately collect blood from the posterior vena cava or cardiac puncture. Dissect the organs of interest.

4 Notes

1. CCl_4 vapors are toxic for humans. The level of CCl_4 exposure should be minimized by use of a fume hood.
2. In our laboratory, we have traditionally used D16022301 from Research Diets, Inc., New Brunswick, NJ. Food should be maintained preferentially under refrigeration (+4 °C); the self-life period is around 6 months.
3. Some publications strongly oppose the use of chow as a control for WD and instead recommend a low-fat, purified ingredient diet for several reasons. First, chow diet is a high-fiber diet composed of agricultural by-product that may include a mixture of corn, oats, alfalfa, wheat, and soybean meal. Additionally, chow diet may have high levels of polyphenols and phytoestrogens, which may affect phenotypic, metabolic, and/or behavioral variables [10, 11]. In contrast, a recent study clearly demonstrated that regular chow diet may be also used as an appropriate control and does not lead to any significant phenotypic, metabolic, and behavioral alterations [12]. Still, the diet content details must be accurately described.
4. The concentration of ethanol significantly impacts the volume and, consequently, the quantity of the daily consumed ethanol. As the concentration of ethanol increased (20% (vol/vol)), the mice drank less volume and intake significantly decreased. At a 5% (vol/vol) concentration, the quantity of consumed EtOH remained relatively low. Only a concentration of 10% (vol/vol) of EtOH resulted in severe pathophysiologic changes [6].
5. The low EtOH intake due to natural aversion in rodents is the main limiting factor for experimental ALD development [13]. In the DUAL model, the sweetened water successfully masks the taste of alcohol, thus increasing alcohol intake. DUAL-fed mice consume in average 30 g/kg of alcohol per day. This resembles approximately 7 L of vodka (40% (vol/vol)) in a 75 kg person daily. In line with this, the relative risk of alcohol-associated cirrhosis increases in humans who drink more than 25–30 g/day [14, 15]. However, due to the small body size, the rate of alcohol catabolism is up to five times faster in mice than humans [16]. Still, in comparison, the liquid Lieber-DeCarli alcohol diet produces EtOH intake of only 14–16 g/kg/day, which is surely not enough to trigger hepatic fibrosis [17]. Moreover, D-glucose in the drinking water potentiates absorption of fructose from the diet, whereas fructose catalyzes glucose uptake and hepatic storage, leading together to greater glucose delivery to the liver [18].
6. Animals showing abnormalities are excluded from the study.

7. Handling (i.e., touching/grabbing) mice is already stressful for the animals. During treatment, animals should be placed on a heating pad at 37 °C to prevent hypothermia.
8. Intraperitoneal (i.p.) CCl_4 injection can lead to local irritation of the skin, and CCl_4 can cause mild peritoneal inflammation.
9. Animal burden/side effects: Usually mice treated i.p. with CCl_4 do not develop severe long term complications. However, i.p. CCl_4 application is clearly associated with transient spasm, abdominal adhesions, and hepatic inflammation.
10. It is critical that investigators conduct the feeding by selecting animals of the appropriate strain, age, and gender and choose the suitable time periods of feeding. The following details must be reported accurately:

    Age: Age has a strong influence in the experimental results. For example, adolescent mice have differential ethanol intake behavioral patterns compared to adults [19]. On the other hand, the prevalence of the metabolic syndrome and MAFLD increases dramatically with aging [20]. Thus, it is recommended to perform DUAL feeding with 10-week-old mice with a body weight over 18–19 g for females and 20 g for males. It is also recommendable to include mice of similar weight (±5%) into groups because their caloric requirements will be similar.

    Gender: The liver is a sexually dimorphic organ, exhibiting major sex differences in the profile of steroids and drug metabolism as well as the expression of several genes and protein [21]. Concerning DUAL feeding, male animals displayed increased liver damage compared to females. Male rodents are more susceptible to metabolic syndrome (MS) and demonstrate increased presence of pro-inflammatory cells and cytokines, significantly elevated extracellular matrix (ECM) deposition, and cell death in the liver. Therefore, only animals of the same gender should be used and compared during the study.

    Mouse strain: There are marked variations in alcohol metabolism between different strains of mice. Therefore, the genetic background can be a major and often unpredictable determinant of the outcome in transgenic and targeted-gene mutant mice. C57BL/6 is the preferred strain to use for DUAL feeding. 129sOla and BALB/c animals exhibit significant protection from detrimental dietary effects. Moreover, the substrains within a single strain may also exhibit variations. There are several studies characterizing the significant behavioral differences between C57BL/6 J and C57BL/6NCrl in ethanol preference. Mice of the C57BL/6 J substrain have been reported as a relatively resistant genotype, due to increased ADH activity that accelerates the elimination of acetaldehyde, the toxic metabolite of alcohol [22–24]. Consequently, it is highly recommended to backcross genetically modified mice into the C57BL/6NCrl background for few generations [17, 25].

Duration of feeding: Depending on the specific stage of hepatic fibrosis that the researcher is seeking to investigate, the duration of the DUAL feeding can be changed. Ten weeks of treatment with the DUAL diet is suitable to study the initiation of fibrosis with steatosis, slight elevation of transaminases, and moderate inflammation. Stepwise feeding with DUAL diet up to 23 weeks has been shown to induce remarkable escalation of fibrogenesis, strong steatohepatitis, and liver damage. Within 52 weeks of DUAL diet, animals developed extraordinary enlarged liver with extensive collagen deposition and well differentiated micronodules, surrounded by fibrotic connective tissue extending between portal regions, which overall indicates the presence already of hepatic cirrhosis [6].

The number of animals in experimental and control group: In fact, no mortality or signs of sickness were observed in DUAL-fed animals, neither after 23 weeks nor after 53 weeks of feeding. Based on the parameters of the biostatistical power analysis, the minimal calculated required sample size in each group is 6–7. However, genetically modified mice may suffer from higher mortality rates, especially if the manipulated gene of interest directly affects pathways of alcohol or lipid metabolism. Therefore, it is highly advisable to run a pilot experiment with few animals in order to understand how the mouse phenotype manifests in response to the DUAL diet, compare it to the published literature, and determine if extra mice are needed for an appropriately powered study.

11. In case of delivery from commercial supplier or another facility, mice should arrive at least 10–14 days prior to being placed on DUAL diet and acclimated on normal chow for this period. It is highly recommendable to use littermates. This will ensure that not only the genetic background but also the environmental factors are comparable. It is especially important when genetically modified strains are analyzed for quantitative phenotypes [26].
12. The individual housing for DUAL-fed mice is expedient. Joint alcohol feeding of a group of mice might affect both the variation of research data and animal welfare issues, as equal ethanol uptake for each animal cannot be guaranteed due to the hierarchical structure of a mouse group [27]. However, the acclimation of mice to single housing in a novel

environment for 14 days prior the beginning of feeding is beneficial [28]. For the control group, no more than three mice per cage can be placed together. Still, mice do not exhibit aggressive behavior toward their cage mates and do not form a social hierarchy. Maintenance of mice in the same housing groups for the duration of the study is recommended. Bedding should not be edible as it will reduce the amount of diet consumed. For the same reason, chew sticks are not recommended.

13. In order to avoid EtOH evaporation and keep the constant concentration during the experimental period, change the drinking bottle biweekly. The time of the day when the food and the water are changed has to be kept consistent through the experiment.
14. Depending on the parameters of interest, animals should be sacrificed consistently in either a fed or fasting state. Fasting is primarily used as a way to standardize and to reduce variability in some metabolic parameters such as levels of insulin, glucose, parameters of lipid metabolism, as well as some metabolic-related hormones. Moreover, an empty stomach, small intestine, and/or large intestine might be required in some gastrointestinal studies. The short 6-h (7:30 AM–1: 30 PM) or long 12-h (7:00 PM–7: 00 AM) fasting can be performed. During the fasting period, the food has to be carefully removed from the cage and the animals should have the access only to tap water ad libitum (water bottles filled only with ordinary tap water without adding glucose).

References

1. Liedtke C, Luedde T, Sauerbruch T et al (2013) Experimental liver fibrosis research: update on animal models, legal issues and translational aspects. Fibrogenesis Tissue Repair 6:19. https://doi.org/10.1186/1755-1536-6-19
2. Tsukamoto H, Mkrtchyan H, Dynnyk A (2008) Intragastric ethanol infusion model in rodents. Methods Mol Biol 447:33–48. https://doi.org/10.1007/978-1-59745-242-7_3
3. Wallace MC, Hamesch K, Lunova M et al (2015) Standard operating procedures in experimental liver research: thioacetamide model in mice and rats. Lab Anim 49(1 Suppl):21–29. https://doi.org/10.1177/0023677215573040
4. Ramadori P, Weiskirchen R, Trebicka J et al (2015) Mouse models of metabolic liver injury. Lab Anim 49(1 Suppl):47–58. https://doi.org/10.1177/0023677215570078
5. Nevzorova YA, Boyer-Diaz Z, Cubero FJ et al (2020) Animal models for liver disease – a practical approach for translational research. J Hepatol 73(2):423–440. https://doi.org/10.1016/j.jhep.2020.04.011
6. Benede-Ubieto R, Estevez-Vazquez O, Guo F et al (2021) An experimental DUAL model of advanced liver damage. Hepatol Commun 5:1051–1068. https://doi.org/10.1002/hep4.1698
7. Tag CG, Sauer-Lehnen S, Weiskirchen S et al (2015) Bile duct ligation in mice: induction of inflammatory liver injury and fibrosis by obstructive cholestasis. J Vis Exp 96:52438. https://doi.org/10.3791/52438
8. Tag CG, Weiskirchen S, Hittatiya K et al (2015) Induction of experimental obstructive cholestasis in mice. Lab Anim

49(1 Suppl):70–80. https://doi.org/10.1177/0023677214567748
9. Torres S, Abdullah Z, Brol MJ et al (2020) Recent advances in practical methods for liver cell biology: a short overview. Int J Mol Sci 21(6):2027. https://doi.org/10.3390/ijms21062027
10. Warden CH, FislerJS (2008) Comparisons of diets used in animal models of high-fat feeding. Cell Metab 7:277. https://doi.org/10.1016/j.cmet.2008.03.014
11. Pellizzon MA, Ricci MR (2018) The common use of improper control diets in diet-induced metabolic disease research confounds data interpretation: the fiber factor. Nutr Metab (Lond) 15:3. https://doi.org/10.1186/s12986-018-0243-5
12. Almeida-Suhett CP, Scott JM, Graham A et al (2019) Control diet in a high-fat diet study in mice: regular chow and purified low-fat diet have similar effects on phenotypic, metabolic, and behavioral outcomes. Nutr Neurosci 22:19–28. https://doi.org/10.1080/1028415X.2017.1349359
13. Nevzorova YA, Boyer-Diaz Z, Cubero FJ et al (2020) Animal models for liver disease – a practical approach for translational research. J Hepatol 73:423–440. https://doi.org/10.1016/j.jhep.2020.04.011
14. Mathurin P, Bataller R (2015) Trends in the management and burden of alcoholic liver disease. J Hepatol 62:S38–S46. https://doi.org/10.1016/j.jhep.2015.03.006
15. Bellentani S, Saccoccio G, Costa G et al (1997) Drinking habits as cofactors of risk for alcohol induced liver damage. The dionysos study group. Gut 41:845–850. https://doi.org/10.1136/gut.41.6.845
16. Brandon-Warner E, Schrum LW, Schmidt CM et al (2012) Rodent models of alcoholic liver disease: of mice and men. Alcohol 46:715–725. https://doi.org/10.1016/j.alcohol.2012.08.004
17. Guo F, Zheng K, Benede-Ubieto R et al (2018) The Lieber-deCarli diet-a flagship model for experimental alcoholic liver disease. Alcohol Clin Exp Res 42:1828–1840. https://doi.org/10.1111/acer.13840
18. Laughlin MR (2014) Normal roles for dietary fructose in carbohydrate metabolism. Nutrients 6:3117–3129. https://doi.org/10.3390/nu6083117
19. Moore EM, Mariani JN, Linsenbardt DN et al (2010) Adolescent C57Bl/6J (but not dba/2j) mice consume greater amounts of limited-access ethanol compared to adults and display continued elevated ethanol intake into adulthood. Alcohol Clin Exp Res 34:734–742. https://doi.org/10.1111/j.1530-0277.2009.01143.x
20. Fontana L, Zhao E, Amir M et al (2013) Aging promotes the development of diet-induced murine steatohepatitis but not steatosis. Hepatology 57:995–1004. https://doi.org/10.1002/hep.26099
21. Lonardo A, Suzuki A (2020) Sexual dimorphism of nafld in adults. Focus on clinical aspects and implications for practice and translational research. J Clin Med 9:1278. https://doi.org/10.3390/jcm9051278
22. Wei VL, Singh SM (1988) Genetically determined response of hepatic aldehyde dehydrogenase activity to ethanol exposures may be associated with alcohol sensitivity in mouse genotypes. Alcohol Clin Exp Res 12:39–45. https://doi.org/10.1111/j.1530-0277.1988.tb00130.x
23. Khisti RT, Wolstenholme J, Shelton KL et al (2006) Characterization of the ethanol-deprivation effect in substrains of C57BL/6 mice. Alcohol 40:119–126. https://doi.org/10.1016/j.alcohol.2006.12.003
24. Matsuo N, Takao K, Nakanishi K et al (2010) Behavioral profiles of three c57bl/6 substrains. Front Behav Neurosci 4:29. https://doi.org/10.3389/fnbeh.2010.00029
25. Bertola A, Mathews S, Ki SH et al (2013) Mouse model of chronic and binge ethanol feeding (the niaaa model). Nat Protoc 8:627–637. https://doi.org/10.1038/nprot.2013.032
26. Holmdahl R, Malissen B (2012) The need for littermate controls. Eur J Immunol 42:45–47. https://doi.org/10.1002/eji.201142048
27. Kappel S, Hawkins P, Mendl MT (2017) To group or not to group? Good practice for housing male laboratory mice. Animals (Basel) 7:88. https://doi.org/10.3390/ani7120088
28. Hebda-Bauer EK, Dokas LA, Watson SJ et al (2019) Adaptation to single housing is dynamic: changes in hormone levels, gene expression, signaling in the brain, and anxiety-like behavior in adult male c57bl/6j mice. Horm Behav 114:104541. https://doi.org/10.1016/j.yhbeh.2019.06.005

Chapter 11

Generation and Culture of Primary Mouse Hepatocyte–Hepatic Stellate Cell Spheroids

Inge Mannaerts, Nathalie Eysackers, and Leo A. van Grunsven

Abstract

In vitro models of liver fibrosis have evolved from mono-cultures of primary rodent hepatic stellate cells and stellate cell lines, to more complex co-cultures of primary or stem cell-derived liver cells. Great progress has been made in the development of stem cell-derived liver cultures; however, the liver cells obtained from stem cells do not yet fully recapitulate the phenotype of their in vivo counterparts. Freshly isolated rodent cells remain the most representative cell type to use for in vitro culture. To study liver injury-induced fibrosis, co-cultures of hepatocytes and stellate cells are an informative minimal model. Here, we describe a robust protocol to isolate hepatocytes and hepatic stellate cells from one mouse and a method for the subsequent seeding and culture as free-floating spheroids.

Key words Fibrosis, 3D, In vivo perfusion, In vitro, Chronic liver disease modeling

1 Introduction

Chronic liver disease is the major cause of progressive liver fibrosis which, in turn, can lead to cirrhosis of the liver. Complications from liver cirrhosis account for more than 170,000 deaths per year in Europe [1]. Moreover, hepatocellular cancer, a rapidly fatal tumor and the third most common cause of cancer-related death worldwide, is closely linked to cirrhosis and is often the direct cause of death [1]. Besides causal treatment like anti-viral therapies or change of lifestyle, which can prevent or retard fibrosis progression, no anti-fibrotic liver therapies are currently available [2].

One major obstacle in the development of efficient therapies is the lack of robust and representative in vitro models of human liver fibrosis to aid in understanding the basic mechanisms of the disease and in the development—of pharmaceuticals. While animal testing is still the most popular preclinical assessment modality, its value in

Inge Mannaerts and Nathalie Eysackers contributed equally with all other contributors.

Ralf Weiskirchen and Scott L. Friedman (eds.), *Hepatic Stellate Cells: Methods and Protocols*, Methods in Molecular Biology, vol. 2669, https://doi.org/10.1007/978-1-0716-3207-9_11,

predicting human physiological response in terms of both drug efficacy and toxicity is frequently poor [1, 3, 4]. Furthermore, due to ethical concerns, there is an urgent need to replace animal models following the 3R principles: "Reduction, Refinement and Replacement." A particularly significant indication of this direction was demonstrated by the European Union's decision in 2013 to ban animal testing in safety evaluation of cosmetic products.

The most widely used in vitro liver fibrosis model is the culture-induced activation of primary rodent hepatic stellate cells (HSCs) [5]. Following liver injury, HSCs, which are the dominant contributors to excessive scar tissue formation in liver fibrosis, "activate" by transitioning from a quiescent vitamin A-storing HSC to a proliferating fibrogenic myofibroblast-like cell [6]. While these mono-layer cultures have contributed greatly to our current understanding of HSC activation, this model system has failed to recapitulate the complexity of this regenerative response. While some typical hallmarks of HSC activation are increased during both in vivo and in vitro activation of HSCs [7], RNA profiling studies have demonstrated that there is only ~25% overlap in differential gene expression between these two models, clearly indicating that current approaches to induce HSC activation in vitro do not fully recapitulate in vivo HSC activation [8, 9]. A key reason for these important discrepancies is the loss of the microenvironmental context in which HSCs reside: in vivo, HSCs interact with the supportive and instructive extracellular matrix in a complex 3D organization, as well as with other cell types such as (damaged) hepatocytes. Some of these key aspects of liver fibrosis, which are missing in 2D mono-cultures of HSCs, should be better recapitulated in co-cultures of primary hepatocytes and HSCs [10]. These cultures are of course a very minimalistic approach to model the complex mechanism involved in liver fibrosis, but some features of HSC activation such as the increase in collagen 6 are present in injured spheroid cultures of hepatocytes and HSCs, while in vitro activated mono-cultures of HSCs do not show this [11].

The co-culture of primary mouse hepatocytes and HSCs demands a robust isolation protocol. The cells obtained with the isolation methods described in this chapter, a Percoll gradient-based isolation of hepatocytes and a flow cytometry-based HSC isolation, can be used for mono-layer cultures of HSCs [7, 8, 12, 13] or hepatocytes [14, 15] or for co-cultures of HSCs and hepatocytes using culture inserts or spheroid cultures enabling direct contact between the cells [11, 16]. Furthermore, we describe here a method to seed and culture these cells for spheroid co-cultures.

2 Materials

2.1 Animals

For a good yield of HSCs, mice of 16–22 weeks of age are optimal. The HSC yield depends on the mouse strain, and C57/Bl6 mice have a lower yield than Balb/c mice (in-house data). Ethical approval is needed to carry out liver perfusions. Please contact your ethical committee for the procedures to obtain approval for this experiment. Liver perfusion is considered a mild or non-recovery procedure.

2.2 Instruments and Surgical Equipment

All equipment, such as scissors, forceps, tubing, and needle holders, should be sterilized by autoclaving for at least 10 min at 121 °C.

1. Titegrip needle holder.
2. Vyclic three-way tap extension.
3. Dissecting scissors.
4. 35 mM clamp.
5. Peristaltic pump, such as Cole-Parmer Masterflex L/S.
6. Silicon Tubing, Masterflex precision pump tubing size 14.
7. Toothed and bent dissecting forceps.
8. Stainless steel operation tray.
9. Sieve.
10. Orbital shaker, Celltron benchtop shaker for CO2 incubator.
11. Recommended: Automated pipetting assistant, Viaflo Assist (Integra Biosciences).
12. Electronic multichannel pipette such as Integra Viaflo 8channel P300 or Rainin Electronic Multichannel pipette for serial aliquoting E8-300XLS+.
13. Refrigerated benchtop centrifuge.
14. Magnetic stirrer with heating.
15. Warm water bath.
16. Cell sorter, such as the BD FACS Aria II, equipped with a UV laser.

2.3 Glassware and Disposables

1. 250 mL beaker.
2. 250 mL Schott glass bottles.
3. 1 cm magnetic stirrer.
4. 1 and 10 mL syringe.
5. Hypodermic needles (1 inch 26 G).
6. Scotch tape.
7. Tissue paper.
8. 15 mL and 50 mL Falcon tubes.

9. Tissue culture pipettes (2, 5, and 10 mL).
10. 96 well ultra-low attachment plates, such as Thermo Nunclon™ Sphera™ 96-well, U-bottom microplates, or Greiner microplate 96 well (polystyrol, clear, sterile, U-bottom, and cell-repellent surface).
11. Micropipettes and filter tips.
12. Minisart Syringe filter 0.45 μm filter.

2.4 Solutions and Chemicals

1. Perfusion buffer 1 (pH 7.3): 137 mM NaCl, 5.4 mM KCl, 0.63 mM NaH_2PO_4, 0.85 mM Na_2HPO_4, 10 mM HEPES, 4.2 mM $NaHCO_3$, 0.5 mM EGTA, 5 mM D-(+)-Glucose, and 0.016 mM Phenol red.
2. Perfusion buffer 2 (pH 7.3): 137 mM NaCl, 5.4 mM KCl, 0.63 mM NaH_2PO_4, 0.85 mM Na_2HPO_4, 10 mM HEPES, 4.2 mM $NaHCO_3$, 3.81 mM $CaCl_2$ x $2H_2O$, and 0.016 mM Phenol red.
3. Seeding medium: Williams E supplemented with 10% fetal bovine serum 50 mg/mL kanamycin sulfate, 10 mg/mL sodium ampicillin, 100 U/mL Penicillin-Streptomycin, and 292 mg/mL L-glutamine and 7 ng/mL Glucagon.
4. Culture medium: Williams E supplemented with 50 μg/mL kanamycin sulfate, 10 μg/mL sodium ampicillin, 100 U/mL Penicillin-Streptomycin, 292 mg/ml L-glutamine, 7 ng/ml Glucagon, 0.5 μg/mL insulin, and 25 μg/mL hydrocortisone sodium succinate.
5. 0.25 mg/mL Collagenase P in perfusion buffer 2.
6. 0.5 mg/mL Pronase E in perfusion buffer 2.
7. 0.4 mg/mL Pronase E and 0.001% DNAse in perfusion buffer 2 (beaker step).
8. Phosphate-buffered saline (PBS): 8.0 g NaCl, 0.2 KCl, and 1.42 g Na_2HPO_4 are solved in 1 L sterile water and the solution is autoclaved.
9. 25% Percoll in PBS.
10. 1 mg/mL Propidium iodide in water.
11. 200 mg/mL Dolethal solution for injection.
12. 70% ethanol.

3 Methods

For dissociation of the liver and the isolation of primary mouse liver cells, a two-step digestion is used (*see* **Note 1**):

Step 1: Perfusion of the liver with enzyme solutions through the vena porta.

Step 2: Digestion of the perfused liver on a magnetic stirrer (= beaker step).

3.1 Solution Preparations

1. Prepare perfusion buffers 1 and 2, verify their pH, and thaw an aliquot of a 2 mg/mL DNAse solution.
2. Make sure there are clean and sterile glassware (Schott bottles for enzymes and beakers with a magnetic stirrer for the collected liver(s)) and surgical material (tubings, sieve with forceps, scissors) available.
3. Weigh enzymes:
   3.1. 20 mg Pronase E for perfusion of one mouse.
   3.2. 10 mg Collagenase P for perfusion of one mouse.
   3.3. For the beaker step weigh 20 mg Collagenase P for 50 mL buffer.
4. Dissolve the enzymes in 10 mL of perfusion buffer 2, and filter the solution using a 10 mL syringe and 0.45 μm filter.
5. Transfer filtered enzyme solution to a Schott bottle, and add perfusion buffer 2 to its final volume (40 mL per mouse for perfusion) and immediately in 50 mL of perfusion buffer 2 in a beaker).
6. Keep the beaker with Collagenase on ice.
7. Prepare a Schott bottle with perfusion buffer 1 to rinse the liver and remove blood (40 mL per mouse).
8. Switch on a water bath for perfusion of the liver (42 °C) and pre-warm a container with water on a heating magnetic stirrer for the beaker step (37 °C).
9. Place Schott bottles with Pronase E and Collagenase P in perfusion buffer 2, in the perfusion water bath for >30 min before start of the surgery.

3.2 Surgery Area Preparation

1. Cover a dissecting board (Styrofoam) with a layer of aluminum foil and fix it in an operation tray to collect draining liquid.
2. Connect two larger pieces of tubing with a Vyclic three-way and connect a third, small piece of tubing to the third (side) way. Close the third tubing with a Titegrip needle holder. This part is intended to work as a bubble trap during the perfusion steps.
3. Fix the first piece of tubing to the peristaltic pump. Break off the filter of a 2 mL pipette and attach it reversed to the front side of the tubing. Put a 26G needle at the end side. Make sure the needle opening is upward; this prevents easy perforation of

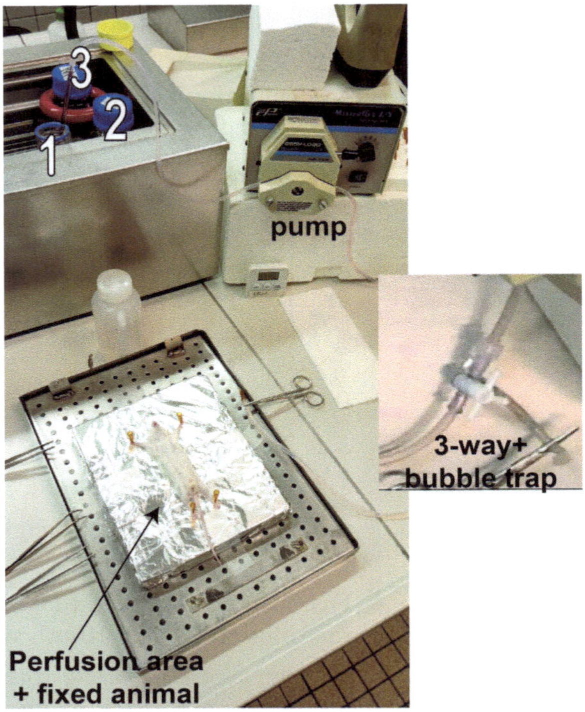

Fig. 1 Surgery and perfusion setup. The image shows the experimental setup with the water bath with 3 solutions: **1.** Perfusion buffer 1, **2.** Pronase E in perfusion buffer 2, and **3.** Collagenase P in perfusion buffer 2. Depicted is the peristaltic pump and the surgical field with the anesthetized and fixed animal. In the insert, a detailed view of the tubing with three-way connector is shown; this part is used to remove air bubbles by repositioning the Titegrip needle holder from the open-end tubing to the closed-end tubing thereby blocking the fluid from bringing air bubbles into the mouse circuit

the vein. Insert the end of the 2 mL pipet in the bottle with perfusion buffer 1 (Fig. 1).

4. Check/adjust the flow rate of the peristaltic pump to 7.5 mL/min.

3.3 Liver Perfusion

1. Anesthetize the mouse by injecting a lethal dose of 300 μg/g body weight Dolethal.
2. Wait until the mouse is fully anesthetized and check unresponsiveness to external stimuli such as a toe-pinch withdrawal test and eye reflex.
3. Pin the mouse with thumbtacks to the aluminum foil-covered Styrofoam board.
4. Disinfect the abdominal skin with 70% ethanol.
5. Open the abdomen using scissors according to the steps indicated in Fig. 2a.

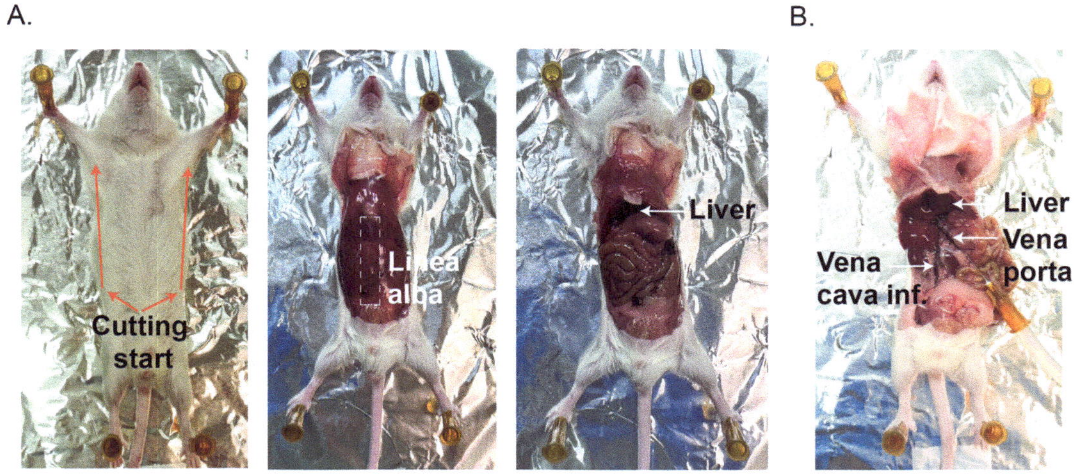

Fig. 2 Opening of mouse abdomen to expose the liver and vena porta. (**a**) The cutting lines and directions are indicated in red, and by opening the abdomen as much as possible, space is created for the needle and tubing and for the liver to increase in size during the perfusion. (**b**) After the intestines are carefully moved to the side, the *vena porta* and vena cava inferior are now visible. It is important to place the needle, lumen up, and to align it with the *vena porta* to avoid perforation of the vein

6. Open the peritoneum and carefully move the intestines to your right side to get a clear view of the liver (Fig. 2b).
7. Carefully place the needle in the portal vein and fix using the clamp, and ensure the needle lumen is faced upwards.
8. Start the perfusion with perfusion buffer 1 by switching on the pump.
9. As soon as the liver starts to turn pale, Fig. 3a, cut the vena cava inferior, and this leads to full discoloration of the liver. Start a timer for 5 min.
10. Switch off the pump after 5 min, switch the pipet-tubing to enzyme bottle 1 (Pronase in buffer 2, Fig. 3b), switch the pump back on, and perfuse for 5 min (repeat with enzyme bottle 2, Fig. 3c).
11. After perfusion, carefully remove the clamp and needle from the *vena porta*.
12. Excise the liver without damaging the lobes.
13. Lift the liver at the ligament using forceps as indicated in the image (Fig. 3d).
14. Cut where the liver is connected (behind the liver = close to diaphragm + just above the kidney), and remove the gallbladder if possible.
15. Put the liver in the beaker with 50 mL perfusion buffer 2, collagenase P, and stirring bar. Depending on downstream needs, 1–3 livers can be transferred to a single beaker.

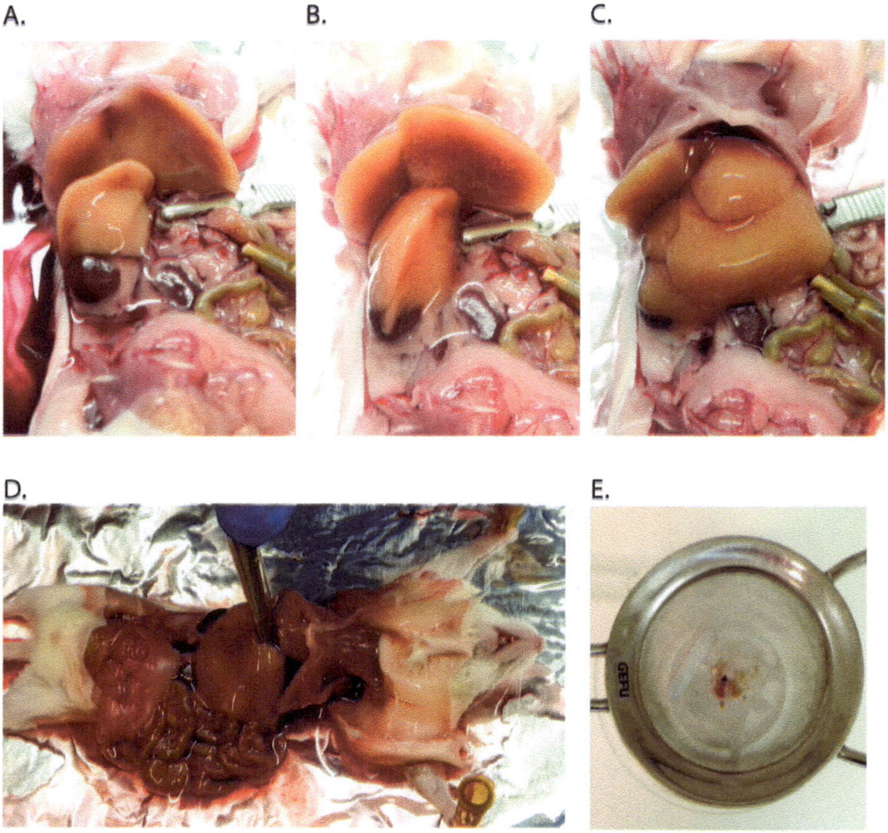

Fig. 3 Visual support for the liver digestion protocol. (**a**) Successful perfusion with perfusion buffer 1 results in blood removal and consequently discoloration of the liver. (**b**) During the Pronase E perfusion, after 2–3 min, the liver starts bloating and softening. (**c**) After the final perfusion, the liver has collapsed again and is now soft and well digested. (**d**) To avoid damage to the cells, the liver is extracted using forceps placed on the ligament between the lobes. Finally, after the in vitro digestion on a magnetic stirrer, most of the liver tissue is digested as shown by the remnants in the sieve (**e**). Cells are now ready to be processed for separating the wanted fractions

3.4 Liver Digestion Until Single Cell Suspension

From this moment on, working in a sterile environment is required.

1. Add 1 mL of pre-warmed DNAse solution to the 50 mL perfusion buffer 2 containing the perfused liver.
2. Gently tear the liver in small pieces with two sterile forceps, and keep the liver in the buffer solution as much as possible.
3. Put the beaker with stirring bar in a pre-warmed container on the heating plate with magnetic stirrer.
4. Stir for 15 min and add one drop of 1 M NaOH at t = 5 min.
5. Pour the crude cell suspension through a sterile sieve into a clean 50 mL Falcon tube (Fig. 3e).

The obtained single cell suspension can be further processed for non-parenchymal cell and hepatocyte purification.

3.5 Hepatocytes Enrichment by Percoll Gradient Centrifugation

The cell suspension obtained in Subheading 3.4 contains all liver cells. From this single cell suspension, hepatocytes and non-parenchymal cells are first separated using a low-speed centrifugation step.

1. Centrifuge cell suspension: 2 min, 50 × g at 4 °C, Fig. 4a. After centrifugation, the supernatant contains the non-parenchymal fraction (NPF) from which HSC will be further purified (see Subheading 3.6). The pellet contains the parenchymal fraction consisting of hepatocytes.
2. During the centrifugation step, prepare a 25% Percoll solution in a 50 mL Falcon tube and mix thoroughly.
3. Discard supernatant and gently resuspend the pellet in 5 mL perfusion buffer 2.

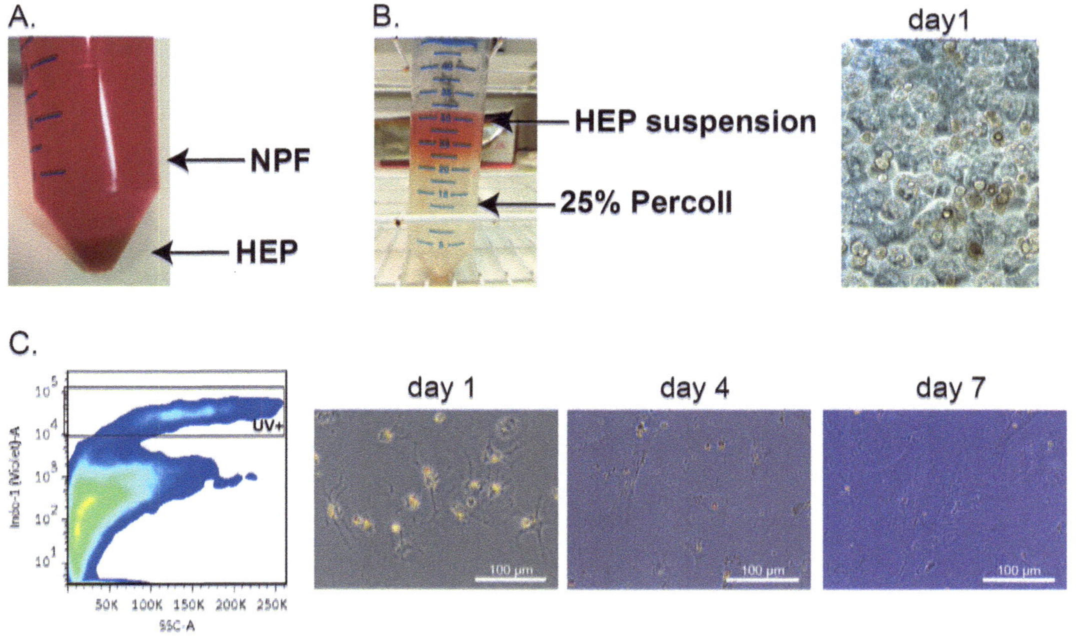

Fig. 4 Cell separation approach and cell type morphology. (**a**) After digestion, NPF and HEP are separated by a 2 min centrifugation at low speed (50 × g). (**b**) Viable hepatocytes are further selected by a Percoll gradient centrifugation. 24 h after seeding on collagen coated dishes, hepatocytes are attached and show their typical cobblestone, binucleated morphology. (**c**) HSCs are purified from the NPF suspension by UV-based cell sorting. On the first day of 2D culture, HSCs are small star-shaped cells containing lipid droplets. After 4 days, cells are activating, they stretch, become larger, and loose the large lipid droplets. HSC, hepatic stellate cell; NPF, non-parenchymal fraction; HEP, hepatocytes

4. Carefully apply the cell suspension on top of the Percoll solution while avoiding mixing the two layers (Fig. 4b).
5. Centrifuge for 20 min at 50 × g with slow acceleration and a minimal speed setting for the brake (4 °C).
6. Discard the upper layer leaving ±5 mL of Percoll solution at the bottom of the Falcon containing the hepatocytes.
7. Gently resuspend hepatocytes, by first tapping the tube and next by gentle pipetting using 5 mL of cold PBS. Finally, fill the Falcon with PBS up to 30 mL.
8. Centrifuge for 5 min at 100 × g (4 °C) and resuspend the cell pellet in seeding medium up to 30 mL.
9. Centrifuge for 5 min at 50 × g (4 °C) and resuspend the cell pellet in 10 mL seeding medium.
10. Count the hepatocytes using a 1/10 dilution in 0.05% trypan blue solution.
11. The hepatocytes can now be seeded in regular tissue culture dishes or plates for 2 D cultures or sandwich cultures. For 2D cultures, plate cells at a density of 57,000 cells per cm^2 on a collagen coated surface. Figure 4b shows a representative image of 2D cultured hepatocytes at day 1 after seeding.

3.6 Hepatic Stellate Cell Isolation Using FACS

The supernatant obtained after the first 50 g centrifugation step is a cell suspension that contains all liver cells except for the larger hepatocytes. The following steps allow the isolation of the HSCs from this NPF fraction.

1. Carefully transfer the supernatant (=NPF) of the 50 × g step to a new 50 mL Falcon tube.
2. Centrifuge 2 min, 50 × g at 4 °C to remove the remaining (small) hepatocytes.
3. Carefully transfer the supernatant to a new 50 mL tube and centrifuge for 8 min at 640 × g at 4 °C.
4. Discard the supernatant and resuspend the cell pellet in 3 mL RBC (1/10 in water) and incubate 3 min at room temperature.
5. Wash the pellet in perfusion buffer 2 and centrifuge the cell suspension for 8 min at 640 × g at 4 °C.
   - During wash prepare FACS tubes:
     1. Sample tube – PI.
     2. Sample tube + PI.
     3. Collection tube HSC 1 mL seeding medium (or other culture medium).

6. Dissolve final pellet in perfusion buffer 1 + DNAse solution (3:1) to a cell density of approximately 10 million cells per ml buffer (for most isolations 3–4 mL).
7. Transfer 150 μL to tube 1.
8. Transfer remaining sample to tube 2 and add 10 μL of PI per mL of cell suspension.
9. Sort the HSCs based on the UV+, FCS gate (Fig. 4c). Settings of the FACS depend on the equipment at hand. Please consult your FACS core facility for appropriate settings.
10. Once the desired number of HSCs is sorted into the collection tube, centrifuge the tube for 8 min at 640 g at 4 °C to pellet the HSCs.
11. Resuspend the cell pellet in 1 mL of seeding medium and count the cells.
12. The HSCs can be plated in regular tissue culture dishes or 6–24 well plates in DMEM 10% FBS for in vitro HSC activation studies. Figure 4c shows representative images of 2D cultured HSCs at days 1, 4, and 7 of culture.

3.7 Seeding and Culture of Hepatocytes–HSC Spheroid Cultures

1. Prepare a cell suspension that contains the desired number of cells per well in 20 μL seeding medium × number of wells needed +10–30% excess. We recommend using a ratio of 2 HSC to 1 hepatocytes and a total number of 2000 cells per spheroid [11, 17].
2. Seed 20 μL of the HSC–hepatocyte suspension per well using an automatic micropipette or an automated system such as the Viaflo Assist (*see* **Note 3**).
3. Allow the cells to settle for 30 min—overnight before placing the 96 well plates on an orbital shaker (set at 80 rpm) in the incubator.
4. Day 1: 24 h after seeding, most cells should be aggregated and forming spheroids. Figure 5a shows possible levels of compactness at day 1. If necessary, plates can be briefly centrifugated at low speed (50 × g), without a brake to force spheroid formation.
5. Add 150 μL of culture medium to each well.
6. Take a picture of day 1 spheroid morphology and if desired collect cells for staining.
7. Day 2: Refresh the medium of the spheroids using culture medium. Adjust the volume of the culture to your needs, and typically 100–135 μL is refreshed (*see* **Note 4**).
8. Refresh the media every 48–72 h, until the end of the culture. Maintain the cultures on the shaking platform during the entire culture time (*see* **Note 5**).

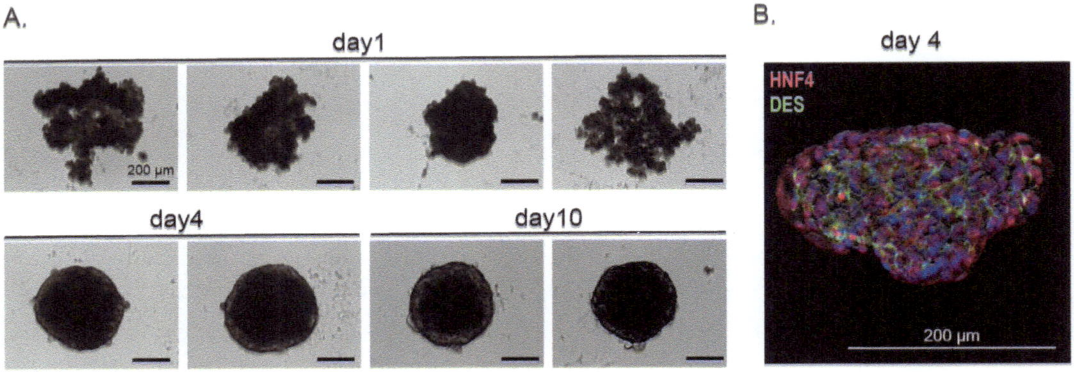

Fig. 5 Morphology of forming and formed mouse liver spheroids. (**a**) After isolation, mouse cells are mixed in the optimized ratio of 1 HEP and 2 HSCs. Per well 2000 cells are dispensed at low speed using an automated pipette in 20 μL of seeding medium. After confirmation that cells start to group together (roundness is not essential at this point), 150 μL culture medium is carefully added to each well, approximately 24 h after seeding. By day 2, spheroids are nicely formed and get more compact throughout the culture, shown here by images of days 4 and 10 spheroid cultures. (**b**) The presence of HEP and HSCs in the cultures can be confirmed by immunofluorescent staining on paraffin embedded spheroid sections, using cell type specific markers such as HNF4α (hepatocyte) or desmin (HSC). HSC, hepatic stellate cell

9. Spheroids can be cultured up to 2 weeks. Figure 5a shows HSC-hepatocyte cultures at days 1, 4, and 7 of culture and IF staining (Fig. 5b) showing HNF4α and desmin at day 4 of the culture (*see* **Note 6**).

4 Notes

1. Since hepatocytes and HSC are needed, perfuse with Pronase E and Collagenase P, but use only Collagenase P in the beaker.
2. A cell viability of >90% is recommended for successful spheroid cultures.
3. Due to issues of evaporation, we recommend not to use the outer wells of the ultra-low attachment 96 well plates for cell seeding. This means one can only culture 60 spheroids per plate. Fill the outer wells with 100 μL PBS + 10% PenStrep. In addition, when seeding the cells in the 96 well plates, make sure the pipette tip is not scraping the bottom of the well, and dispense the medium in the center of the well. Scraping will damage the coating of the well, which leads to cell adhesion and activation of HSCs.
4. Medium refreshing of the spheroid cultures needs to be done with precision to avoid loss of the spheroids. Ideally, medium change is done using a p1000 electronic multichannel. The majority of the medium is removed by placing the pipette tip

at the medium-air interface at low aspiration speed, and fresh medium is added at lowest dispense speed to the side of the well. Be careful to remove the same amount of medium from each well to ascertain equal culture conditions in every well of the culture.

5. Only part of the medium is refreshed. If you wish to remove a component of the medium, make sure to wash several times and incubate the plate on the shaker for 30–60 min in between washes.

6. For mRNA level analysis by RT-qPCR, we recommend using 4–6 spheroids for the isolation of mRNA, and we recommend at least 6–10 spheroids per condition for immunohistochemical or immunofluorescent evaluations.

References

1. Blachier M, Leleu H, Peck-Radosavljevic M, Valla DC, Roudot-Thoraval F (2013) The burden of liver disease in Europe: a review of available epidemiological data. J Hepatol 58(3): 593–608. https://doi.org/10.1016/j.jhep.2012.12.005
2. Schuppan D, Kim YO (2013) Evolving therapies for liver fibrosis. J Clin Invest 123(5): 1887–1901. https://doi.org/10.1172/JCI66028
3. Greek R, Menache A (2013) Systematic reviews of animal models: methodology versus epistemology. Int J Med Sci 10(3):206–221. https://doi.org/10.7150/ijms.5529
4. Schuster D, Laggner C, Langer T (2005) Why drugs fail–a study on side effects in new chemical entities. Curr Pharm Des 11(27): 3545–3559. https://doi.org/10.2174/138161205774414510
5. Mederacke I, Hsu CC, Troeger JS, Huebener P, Mu X, Dapito DH, Pradere JP et al (2013) Fate tracing reveals hepatic stellate cells as dominant contributors to liver fibrosis independent of its aetiology. Nat Commun 4: 2823. https://doi.org/10.1038/ncomms3823
6. Iwaisako K, Brenner DA, Kisseleva T (2012) What's new in liver fibrosis? The origin of myofibroblasts in liver fibrosis. J Gastroenterol Hepatol 27(Suppl 2):65–68. https://doi.org/10.1111/j.1440-1746.2011.07002.x
7. De Smet V, Eysackers N, Merens V, Kazemzadeh Dastjerd M, Halder G, Verhulst S, Mannaerts I et al (2021) Initiation of hepatic stellate cell activation extends into chronic liver disease. Cell Death Dis 12(12):1110. https://doi.org/10.1038/s41419-021-04377-1
8. Mannaerts I, Leite SB, Verhulst S, Claerhout S, Eysackers N, Thoen LF, Hoorens A et al (2015) The Hippo pathway effector YAP controls mouse hepatic stellate cell activation. J Hepatol 63(3):679–688. https://doi.org/10.1016/j.jhep.2015.04.011
9. De Minicis S, Seki E, Uchinami H, Kluwe J, Zhang Y, Brenner DA, Schwabe RF (2007) Gene expression profiles during hepatic stellate cell activation in culture and in vivo. Gastroenterology 132(5):1937–1946. https://doi.org/10.1053/j.gastro.2007.02.033
10. van Grunsven LA (2017) 3D in vitro models of liver fibrosis. Adv Drug Deliv Rev 121:133–146. https://doi.org/10.1016/j.addr.2017.07.004
11. Mannaerts I, Eysackers N, Anne van Os E, Verhulst S, Roosens T, Smout A, Hierlemann A et al (2020) The fibrotic response of primary liver spheroids recapitulates in vivo hepatic stellate cell activation. Biomaterials 261:120335. https://doi.org/10.1016/j.biomaterials.2020.120335
12. Mannaerts I, Thoen LFR, Eysackers N, Cubero FJ, Batista Leite S, Coldham I, Colle I et al (2019) Unfolded protein response is an early, non-critical event during hepatic stellate cell activation. Cell Death Dis 10(2):98. https://doi.org/10.1038/s41419-019-1327-5
13. Lambrecht J, Jan Poortmans P, Verhulst S, Reynaert H, Mannaerts I, van Grunsven LA (2017) Circulating ECV-associated miRNAs as potential clinical biomarkers in early stage HBV and HCV induced liver fibrosis. Front

Pharmacol 8:56. https://doi.org/10.3389/fphar.2017.00056
14. Mannaerts I, Schroyen B, Verhulst S, Van Lommel L, Schuit F, Nyssen M, van Grunsven LA (2013) Gene expression profiling of early hepatic stellate cell activation reveals a role for Igfbp3 in cell migration. PLoS One 8(12): e84071. https://doi.org/10.1371/journal.pone.0084071
15. Verbeke L, Mannaerts I, Schierwagen R, Govaere O, Klein S, Vander Elst I, Windmolders P et al (2016) FXR agonist obeticholic acid reduces hepatic inflammation and fibrosis in a rat model of toxic cirrhosis. Sci Rep 6: 33453. https://doi.org/10.1038/srep33453
16. Stradiot L, Verhulst S, Roosens T, Oie CI, Moya IM, Halder G, Mannaerts I et al (2017) Functionality based method for simultaneous isolation of rodent hepatic sinusoidal cells. Biomaterials 139:91–101. https://doi.org/10.1016/j.biomaterials.2017.05.047
17. Leite SB, Roosens T, El Taghdouini A, Mannaerts I, Smout AJ, Najimi M, Sokal E et al (2016) Novel human hepatic organoid model enables testing of drug-induced liver fibrosis in vitro. Biomaterials 78:1–10. https://doi.org/10.1016/j.biomaterials.2015.11.026

Chapter 12

Hepatic Stellate Cell Depletion and Genetic Manipulation

Qiuyan Sun and Robert F. Schwabe

Abstract

Hepatic stellate cells (HSCs) exert key roles in the development of liver disease. Cell-specific genetic labeling, gene knockout and depletion are important for the understanding of the HSC in homeostasis and a wide range of diseases ranging from acute liver injury and liver regeneration to nonalcoholic liver disease and cancer. Here, we will review and compare different Cre-dependent and Cre-independent methods for genetic labeling, gene knockout, HSC tracing and depletion, and their applications to different disease models. We provide detailed protocols for each method including methods to confirm successful and efficient targeting of HSCs.

Key words Hepatic stellate cell depletion, Cre-lox, Transgenic mice, Diphtheria toxin, Thymidine kinase

1 Introduction

Hepatic stellate cells (HSCs) constitute about 10% of resident cells in the liver [1]. Beyond their role as the main fibrogenic cell type, HSCs have emerged as a major hub of cell-cell communication with roles in a wide range of disease processes [2]. Accordingly, single cell RNA sequencing based ligand-receptor analyses have suggested HSCs as the cell type in the liver that interacts the most with other cell types in various disease conditions [2–6]. Hence, it is believed that HSCs may serve functions far beyond fibrogenesis, for which they are best known [1, 7, 8]. Their functions range from their role as main stores of vitamin A in homeostatic conditions [1] to contributors to liver regeneration following partial hepatectomy or acute liver injury [9, 10] to tumor-promoting as well as tumor-restricting roles in primary liver cancer and liver metastasis [3, 4, 11]. Accordingly, HSCs express a wide range of ligands and receptors that allow them to interact with epithelial, endothelial, and immune cells [2, 12]. Besides type I collagen, HSCs secrete chemokines and cytokines, growth factors, and a wide range of col-

Ralf Weiskirchen and Scott L. Friedman (eds.), *Hepatic Stellate Cells: Methods and Protocols*, Methods in Molecular Biology, vol. 2669, https://doi.org/10.1007/978-1-0716-3207-9_12,

lagens and non-collagenous extracellular matrix components [2, 7, 12]. While initial studies have largely focused on culturing HSC to understand their functions, there is increasing recognition that in vivo studies and genetic manipulations are needed to better capture their complex functions, including the many interactions with other cell types and the underlying ligand-receptor systems, in the healthy and diseased liver.

Transgenic mice are a powerful method to genetically label and manipulate many different cell types, including hepatic stellate cells, by using specific promoters to drive the expression of Cre recombinase, reporter genes, or genes that sensitize cells to cell death such as the diphtheria toxin receptor. Widely used promoters to drive gene expression in HSCs include *Gfap* [13], *Pdgfrb* [14], *Lrat* [15], *Acta2*, and *Col1a1* [16]. Each of these systems may have distinct advantages and disadvantages, summarized in Table 1. Glial fibrillary acidic protein (GFAP)-based drivers were among the first systems used [13], but GFAP may also label other cell types in the liver, and absent labeling of HSC in mice GFAP-Cre has also been reported [15]. Moreover, GFAP is abundant in the brain and the cells of the intestinal tract. Pdgfrb, Col1a1, and Acta2 are expressed by a wide range of pericytes or myofibroblasts throughout the body and may thus affect myofibroblast in organs beyond the liver. Lrat, although highly enriched in HSC and specific for HSC within the liver, is also expressed during development, and Lrat-Cre-transgenic mice therefore show positive labeling in other organs. At least for Lrat, the use of inducible Cre systems may alleviate some of these concerns as Cre activation in the adult liver is likely to be more HSC-selective.

Besides the choice of the specific promoter driving expression, its length is also important in controlling Cre expression. Knock-in and BAC transgenic mice often achieve similar expression patterns as the endogenous gene, as they incorporate full promoter and enhancer sequences, whereas shorter constructs often lack elements important for controlling expression.

Transgenic mice allow fluorescent labeling to trace HSC. Tracing of HSC is best achieved by Cre-transgenic mice as Cre-mediated deletion, e.g., of a "lox-stop-lox" (LSL) cassette flanking a fluorescent reporter gene, permanently labels a cell and its progeny as opposed to the use of reporter genes, whose expression can fluctuate. Previous studies have also combined Cre-transgenic mice with an LSL-TdTomato and a Col1a1-GFP reporter to determine that number of Lrat-Cre-labeled cells among all Col1a1-GFP positive collagen-producing cells in the liver [15]. The incorporation of an inducible Cre, e.g., either tamoxifen- or tetracycline-inducible, allows to trace the fate of cells that were labeled at one time point.

Table 1
Cre-transgenic mice for recombination in HSCs

	Recombination efficacy in HSCs	Considerations	Available from	References
Gfap-Cre	Various	May recombine in cholangiocytes and other organs including brain and intestine	JAX strains #024098 and #012886	[13, 15]
Pdgfrb-Cre	High (by reporter gene fluorescence in liver sections)	Recombines in all Pdgfrb^{+} cells, i.e., many fibroblasts from other organs	Not publicly available	[14, 38]
Pdgfrb-CreERT	90% (by reporter gene fluorescence in isolated HSC)	Recombines in all Pdgfrb^{+} cells, i.e., many fibroblasts from other organs	JAX strain #030201	Unpublished (personal communication by Ingmar Mederacke)
Lrat-Cre	99% (by reporter gene fluorescence in isolated HSC)	Recombines in all Lrat^{+} cells, i.e., also in other organs during development	MMRRC strain #069595-JAX	[3, 4, 11, 15, 39]
Lrat-rTA x TRE-Cre	Not investigated in detail	Recombines in all Lrat^{+} cells, but treatment with doxycycline in adulthood likely to reduce background strongly	Not publicly available	[17]
Col1a1-Cre	Not investigated in detail	Recombines in many type I collagen-producing cells and has been used for liver fibrosis and HSC research	Not publicly available	[20, 40]
Col1a1-CreERT2	Not well established	Recombines in many type I collagen-producing cells and has been used for liver fibrosis research	JAX strain #016241	[18, 41]
Col1a2-CreERT2	Not well established	Recombines in many type I collagen-producing cells	JAX strain #029567	[19, 42]

The deletion of floxed genes in Cre-transgenic mice is a powerful method to study the role of mediators or pathways in HSCs. Cre-mediated deletion has been applied to a wide range of gene targets, including ECM genes, growth factors, regulatory genes, and transcription factors in both healthy and a wide range of disease models, ranging from fibrosis to different tumor models. Gfap-Cre, Pdgfrb-Cre, and Lrat-Cre are among the most commonly used mouse strains [13–15]. Recently, a tetracycline-inducible Lrat-Cre system has been developed [17], which may represent the most HSC-selective approach for gene deletion on a whole-body scale. There are also several mice in which Cre or CreERT is driven by the Col1a1 or Col1a2 promoter [18–20]. However, not every single strain shows strong recombination in HSC, as many transgenic mice do not use the full endogenous promoter to drive Cre or CreERT expression. For all mice, it is important to quantify deletion of floxed genes by qPCR in isolated HSC and, if possible, by immunohistochemical staining.

Cre-transgenic mice can also be used to overexpress genes that are flanked by an LSL cassette. This system has been used to overexpress Cre-inducible diphtheria toxin receptor (iDTR), which is commonly used to genetically deplete specific populations. Lrat-Cre-mediated expression of iDTR is highly efficient for HSC ablation in different settings [3, 4, 11, 15].

Additional methods for HSC depletion included transgenic mice in which DTR expression is driven by a GFAP promoter [10]. It is also conceivable to express antigens such as GFP that are recognized by CAR T cells [21]. For all depletion studies, it is important to quantify HSC depletion and to be aware that killing of HSC for depletion leads to temporary increases in inflammation due to the recruitment of different immune cell population that engulf and remove dead HSC.

2 Conditional Gene Deletion in HSCs in the Normal and Diseased Liver

2.1 General Notes

Conditional deletion of genes in HSCs is a powerful approach to understand the role of HSCs and may shed light on the functions of specific pathways in HSCs or HSC-derived mediators in homeostasis or liver disease. There are several well-established approaches for conditional gene deletion in HSCs, but it is important to carefully validate gene deletion and also assess the mouse phenotype, as loss of HSC genes could disrupt liver and whole-body homeostasis as Cre-transgenic mice often recombine in organs other than the liver.

2.2 Breeding Mice

1. Different mice for conditional deletion in HSCs are listed in Table 1.
   (a) Lrat-Cre-transgenic mice can be obtained from the Jackson Laboratory (#069595-JAX) or a similar Lrat-P2A-iCre strain from GemPharmatech (T006205). A tetracycline-inducible Lrat-Cre strain has been described [17].
   (b) For deletion in PDGFR-beta positive fibroblasts including HSCs, Pdgfrb-Cre provides another powerful approach [14].

(c) GFAP-Cre-transgenic mice have been used for conditional deletion in HSCs [13], but some studies have reported recombination in biliary cells instead of HSCs [15].

(d) Col1a1-Cre, Col1a1-CreERT2 (#016241-JAX), and Col1a2-CreERT2 (#029567-JAX) stains are also available for deletion in Col1a1 or Col1a2 positive fibroblast and have been used for deletion in the context of liver fibrosis [18–20].

2. Mice are housed and bred at a density of three to five per cage in a facility according to the Institutional Animal Care and Use Committee (IACUC) guidelines. For breeding of the Lrat-Cre strain, Lrat-Cre-positive female mice need to be bred with Lrat-Cre negative male mice as Lrat-Cre^{pos} male mice may have Cre activity or inactivation in the germ line, leading to whole-body knockout/activation of reporter genes or insufficient activity, respectively.

3. The primers used for genotyping Lrat-Cre mice are CCTTTC TTTGACCCCCTGCAC (forward primer) and GACCGG CAAACGGACAGAAG (reverse primer). Annealing temperature is optimal between 57.5 °C and 62 °C. The expected PCR product is at around 315 bp. Primers for other Cre-transgenic mice are described in the literature [13–15, 17].

2.3 Appropriate Controls

Appropriate controls are essential to ensure the validity and reliability of in vivo studies. The most common control is the use of gender-matched Lrat-Cre^{neg} floxed littermates. Furthermore, one can additionally consider the use of Lrat-Cre^{pos} wild-type (WT) littermates as additional negative controls.

2.4 Determining Deletion

HSC-specific gene deletion efficiency should be confirmed prior to experimental studies. HSC isolation in combination with FACS sorting is recommended for this purpose; administration of CCl_4 is optional but increases the yield of HSCs.

1. Optional: Inject four times CCl_4 (first time, i.p. 0.5 mL kg^{-1}, diluted 1:8 in corn oil; second to fourth time, i.p. 0.5 mL kg^{-1}, diluted 1:4 in corn oil) every 3 days in both Lrat-Cre^{pos} and Lrat-Cre^{neg} floxed gene mice (*see* **Notes 1** and **2**).
2. Perform HSC isolation in untreated mice or mice 2 days after the last injection of CCl_4, according to the protocol described in the chapter "Isolation and Culture of Primary Murine Hepatic Stellate Cells" Subheadings 3.1, 3.2, 3.3, 3.4 and 3.5 or our previously described protocol [22].
3. Resuspend the cells in phenol red-free medium containing 1% (vol/vol) FBS.

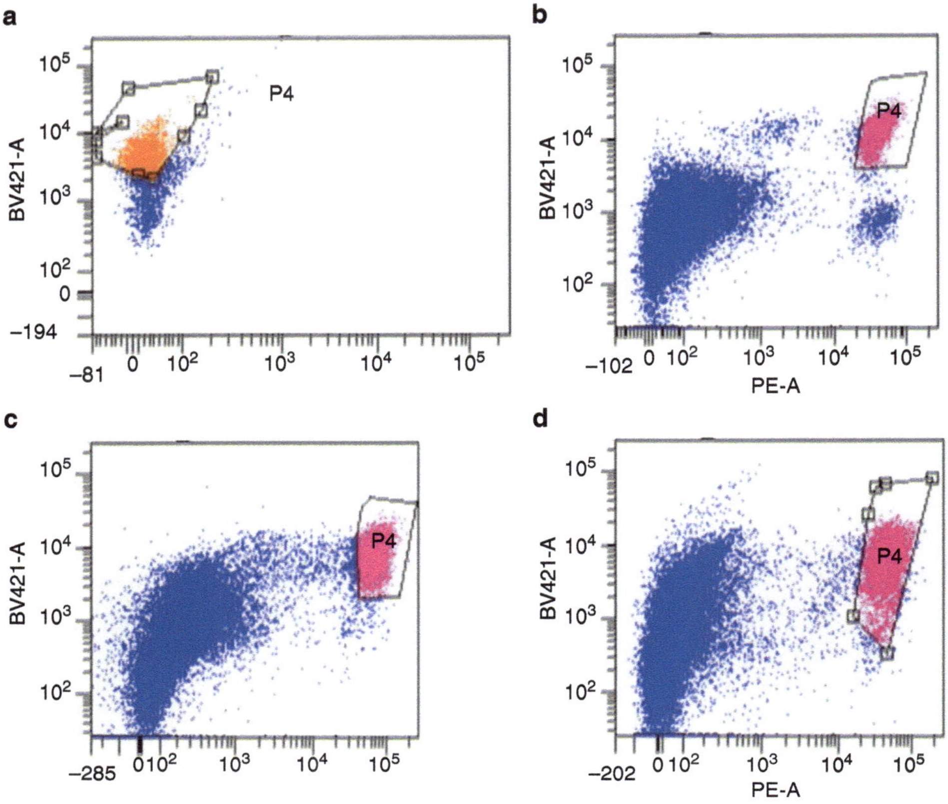

Fig. 1 Retinoid- and reporter-based FACS sorting for HSC isolation. (**a**) Retinoid-based FACS sorting of 11% Nycodenz gradient-purified HSCs from 4× CCl_4 treated mice. (**b, c, d**) Retinoid- and TdTomato-based FACS sorting of 17% Nycodenz gradient-purified HSC from untreated, 4× CCl_4 treated, and 12× CCl_4 treated mice

4. FACS-sort the cells for HSC-specific vitamin A fluorescence (Fig. 1a). Use a 405–407 nM laser for excitation and a 450/50 nM band-pass filter for detection as described in [22] or according to the protocol described in the chapter "Isolation and Culture of Primary Murine Hepatic Stellate Cells" (Chap. 1). This should give around 500,000 to 1,000,000 HSCs of 99% purity.
5. Extract RNA from the sorted HSCs, e.g., using RNeasy Micro Kit (Qiagen, #74004). We recommend using syringe to resuspend the cells in cell lysis buffer eight to ten times to facilitate the cell lysis process. DNAse treatment, e.g., on column, is highly recommended.
6. Perform RT-qPCR experiment and downstream analysis. We use probe-based TaqMan reverse transcription reagents, but other detection methods, using SYBR green-based, work as well. It is important to choose primers so that at least one of the primers lies within the deleted exon. We use relative standard curve to quantify absolute values and normalize to 18S

expression. Fold changes in gene expression in Lrat-Cre^{pos} mice are calculated with respect to Lrat-Cre^{neg} mice. Target gene deletion in HSC is expected to be 95–99%.

2.5 Applying HSC-Specific Deletion in Mouse Models of Liver Fibrosis and Cancer

HSC-specific gene deletion is an important and necessary method to study the role of candidate genes in liver fibrosis and liver cancer. Lrat-Cre x floxed gene model described above can be used in combination with different mouse models of liver fibrosis and cancer for this purpose. Timeline and methods for the models are summarized in this section. Generally, one piece of tissue should be taken from each lobe of the liver for paraffin and frozen sections and RNA. The rest of the tissue can be saved for additional needs.

1. CCl_4-induced liver fibrosis model: Inject six times CCl_4 (first time, i.p. 0.5 mL kg^{-1} CCl_4, diluted 1:8 in corn oil; two to six times, i.p. 0.5 mL kg^{-1} CCl_4, diluted 1:4 in corn oil) every 3 days in 8- to 10-week-old mice. Sacrifice the mice 2 days after the last dose of CCl_4.
2. NASH-induced fibrosis model: Put 8–10-week-old mice on one of the following profibrogenic high fat diets or the diet of your choice:
   (a) L-amino acid diet with 60 kcal% fat and 0.1% methionine and no added choline (HF-CDAA diet, Research Diet, A06071302).
   (b) High glucose (23 g L^{-1} in water), high fructose (19 g L^{-1} in water), and high-fat diet (FPC-NASH diet, Envigo, TD.160785).
   (c) Sacrifice mice 6 weeks after the start of diet.
3. Biliary type liver fibrosis model: Subject 8–10-week-old mice to the following:
   (a) 0.1% 3,5-diethoxycarbonyl-1,4-dihydrocollidine (DDC) diet [23].
   (b) Surgical ligation of the common bile duct (BDL) [24].
   (c) Sacrifice the mice 3 weeks after DDC diet or around 16 days after BDL surgery.
   (d) In addition to diet or surgery, multidrug resistance-associated protein 2-deficient (Mdr2^{KO}) mice provide a spontaneous liver/biliary fibrosis model [25, 26]. Sacrifice Mdr2^{KO} mice 8–12 weeks after birth for studies in liver fibrosis.
4. DEN-CCl_4 model for HCC.
   (a) Moderate fibrosis model [4]: Inject N-nitrosodiethylamine (DEN, Sigma, N0258, i.p. 25 mg kg^{-1}) on day 15 postpartum, followed by around 14 times CCl_4 injections (first time, i.p. 0.5 mL kg^{-1} CCl_4, diluted 1:8 in corn oil; two to 14 times, i.p. 0.5 mL kg^{-1}, diluted 1:4 in corn oil) every week starting at week 6 after birth.

(b) Profound fibrosis model [4]: Inject DEN (i.p. 80 mg kg^{-1}) at 4–5 weeks after birth, followed by around 44 times increased doses of CCl_4 injections starting at 8–9 weeks after birth (i.p. first week, 1× 0.5 mL kg^{-1}, diluted 1:8 in corn oil; second week, 3× 0.5 mL kg^{-1}, diluted 1:4 in corn oil; third week, 3× 1 mL kg^{-1}, diluted 1:4 in corn oil; fourth week, 3× 1.5 mL kg^{-1}, diluted 1:4 in corn oil; from the fifth to 20th week, 2× 1.5 mL kg^{-1}, diluted 1:4 in corn oil) (*see* **Notes 3** and **4**).

5. NASH-HCC model: Put mice on HF-CDAA diet for 8 months [27] or high fat choline-deficient diet for 12 months [28] starting at around 8–12 weeks of age. Sacrifice the mice at the end of the diet.
6. $Mdr2^{KO}$-HCC model: $Mdr2^{KO}$ provides a spontaneous HCC model [29]. Sacrifice the mice at around 15 months of age for studies in HCC.
7. Oncogene-driven HCC/ICC model: HCC/ICC can be induced by hydrodynamic tail vain injection of various oncogene(s) and sleeping beauty transposase (normally at 20 μg:5 μg ratio per 25 g of body weight) alone or in combination with CCl_4 injection or diet at 7–9 weeks of age. Known models for HCC include the following:

(a) TAZ-S89A alone and with 4 months of FPC diet or 16 times CCl_4 injections (i.p. 0.5 mL kg^{-1}, diluted 1:4 in corn oil) 2 times a week [4].

(b) cMet in combination with CTNNB1-S45Y for 8 weeks [30].

(c) Myc alone for 5 weeks (this model works in mice with FVB/N background) [31].

(d) Myc in combination with AKT or HRASV12 for 7–8 weeks [32].

(e) NRASV12 in combination with AKT for 4–5 weeks [33].

(f) Known models for ICC include myr-Akt in combination with YAPS127A, FBXW7DF or NICD1, and KRASG12D in combination with CRISPR/Cas9 sgRNA-p19 [3, 34–37].

3 In Vivo Labeling and Tracing of HSCs

3.1 General Notes

Constitutively expressing Lrat-Cre or tetracycline-inducible Lrat-Cre in combination with a Cre-inducible reporter can be used for in vivo tracing of HSCs. Cre-activated reporters will be permanently turned on after Cre-mediated recombination of the LSL cassette and remain positive in all of the cell's progeny. While constitutive Lrat-Cre marks Lrat-positive cells at all stages including cells that were only positive at one time point, e.g., during development, tetracycline-inducible Lrat-Cre tracing is more selective, allowing precise control of Cre activity depending on the timing of doxycycline delivery.

3.2 Breeding Mice

Two widely used models for in vivo tracing of HSCs include Lrat-Cre (Subheading 2.2.1) and tetracycline-inducible Lrat-Cre (Subheading 2.2.1) in combination with transgenic mice that express reporter genes with an LSL cassette:

1. LSL-reporter mice are available from Jackson Laboratory including TdTomato (#007908-JAX), ZsGreen (#007906-JAX), YFP (#006148-JAX), and mTom/mGFP (#007576-JAX).
2. Breeding of Lrat-Cre and tetracycline-inducible Lrat-Cre mice is described in Subheading 2.2.
3. Determination of recombination efficiency can be done by FACS for Cre-positive mice that have been crossed with a fluorescent reporter:
   (a) Isolate HSCs and FACS sort the cells as described in Subheading 2.4.
   (b) During sorting, use both HSC-specific vitamin A fluorescence channel and the reporter-specific channel.
   (c) Recombination efficiency can be determined as the percentage of reporter positive cells versus vitamin A-positive cells.

3.3 Detecting HSC by Fluorescence

1. In general, frozen sections from the liver tissue are taken to visualize the reporter. This can be combined with other markers (e.g., Col-GFP) or immunofluorescence staining (e.g., CK19) for tracing and to learn about the spatial relationship between HSCs and the other cell types in the liver. However, formalin-fixed paraffin-embedded sections can also be used and will require immunohistochemical staining with an antibody specific to the employed reporter gene. This is commonly done for reporters that have lower endogenous fluorescence, e.g., YFP, or when co-staining with other markers in paraffin sections is desired.
2. FACS sorting as described in Subheading 2.4 using the reporter-specific channel in addition to the vitamin A specific fluorescence is also used for quantification and for more sensitive downstream analysis (e.g., bulk-RNA-seq) that requires cell purity (Fig. 1b–d) (*see* **Note 5**).

4 In Vivo Genetic Depletion of HSCs to Understand their Role in the Normal and Diseased Liver

4.1 General Notes

Depletion of HSCs provides powerful insight into their roles in the healthy and diseased liver. As HSC depletion is achieved via the induction of cell death, one needs to be aware of the side effects of such procedures. These include the induction of cell death and ensuing inflammation, the potential induction of cell death outside the liver if the employed Cre-transgenic mouse recombines outside the liver, as well as the potentially massive changes in liver homeostasis and architecture associated with the extensive reduction of HSCs often throughout the entire liver, which could be the topic of specific studies but could also affect the interpretation of other studies, especially when long-term HSC depletion is performed. Hence, complementary experimental approaches may be appropriate for specific research questions.

4.2 Breeding Mice

Two widely used models of in vivo genetic depletion of HSCs include Lrat-Cre x iDTR model (4.2.1–4.2.3) and αSMA-TK model (4.2.4):

1. iDTR mice can be obtained from Jackson Laboratory (JAX, #007900). Details on Lrat-Cre-transgenic mice and their breeding are included in Subheading 2.2 in this chapter.
2. For Lrat-Cre x iDTR model, we recommend to include a reporter (e.g., LSL-TdTomato) for HSC depletion tracing.
3. Breeding should be set up to include gender-matched littermate Lrat-Cre^{pos}TdTom^{pos}DTR^{pos} mice and Lrat-Cre^{pos}TdTom^{pos}DTR^{neg} mice as control.
4. αSMA-TK mice can be obtained from Jackson Laboratory (JAX, #029921). These mice are on FVB background. If C57Bl/6 background is needed, then additional backcrosses are required.

4.3 Determining HSC Depletion

Successful HSC depletion should be verified prior to experimental studies. Reporter tracing (3.3.1–3.3.2), RT-qPCR (3.3.3–3.2.5), and IHC staining for Sirius Red (3.3.6–3.2.7) and αSMA (3.3.8–3.2.9) are recommended for this purpose.

1. For study using Lrat-Cre x iDTR x LSL-TdTom model to deplete HSCs, take one piece of tissue from each lobe of the liver for frozen sections.
2. Quantify TdTom^{pos} area using fluorescence microscope in experimental (Lrat-Cre^{pos}TdTom^{pos}DTR^{pos}) and control group (Lrat-Cre^{pos}TdTom^{pos}DTR^{neg}). The level of depletion as indicated by reduction in TdTom^{pos} area is expected to be

more than 90% in uninjured livers (unpublished data) and around 50% to more than 90% in chronic liver injury and liver cancer [4].

3. For studies using either model, take one piece of tissue from each lobe of liver for paraffin sections and RNA extraction.
4. Isolate RNA from liver tissue collected, e.g., using TRIzol and a High Pure Tissue RNA Isolation Kit (Roche, #11828665001) or similar.
5. Perform RT-qPCR experiment and downstream analysis as described in Subheading 2.4.6. using primer-probe sets (ThermoFisher) for HSC marker and activation genes (Acta2 (Mm01546133_m1), Col1a1 (Mm00801666_g1), and Lrat (Mm00469972_m1) or SYBR green detection-based qPCR. A significant reduction of the HSC marker gene expression should be detected in HSC depleted mice liver.
6. Perform Sirius Red staining on the paraffin sections for fibrosis detection.
7. For Sirius Red staining, take one or more representative pictures of each lobe of the liver. Using a polarized light filter reduces background signal for quantification of Sirius Red staining. We use Adobe Photoshop (v.11.0) to quantify Sirius Red-positive area, but there is a wide range of quantification software. A significant reduction in fibrosis as indicated by Sirius Red-positive area should be detected in HSC depleted mice liver in most liver disease model.
8. Perform αSMA staining on paraffin sections, e.g., using anti-α SMA-FITC (Sigma, F3777; 1:2000), anti-FITC (Abcam, ab6655; 1:250), and a Vectastain Elite ABC-HRP Kit (Vector Laboratories) with a DAB Peroxidase Substrate Kit (Vector Laboratories).
9. For αSMA staining, we scan the slides (e.g., using a Leica SCN400 slide scanner) and quantify DAB-positive area (e.g., using LEICA Digital Image Hub 4.0 image server). A significant reduction in HSC activation as indicated by αSMA-positive area should be detected in HSC depleted mice liver in most liver disease model.

4.4 Timing of Depletion in Disease Models

The specific time to induce HSC depletion depends on the goal or question underlying the study. Below we describe the generic methods for HSC depletion in the two models. High level depletion can be achieved in acute models, where depletion is short term and usually stopped 7–10 days before starting the experiment and obtaining liver tissue. This may apply to study the role of HSCs in homeostasis, regeneration, and acute liver injury. In chronic disease models such as nonalcoholic fatty liver disease, in particular nonalcoholic steatohepatitis, long-term fibrosis models, and liver cancer, repeated injections may be helpful to maintain efficient HSC depletion.

1. iDTR model: Inject diphtheria toxin (DT, Sigma 322,326, 0.25 ng g^{-1}) intraperitoneally once per week or as indicated.
2. αSMA-TK model: Inject GCV (ganciclovir, InvivoGen, i.p. 10 mg/kg) intraperitoneally three times per week or as indicated.
3. For study of the role of HSC in early tumorigenesis, i.p. injection of DT or ganciclovir could be started after first dose of CCl_4 injection or after oncogene delivery.
4. For study of the role of HSC in late tumorigenesis or established tumors, i.p. injection of DT or ganciclovir could be started several weeks and last till the end of the experiments.
5. Note: Inflammation occurs after HSC depletion, particularly when HSC death is massive, e.g., after the first depletion. Therefore, appropriate timing for tissue collection should be considered.

5 Notes

1. This step is optional. It is recommended if a gene is not expressed or only weakly expressed in quiescent HSCs but upregulated during HSC activation.
2. For greater number of HSC yield, we recommend using mice older than 15 weeks.
3. The exact times of CCl_4 injections in both moderate and profound fibrosis model may need to be adjusted for some strains, as faster or slower tumor development can occur due to genetic background, genetic drift, or differences in mouse facilities.
4. Liver tissues are generally collected 2 days after last CCl_4 injection for moderate fibrosis model and 1 week after last CCl_4 injection for a profound fibrosis model.
5. For sorting HSCs in chronic injury or a liver cancer model, an HSC-specific reporter is needed because vitamin A is depleted when HSCs are activated (Fig. 1d).

References

1. Tsuchida T, Friedman SL (2017) Mechanisms of hepatic stellate cell activation. Nat Rev Gastroenterol Hepatol 14:397–411
2. Wallace SJ, Tacke F, Schwabe RF et al (2022) Understanding the cellular interactome of non-alcoholic fatty liver disease. JHEP Rep 4: 100524

3. Affo S, Nair A, Brundu F et al (2021) Promotion of cholangiocarcinoma growth by diverse cancer-associated fibroblast subpopulations. Cancer Cell 39:866–882 e11
4. Filliol A, Saito Y, Nair A et al (2022) Opposing roles of hepatic stellate cell subpopulations in hepatocarcinogenesis. Nature 610:356–365
5. Wang ZY, Keogh A, Waldt A et al (2021) Single-cell and bulk transcriptomics of the liver reveals potential targets of NASH with fibrosis. Sci Rep 11:19396
6. Xiong X, Kuang H, Ansari S et al (2019) Landscape of intercellular crosstalk in healthy and NASH liver revealed by single-cell secretome gene analysis. Mol Cell 75(644–660):e5
7. Friedman SL (2008) Hepatic stellate cells: protean, multifunctional, and enigmatic cells of the liver. Physiol Rev 88:125–172
8. Wells RG, Schwabe RF (2015) Origin and function of myofibroblasts in the liver. Semin Liver Dis 35:e1
9. Kitto LJ, Henderson NC (2021) Hepatic stellate cell regulation of liver regeneration and repair. Hepatol Commun 5:358–370
10. Stewart RK, Dangi A, Huang C et al (2014) A novel mouse model of depletion of stellate cells clarifies their role in ischemia/reperfusion- and endotoxin-induced acute liver injury. J Hepatol 60:298–305
11. Bhattacharjee S, Hamberger F, Ravichandra A et al (2021) Tumor restriction by type I collagen opposes tumor-promoting effects of cancer-associated fibroblasts. J Clin Invest 131(11):e146987
12. Carter JK, Friedman SL (2022) Hepatic stellate cell-immune interactions in NASH. Front Endocrinol (Lausanne) 13:867940
13. Yang L, Jung Y, Omenetti A et al (2008) Fate-mapping evidence that hepatic stellate cells are epithelial progenitors in adult mouse livers. Stem Cells 26:2104–2113
14. Henderson NC, Arnold TD, Katamura Y et al (2013) Targeting of alphav integrin identifies a core molecular pathway that regulates fibrosis in several organs. Nat Med 19:1617–1624
15. Mederacke I, Hsu CC, Troeger JS et al (2013) Fate tracing reveals hepatic stellate cells as dominant contributors to liver fibrosis independent of its aetiology. Nat Commun 4:2823
16. Magness ST, Bataller R, Yang L et al (2004) A dual reporter gene transgenic mouse demonstrates heterogeneity in hepatic fibrogenic cell populations. Hepatology 40:1151–1159
17. Wang S, Zhu Q, Liang G et al (2021) Cannabinoid receptor 1 signaling in hepatocytes and stellate cells does not contribute to NAFLD. J Clin Invest 131(22):e152242
18. He X, Tolosa MF, Zhang T et al (2022) Myofibroblast YAP/TAZ activation is a key step in organ fibrogenesis. JCI Insight 7(4):e146243
19. Hung CT, Su TH, Chen YT et al (2022) Targeting ER protein TXNDC5 in hepatic stellate cell mitigates liver fibrosis by repressing non-canonical TGFbeta signalling. Gut 71:1876–1891
20. Kisseleva T, Cong M, Paik Y et al (2012) Myofibroblasts revert to an inactive phenotype during regression of liver fibrosis. Proc Natl Acad Sci USA 109:9448–9453
21. Agudo J, Ruzo A, Park ES et al (2015) GFP-specific CD8 T cells enable targeted cell depletion and visualization of T-cell interactions. Nat Biotechnol 33:1287–1292
22. Mederacke I, Dapito DH, Affo S et al (2015) High-yield and high-purity isolation of hepatic stellate cells from normal and fibrotic mouse livers. Nat Protoc 10:305–315
23. Pose E, Sancho-Bru P, Coll M (2019) 3,5-Diethoxycarbonyl-1,4-Dihydrocollidine diet: a rodent model in cholestasis research. Methods Mol Biol 1981:249–257
24. Tag CG, Sauer-Lehnen S, Weiskirchen S et al (2015) Bile duct ligation in mice: induction of inflammatory liver injury and fibrosis by obstructive cholestasis. J Vis Exp 96:52438. https://doi.org/10.3791/52438
25. Fickert P, Zollner G, Fuchsbichler A et al (2002) Ursodeoxycholic acid aggravates bile infarcts in bile duct-ligated and Mdr2 knockout mice via disruption of cholangioles. Gastroenterology 123:1238–1251
26. Popov Y, Patsenker E, Fickert P et al (2005) Mdr2 (Abcb4)-/- mice spontaneously develop severe biliary fibrosis via massive dysregulation of pro- and antifibrogenic genes. J Hepatol 43:1045–1054
27. Wei G, An P, Vaid KA et al (2020) Comparison of murine steatohepatitis models identifies a dietary intervention with robust fibrosis, ductular reaction, and rapid progression to cirrhosis and cancer. Am J Physiol Gastrointest Liver Physiol 318:G174–G188
28. Wolf MJ, Adili A, Piotrowitz K et al (2014) Metabolic activation of intrahepatic CD8+ T cells and NKT cells causes nonalcoholic steatohepatitis and liver cancer via cross-talk with hepatocytes. Cancer Cell 26:549–564
29. Katzenellenbogen M, Pappo O, Barash H et al (2006) Multiple adaptive mechanisms to chronic liver disease revealed at early stages of liver carcinogenesis in the Mdr2-knockout mice. Cancer Res 66:4001–4010

30. Tao J, Xu E, Zhao Y et al (2016) Modeling a human hepatocellular carcinoma subset in mice through coexpression of met and point-mutant beta-catenin. Hepatology 64:1587–1605
31. Wang H, Wang P, Xu M et al (2021) Distinct functions of transforming growth factor-beta signaling in c-MYC driven hepatocellular carcinoma initiation and progression. Cell Death Dis 12:200
32. Xin B, Yamamoto M, Fujii K et al (2017) Critical role of Myc activation in mouse hepatocarcinogenesis induced by the activation of AKT and RAS pathways. Oncogene 36:5087–5097
33. Ho C, Wang C, Mattu S et al (2012) AKT (v-akt murine thymoma viral oncogene homolog 1) and N-Ras (neuroblastoma ras viral oncogene homolog) coactivation in the mouse liver promotes rapid carcinogenesis by way of mTOR (mammalian target of rapamycin complex 1), FOXM1 (forkhead box M1)/SKP2, and c-Myc pathways. Hepatology 55: 833–845
34. Fan B, Malato Y, Calvisi DF et al (2012) Cholangiocarcinomas can originate from hepatocytes in mice. J Clin Invest 122:2911–2915
35. Seehawer M, Heinzmann F, D'Artista L et al (2018) Necroptosis microenvironment directs lineage commitment in liver cancer. Nature 562:69–75
36. Wang J, Dong M, Xu Z et al (2018) Notch2 controls hepatocyte-derived cholangiocarcinoma formation in mice. Oncogene 37:3229–3242
37. Wang J, Wang H, Peters M et al (2019) Loss of Fbxw7 synergizes with activated Akt signaling to promote c-Myc dependent cholangiocarcinogenesis. J Hepatol 71:742–752
38. Foo SS, Turner CJ, Adams S et al (2006) Ephrin-B2 controls cell motility and adhesion during blood-vessel-wall assembly. Cell 124: 161–173
39. Mederacke I, Filliol A, Affo S et al (2022) The purinergic P2Y14 receptor links hepatocyte death to hepatic stellate cell activation and fibrogenesis in the liver. Sci Transl Med 14: eabe5795
40. Lai KKY, Kweon SM, Chi F et al (2017) Stearoyl-CoA desaturase promotes liver fibrosis and tumor development in mice via a Wnt positive-signaling loop by stabilization of low-density lipoprotein-receptor-related proteins 5 and 6. Gastroenterology 152:1477–1491
41. Zheng B, Zhang Z, Black CM et al (2002) Ligand-dependent genetic recombination in fibroblasts: a potentially powerful technique for investigating gene function in fibrosis. Am J Pathol 160:1609–1617
42. Kim JE, Nakashima K, de Crombrugghe B (2004) Transgenic mice expressing a ligand-inducible cre recombinase in osteoblasts and odontoblasts: a new tool to examine physiology and disease of postnatal bone and tooth. Am J Pathol 165:1875–1882

Chapter 13

Human Hepatic Stellate Cells: Isolation and Characterization

Xiao Liu, David A. Brenner, and Tatiana Kisseleva

Abstract

Liver fibrosis of different etiologies is characterized by activation of hepatic stellate cells (aHSCs) into collagen type I secreting myofibroblasts, which produce fibrous scar and make the liver fibrotic. aHSCs are the major source of myofibroblasts and, therefore, the primary targets of anti-fibrotic therapy. Despite extensive studies, targeting of aHSCs in patients provides challenges. The progress in anti-fibrotic drug development relies on translational studies but is limited by the availability of primary human HSCs. Here we describe a perfusion/gradient centrifugation-based method of the large-scale isolation of highly purified and viable human HSCs (hHSCs) from normal and diseased human livers and the strategies of hHSC cryopreservation.

Key words Liver fibrosis, Myofibroblasts, Human hepatic stellate cells, Enzymatic digestion, Gradient centrifugation, Whole human livers

1 Introduction

Nonalcoholic fatty liver disease (NAFLD) is a spectrum of liver disease ranging from steatosis (nonalcoholic fatty liver, NAFL) to nonalcoholic steatohepatitis (NASH) with fibrosis [1, 2] and hepatocellular carcinoma (HCC). Hepatic fibrosis is characterized by the replacement of healthy tissue with extracellular matrix (ECM) scar tissue. Hepatic stellate cells (HSCs) are the dominant cell type implicated in fibrosis with negligent contributions from other matrix-producing cells, such as portal fibroblasts and bone marrow derived fibrocytes [3].

Hepatic stellate cells were originally identified as liver "lipocytes" due to their ability to store retinoids in lipid droplets. Under physiological conditions, quiescent HSCs (qHSCs) reside in the space of Disse, store vitamin A, and express neural (Lrat, NGFR1, GFAP) [4, 5] and lipogenic (PPARγ, Adipor1) markers [6]. In response to chronic injury, qHSCs downregulate expression of

Ralf Weiskirchen and Scott L. Friedman (eds.), *Hepatic Stellate Cells: Methods and Protocols*, Methods in Molecular Biology, vol. 2669, https://doi.org/10.1007/978-1-0716-3207-9_13,

neural and lipogenic markers and activate into collagen type I-producing aHSCs/myofibroblasts [7]. Activation of HSCs is triggered by TGF-β1, the major pro-fibrogenic factor secreted by myeloid cells, while HSC proliferation is regulated by PDGFβ. Although other inflammatory and fibrogenic cytokines (IL-6, TNFα, IL-1β [8–12]) contribute to aHSC activation, very few chemokines (leptin, CTGF, IL-6) can directly drive collagen type I production in aHSCs [3]. aHSCs are characterized by increased proliferation and high contractility with expression of pericellular matrix proteins (α-smooth muscle actin (α-SMA), vimentin), activation of TGFβ signaling pathway (TGFβ1, Smad2/3/4, PAI-1, Activin, and others), and secretion of abundant extracellular matrix proteins (fibronectin, collagen type I and III). HSCs release inflammatory, proliferative, and fibrogenic cytokines such as IL-6, PDGF, and TGFβ, through direct contact with their neighboring cells [3]. HSCs may also function as regulatory bystanders and contribute to liver-induced tolerance [13]. HSCs contribute to liver regeneration [14] and mediate sinusoidal blood flow via contraction and regulate microvascular structure and function in liver [15].

New investigations into HSC biology may provide important mechanistic insights into the pathophysiology of liver fibrosis and hold the key to developing therapeutic approaches that block HSC activation and liver fibrosis [16]. Single-cell and single nucleus-based approaches of hHSC isolation and analysis have made it possible to analyze the gene expression profiles and epigenetic landscapes of different subsets of hHSCs and reveal the pathways of their regulation, activation, and interaction with other key fibrogenic cell types in a well-defined context. As an example, one such analysis led to the identification of relevant regulatory pathways such as TGFβ and PDGFβ.

HSCs were first isolated from rat livers of retired breeders [17]. The rat liver was perfused with collagenase to digest the tissue and pronase to kill the hepatocytes. The accumulation of lipid droplets in these HSCs enabled the use of gradient centrifugation to purify the high buoyancy HSCs from the other less buoyant non-parenchymal cells [6, 18]. Subsequently, HSCs were purified from mouse livers in order to take advantage of the many transgenic and knockout mice. The gradient centrifugation protocol was slightly modified for mouse HSCs [6, 19].

Furthermore, HSCs could be purified from both normal mouse livers and fibrotic mouse livers. Earlier studies isolated HSCs from wedge biopsies of normal human liver by mincing the tissue prior to digestion in collagenase and pronase [17]. hHSCs were subsequently cultured and passaged so that they had an activated phenotype [17]. In addition, several immortalized human HSC cell lines were established. For example, LX-2 [20] and hTERT [21] became a useful tool to study hHSC responses and are suitable for certain experiments but exhibit the activated state and do not reflect the full range of cellular phenotypes [22]. Human inducible pluripotent stem cells (iPSCs) can be directly differentiated into HSC-like cells [23]. This enables the development of HSC-like cells from the genetic background of any human. However, the differentiated iPSCs are a mixture of cells with different levels of PDGFRβ expression.

The current protocol enables the isolation of HSCs from human livers that range from normal to moderately fibrotic (fibrosis stage 2–3). Donor livers are prepared for transplantation but declined due to various reasons. Large quantities of HSCs can be isolated from the left liver lobe, resulting in collection of highly purified (>90%) and viable (>85%), well oxygenated and detoxified, and functionally intact HSCs, which have been used in numerous studies [24–32]. In comparison, cell isolation from resected/explanted livers [33] can cause tissue under−/overdigestion and mechanical cell injury, leading to reduced cell viably and low yield and, most importantly, underrepresentation of specific cell types in the final cellular fraction [17].

2 Materials

2.1 Reagents and Chemicals/Solutions

1. Pronase (Roche, cat. no. 10-165-921-001).
2. Collagenase D (Roche, cat. no. 11-088-874-103).
3. Collagenase (VitaCyte, cat. no. 001-2030).
4. Protease (VitaCyte, cat. no. 003-1000).
5. DNase I (Roche, cat. no. 10-104-159-001).
6. Potassium chloride (KCl; Sigma Aldrich, cat. no. P9541-500G).
7. Magnesium chloride hexahydrate ($MgCl_2 \times 6H_2O$; Sigma Aldrich, cat. no. M2670-100G).
8. Magnesium sulfate heptahydrate ($MgSO_4 \times 7H_2O$; Sigma Aldrich, cat. no. M5921-500G).
9. Sodium phosphate dibasic (Na_2HPO_4; Sigma Aldrich, cat. no. S3264-500G).
10. Potassium phosphate monobasic (KH_2PO_4; Sigma Aldrich, cat. no. P9791-100G).
11. D-(+)-Glucose (Sigma Aldrich, cat. no. G8270-1KG).
12. Sodium bicarbonate (NaHCO3; Fisher, cat. no. S233-500G).
13. Calcium chloride dihydrate ($CaCl_2$ x $2H_2O$; Fisher, cat. no. C70500).
14. Dexamethasone (MP Biomedicals, cat. no. 0219456180).
15. Nycodenz AG (Accurate Chemical, cat. no. AN1002424).

16. HEPES (VWR, cat. no. 45000-690).
17. EGTA (Fisher, cat. no. 50-255-957).
18. Gey's Balanced Salt Solution B (GBSS/B; Fisher, cat. no. J67569-K2).
19. 10 X DPBS (VWR. Cat. no. 45000-426).
20. Hanks' Balanced Salt Solution (HBSS; Fisher, cat. no. 14175-095).
21. Eagle's Minimum Essential Medium (EMEM; VWR, cat. no.45000-386).
22. PERCOLL (Sigma Aldrich, cat. no. GE17-0891-09).
23. Fructose (VWR, cat. no. 97061-236).
24. HMM SingleQuot Kit (Lonza, cat. no. CC-4192).
25. Cellbanker 1 (AMSBIO, cat. no. 11888).
26. CS10 (BioLife Solutions, cat. no. 210102).
27. Surgical grade glue (VWR, cat. no. 37001-738).
28. CD11b MicroBeads, human and mouse (Miltenyi Biotec, cat. no. 130-049-601).
29. CD31 MicroBead, human (Miltenyi Biotec, cat. no. 130-091-935).

2.2 Reagents for Cell Culture

1. DMEM (Fisher, cat. no. 11965118).
2. Antibiotic-antimycotic 100× (Gibco, cat. no. 15240062).
3. FBS (Gemini Bio-Products, cat. no. 100-106).

2.3 Buffers

1. Buffer A: HBSS +1.0 mM EGTA.
2. Buffer B: EMEM +25 mM HEPES+ 400 mg/L pronase.
4. Buffer C: EMEM +25 mM HEPES+ 700–900 mg/L collagenase D.
5. DNase I (2 mg/mL): Dissolve 100 mg into 50 mL of GBSS/B buffer, aliquot 1 ml/tube, store −20 °C up to 6 months, and avoid freezing and thawing.
6. Nycodenz I: 5.2 g–5.5 g/total volume 15 mL in GBSS/A followed by 0.2 μM bottle top filter; final concentration, 10.4–11%.
7. Nycodenz II: 3.63 g/total volume 25 mL in GBSS/A followed by 0.2 μM bottle top filter; final concentration, 14.52%.
8. Gey's Balanced Salt Solution A, GBSS/A: Prepare the solution by dissolving the listed components in 1 l of ddH_2O (Table 1). Adjust pH to 7.3–7.4 followed by 0.2 μM bottle top filter. Store at 4 °C up to 6 months.

Table 1
Preparation of GBSS/A solution

Ingredients	Final concentration (mg/L)
KCl	370
$MgCl_2 \times 6H_2O$	210
$MgSO_4 \times 7H_2O$	70
Na_2HPO_4	59.6
KH_2PO_4	30
Glucose	991
$NaHCO_3$	227
$CaCl_2 \times 2H_2O$	225

2.4 Additional Equipment

1. Water bath.
2. Refrigerated benchtop centrifuge with swinging bucket rotor.
3. CryoMed controlled-rate freezer.
4. Perfusion pump and tubes.
5. Heat lamp.
6. Biosafety cabinet.
7. Inverted microscope.
8. Cell culture incubators with 5% CO_2, 37 °C.
9. Glass door refrigerator.
10. −20 °C freezer.
11. −80 °C freezer.
12. Surgical instruments.
13. Precision balance.
14. Cell counter.
15. Pipet aids.
16. Syringe, 10 mL and 60 mL.
17. 2-0 and 3-0 suture.
18. Sterile pipettes: 10 and 25 mL.
19. Falcon tubes: 15 and 50 mL.
20. Bottle top filter, 0.2 μM.
21. 2 mL cryovials.
22. Liquid nitrogen.
23. Nylon mesh 500, 250, and 85 μM.
24. Catheters and tubes.

3 Methods

3.1 General Remarks

3.1.1 Patient Material

The current protocol has been developed for the isolation of human HSCs from normal, NAFL, or NASH donor livers that are obtained through organ procurement organizations. De-identified medical records (including age, gender, BMI, and ALT/AST) are provided by the OPO. Whole livers are prepared for transplantation but declined at the OR (operation room) due to various reasons (such as blunt injury to the liver, fatty liver, steatohepatitis, and NASH). The left lobe is used for paraffin blocks or to snap frozen tissue [26]. The right lobe is used for human HSC isolation using perfusion/gradient centrifugation method.

3.1.2 Tissue Analysis

Tissues are stained for H&E, Trichrome, Sirius Red, desmin, αSMA, and CD68. Steatosis, inflammation, and fibrosis stage should be graded by a pathologist using a double-blinded method [34] and identified as normal, NAFL, or NASH. (a) Inflammatory and fibrogenic (IL-6, TNFα, IL-1β, TGFβ1) cytokines are measured by qRT-PCR [17]. (b) Livers are stained with Sirius Red and Trichrome. (c) Expression of Col1a1, αSMA, TIMP-1, TGFβRI, and CD68 is analyzed using immunohistochemistry and qRT-PCR. (d) Steatosis (Oil Red O) is measured. Upon completion of the tissue analysis (normal, NAFL vs. NASH) and grading of steatosis (grade, 0–3), inflammation (grade, 0–3), hepatocyte ballooning (grade, 0–2), and fibrosis (METAVIR stage, F = 1–4) [34], livers are identified as normal, NAFL, or NASH.

3.1.3 Characterization of Isolated hHSCs

One should anticipate isolating $>1 \times 10^8$ viable human HSCs/lobe. Single-cell suspension of viable vitamin A^+ human HSCs is immunophenotyped (for GFAP, Desmin, TE-7, F4/80, and CD31; Fig. 1a). Normal hHSCs and NASH hHSCs were stimulated ± TGF-β1 (5 ng/mL), SB431542 (10 μM), IL-6 (10 ng/mL), and PDGFβ (10 ng/mL) for 24 h and analyzed by qRT-PCR.

3.2 Preparation of the Liver

1. Sterilize all instruments and the materials in advance to avoid bacterial contamination during the isolation process.
2. Turn on two biological safety cabinets and run for 10 min before using. Place peristaltic pump and filled water bath into the first biological safety cabinet. Turn on water bath and warm to 39 °C.
3. Prepare perfusion solutions (Buffers A–C), and place them in the water bath to warm up 20 min before starting the perfusion (*see* **Note 1**). Load perfusion tubing(s) into the peristaltic pump, and maintain sterile conditions. Place all ends into the first bottle containing warm perfusion Buffer A. Turn on the pump to prime the tubing, and allow the solution to circulate continuously.

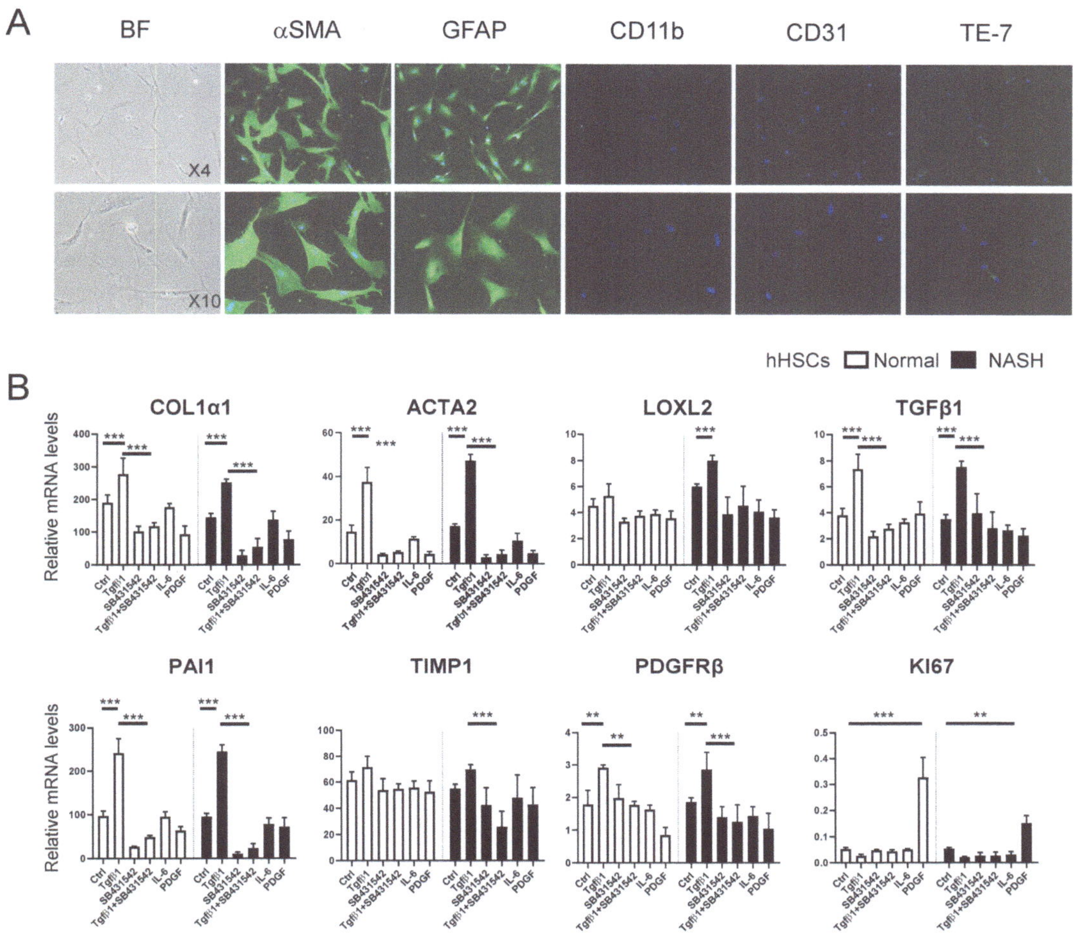

Fig. 1 Characterization of human hepatic stellate cells. (**a**) Primary human HSCs were immunostained for GFAP, αSMA, CD11b, CD31, and elastin (TE-7 Ab). (**b**) Normal hHSCs and Nash hHSCs were stimulated ± TGF-β1 (5 ng/mL), SB431542 (10 μM), IL-6 (10 ng/mL), and PDGFβ (10 ng/mL) for 24 h and analyzed by qRT-PCR, *$p < 0.05$, ** $p < 0.01$, *** $p < 0.001$, student *t*-test, and one-way ANOVA

4. Meanwhile, prepare the workstation for resection of the right liver lobe. These procedures are performed on ice. Fill a plastic pan with ice, and place into a second biological safety cabinet. Place a sterile waterproof placemat over the ice pan. Transfer the liver on top of the sterile field. Place all sterile surgical instruments on another sterile placemat next to the ice pan.
5. Use sterile gloves. Pick up the liver and examine for any lacerations, bruising, and other abnormalities. Resect the left lobe; it should have intact liver capsule and only one resected surface.
6. Insert tubing into the major portal and hepatic veins but do not secure (*see* **Note 2**). Connect a 60 mL syringe filled with cold EMEM medium to each tubing and flush the liver with

medium. The purpose is to identify the blood vessels which provide maximal perfusion to the entire lobe. Flushing with EMEM also removes residual blood. Select one to two blood vessels for optimal perfusion.

7. Secure the catheters into the selected vessels using nylon (2.0 suture material, five to seven stiches per blood vessel). Suture the resected liver surface (which lacks liver capsule) to occlude the open blood vessels, and seal remaining open areas with surgical grade glue. Designed to "restore" the integrity of liver capsule, this step is important to minimize area with the open blood vessels, prevent leakiness and loss of the perfusion solution, and achieve maximal perfusion efficiency. Flush the liver again with cold EMEM to test for any major leaks from the liver. High perfusion efficiency can be estimated by full inflation of the perfused liver piece and liver discoloration (pale color, due to the removal of residual red blood cells).
8. Perfusion with Buffer A. Place liver tissue with catheters into an empty sterile plastic bag. Weigh the liver tissue with the bag closed. Place the bag containing the liver into 39 °C water bath. Secure the bag above the water level. Connect the catheters to a peristaltic pump with flow rate of 20–30 mL/min, and perfuse the liver with Buffer A for 10–15 min (*see* **Note 3**).
9. Following perfusion with Buffer A, the liver is ready for enzymatic digestion. Specifically, liver perfusion with Buffer B (contains pronase to destroy the parenchymal cells) followed by Buffer C (contains collagenase D) is designed to dissociate the non-parenchymal cells from the liver scaffold with the maximal efficiency.

3.3 Human Hepatic Stellate Cell Isolation

1. Perfuse the liver with Buffer B for 15–30 min. Then, perfuse the liver with Buffer C for 15–30 min. Perfusion should be stopped when the liver tissue loses firmness and exhibits visible signs of cell separation from the liver capsule (*see* **Note 4**).
2. Slowly remove the digested liver from the plastic bag, and place into a sterile plastic beaker that contains ice-cold EMEM (to stop enzymatic digestion). Using sterile scissors, cut liver capsule, and gently release liver cells. Remove connective tissue, large clumps, and non-digested particles. Gently mix the cells to obtain a homogenous cell suspension.
3. Filter the cell suspension through sterile three-layer nylon mesh-covered funnels to remove cellular debris and clumps of undigested tissues (*see* **Note 5**).
4. Transfer the filtrate into 50 mL conical tubes. Remove hepatocyte debris from the non-parenchymal fraction by low-speed centrifugation of the filtered cell suspension at 60–80 × g for 5 min at 4 °C.

5. Collect the supernatant into new 50 mL tubes, and centrifuge at 935 × g for 8 min at 18 °C to collect total NPCs.
6. Resuspend each pellet in 10 mL EMEM, and bring the volume in each tube up to 50 mL with EMEM (room temperature). Spin again at 700 × g for 8 min. This step is critical for washing the NPCs from the residual debris. Discard the supernatant, add 10 mL GBSS/B to each tube, and thoroughly resuspend the pellet by inversion.
7. The next step is preparing for gradient centrifugation. Add 15 mL Nycodenz I to each cell suspension (from **step 6**), and bring total volume up to 50 mL with GBSS/B (room temperature). Mix well by inversion (*see* **Notes 6** and 7).
8. Prepare new 50 mL conical tubes with 10 mL Nycodenz II. Next, very slowly layer 30 mL of the cell mixture from **Step** 7 on top of Nycodenz II, and avoid mixing of the layers. Then layer another 10 mL of GBSS/B (room temperature) onto top of the cell mixture for total of three layers. Centrifuge 2000 × g, and break off for 20 min at 18 °C (*see* **Note 8**).
9. Under the first layer of clear GBSS/B solution is a white layer. This layer contains enriched hepatic stellate cells. Using a 1 mL pipet, carefully go through the top layer, and collect the HSC-enriched layer, and transfer it to new 50 mL tubes.
10. Add GBSS/B up to 50 mL (room temperature) to wash cells, and centrifuge 935 × g for 8 min at 18 °C (*see* **Note 9**).
11. Repeat **Step 10**, but centrifuge 700 × g for 8 min at 18 °C.
12. Proceed to downstream applications (*see* **Note 10**).
13. Proceed with Section C to culture hepatic stellate cells and Section D to cryopreserve hepatic stellate cells (*see* **Notes 11–12**).

3.4 Culturing Hepatic Stellate Cells

1. Count cells using a hemacytometer.
2. Culture 0.5 million hepatic stellate cells with DMEM +10% FBS + 1% antibiotic-antimycotic 100× in 10 cm dish.
3. Change media every 3–4 days until cells are 90% confluent on cell culture dish, and continue to applications below.

3.5 Cryopreservation of Hepatic Stellate Cells

1. Count cells using a hemacytometer.
2. Resuspend cell pellet in CELLBANKER® 1.
   (a) For hepatic stellate cells, aliquot 1 mL of cell suspension per freezing tube. Cell number per tube is 1.1×10^6.
3. Place tubes with cell suspension in control-rate step down freezer.

4. Start step down program and set the starting temperature to be 4 °C.
   (a) Step 1 . . . . Wait at 4.0 °C.
   (b) Step 2 . . . . 0.1.0 C/m S to −4.0 °C.
   (c) Step 3 . . . . 0.25.0 C/m C to −40 °C.
   (d) Step 4 . . . . 0.10.0 C/m C to −12.0 °C.
   (e) Step 5 . . . . 0.1.0 C/m C to −0 °C.
   (f) Step 6 . . . . 0.10.0 C/m C to −90 °C.
   (g) Step 7 . . . . End.
5. Transfer tubes into liquid nitrogen vapor phase for long-term storage.

4 Notes

1. The required volumes of solutions depend on the size of the liver. If left lobe is used, 600 mL of perfusion buffers should be prepared.
2. Multiple vessels can be used, but use as few vessels as possible while achieving maximum inflation of liver.
3. EGTA chelates calcium that leads to the separation of cell junctions and helps to remove any residual blood. Flow rate is based on the liver tissue size and number of catheters. If one tubing is used to perfuse the left lobe, the flow rate should not exceed 30 mL per min. Perfusion with enzymes requires the adjustment of the flow rate to 28 mL per min.
4. Stop enzymatic perfusion when liver tissue becomes soft, shown fracturing, and separates from the liver capsule.
5. Three-layer nylon mesh must be from outside to inside with size of 85 μM > 250 μM > 500 μM. Nylon mesh should be autoclaved before use. This step can be repeated as many times as needed to obtain the maximum number of single-cell suspension.
6. Nycodenz I concentration must be adjusted based on the fat content in the liver. This is a critical step; the final concentration should be 10.4–11%.
7. Gradient centrifugation: Before beginning the layering process, wet the sides of the tube with Nycodenz II solution which is already in the tube. The layering is very important, and the layer should not be disrupted or mixed with other layers.
8. Centrifugation without brake is critical.
9. Avoid pipetting cells with too much force to prevent mechanical stress when collecting and washing isolated HSCs.

10. Contamination of HSCs with fatty hepatocytes: To remove contaminating hepatocytes, the suspension of HSCs can be transferred into new 50 mL conical tubes. Hepatocytes can be removed from HSCs by low-speed centrifugation at 60–80 × g for 5 min at 4 °C. Centrifugation speed depends on the fat content in the hepatocytes.
11. The purified HSCs can be studied directly, after culturing, or frozen for subsequent use. The frozen HSCs have been thawed and used in 2D cultures or mixed with other cells to form spheroids or organoids. The standard 2D cultures of the HSCs are used for studies in signal transduction and gene expression. The HSCs are genotyped so that the effect of specific single nucleotide polymorphisms (SNPs), such as PNPLA3, on HSC function can be assessed [32, 35]. Our standard approach is to serum starve the HSCs overnight, stimulate with agonists such as TGF-β [17] (10 ng/mL), and then measure gene expression and protein secretion [36]. The HSCs are transfectable with transgenes or siRNA to assess the role of specific genes in HSC activation and signaling [17].
12. This protocol relies on pronase to eliminate hepatocytes and on gradient centrifugation for separation of the HSCs from the other cells. Therefore, there may be residual steatotic hepatocytes with light buoyancy that contaminate the HSC population [26]. These hepatocytes are eliminated by short-term culturing the HSC preparation [26]. However, culturing the HSC preparation results in HSC activation with a change of gene expression. Alternatively, the contaminating cells are identified and excluded by computational approaches in some studies such as single-cell RNA-seq.

References

1. Loomba R, Sanyal AJ (2013) The global NAFLD epidemic. Nat Rev Gastroenterol Hepatol 10:686–690
2. Friedman SL, Neuschwander-Tetri BA, Rinella M, Sanyal AJ (2018) Mechanisms of NAFLD development and therapeutic strategies. Nat Med 24:908–922
3. Kisseleva T, Brenner D (2021) Molecular and cellular mechanisms of liver fibrosis and its regression. Nat Rev Gastroenterol Hepatol 18:151–166
4. Kendall TJ et al (2009) p75 Neurotrophin receptor signaling regulates hepatic myofibroblast proliferation and apoptosis in recovery from rodent liver fibrosis. Hepatology 49: 901–910
5. Sachs BD et al (2007) p75 neurotrophin receptor regulates tissue fibrosis through inhibition of plasminogen activation via a PDE4/cAMP/PKA pathway. J Cell Biol 177:1119–1132
6. Friedman SL, Roll FJ, Boyles J, Bissell DM (1985) Hepatic lipocytes: the principal collagen-producing cells of normal rat liver. Proc Natl Acad Sci USA 82:8681–8685
7. Bataller R, Brenner DA (2005) Liver fibrosis. J Clin Invest 115:209–218
8. Naugler WE et al (2007) Gender disparity in liver cancer due to sex differences in MyD88-dependent IL-6 production. Science 317:121–124
9. Gu FM et al (2011) IL-17 induces AKT-dependent IL-6/JAK2/STAT3 activation and tumor progression in hepatocellular carcinoma. Mol Cancer 10:150
10. Li J et al (2011) Interleukin 17A promotes hepatocellular carcinoma metastasis via NF-kB

induced matrix metalloproteinases 2 and 9 expression. PLoS One 6:e21816

11. Di Rosa M, Malaguarnera L (2012) Genetic variants in candidate genes influencing NAFLD progression. J Mol Med (Berl) 90: 105–118
12. Generon (Shanghai) Corporation Ltd (2012) Generon initiates a phase I clinical study for F-652 in Australia. Business Wire, A Berkshire Hathaway
13. Ichikawa S, Mucida D, Tyznik AJ, Kronenberg M, Cheroutre H (2011) Hepatic stellate cells function as regulatory bystanders. J Immunol 186:5549–5555
14. Saito Y, Morine Y, Shimada M (2017) Mechanism of impairment on liver regeneration in elderly patients: role of hepatic stellate cell function. Hepatol Res 47:505–513
15. Lee JS, Semela D, Iredale J, Shah VH (2007) Sinusoidal remodeling and angiogenesis: a new function for the liver-specific pericyte? Hepatology 45:817–825
16. Hernandez-Gea V, Friedman SL (2011) Pathogenesis of liver fibrosis. Annu Rev Pathol 6: 425–456
17. Shang L, Hosseini M, Liu X, Kisseleva T, Brenner DA (2018) Human hepatic stellate cell isolation and characterization. J Gastroenterol 53:6–17
18. Knook DL, Seffelaar AM, de Leeuw AM (1982) Fat-storing cells of the rat liver. Their isolation and purification. Exp Cell Res 139: 468–471
19. Mederacke I, Dapito DH, Affo S, Uchinami H, Schwabe RF (2015) High-yield and high-purity isolation of hepatic stellate cells from normal and fibrotic mouse livers. Nat Protoc 10:305–315
20. Xu L et al (2005) Human hepatic stellate cell lines, LX-1 and LX-2: new tools for analysis of hepatic fibrosis. Gut 54:142–151
21. Schnabl B, Purbeck CA, Choi YH, Hagedorn CH, Brenner D (2003) Replicative senescence of activated human hepatic stellate cells is accompanied by a pronounced inflammatory but less fibrogenic phenotype. Hepatology 37:653–664
22. Herrmann J, Gressner AM, Weiskirchen R (2007) Immortal hepatic stellate cell lines: useful tools to study hepatic stellate cell biology and function? J Cell Mol Med 11:704–722
23. Coll M et al (2018) Generation of hepatic stellate cells from human pluripotent stem cells enables in vitro modeling of liver fibrosis. Cell Stem Cell 23(101–113):e107
24. Baeza-Raja B et al (2020) Pharmacological inhibition of P2RX7 ameliorates liver injury by reducing inflammation and fibrosis. PLoS One 15:e0234038
25. Chinnadurai R et al (2019) Molecular genetic and immune functional responses distinguish bone marrow mesenchymal stromal cells from hepatic stellate cells. Stem Cells 37:1075–1082
26. Liu X et al (2020) Primary alcohol-activated human and mouse hepatic stellate cells share similarities in gene-expression profiles. Hepatol Commun 4:606–626
27. Liu X et al (2020) Identification of lineage-specific transcription factors that prevent activation of hepatic stellate cells and promote fibrosis resolution. Gastroenterology 158(6): 1728–1744.e14
28. Povero D et al (2019) Human induced pluripotent stem cell-derived extracellular vesicles reduce hepatic stellate cell activation and liver fibrosis. JCI Insight 5(14):e125652
29. Drinane MC et al (2017) Synectin promotes fibrogenesis by regulating PDGFR isoforms through distinct mechanisms. JCI Insight 2(24):e92821
30. Xu J et al (2017) The role of human cytochrome P450 2E1 in liver inflammation and fibrosis. Hepatol Commun 1:1043–1057
31. Baglieri J, Brenner DA, Kisseleva T (2019) The role of fibrosis and liver-associated fibroblasts in the pathogenesis of hepatocellular carcinoma. Int J Mol Sci 20(7):1723
32. Rady B et al (2021) PNPLA3 downregulation exacerbates the fibrotic response in human hepatic stellate cells. PLoS One 16:e0260721
33. Kegel V et al (2016) Protocol for isolation of primary human hepatocytes and corresponding major populations of non-parenchymal liver cells. J Vis Exp 109:e53069
34. Kleiner DE et al (2005) Design and validation of a histological scoring system for nonalcoholic fatty liver disease. Hepatology 41:1313–1321
35. Romeo S et al (2008) Genetic variation in PNPLA3 confers susceptibility to nonalcoholic fatty liver disease. Nat Genet 40:1461–1465
36. Seki E et al (2007) TLR4 enhances TGF-beta signaling and hepatic fibrosis. Nat Med 13: 1324–1332

Chapter 14

Decellularization of the Human Liver to Generate Native Extracellular Matrix for Use in Automated Functional Assays with Stellate Cells

Emma L. Shepherd, Ellie Northall, Pantelitsa Papakyriacou, Karolina Safranska, Karen K. Sorensen, and Patricia F. Lalor

Abstract

With the incidence of liver disease on the rise globally, increasing numbers of patients are presenting with advanced hepatic fibrosis and significant mortality risk. The demand far outstrips possible transplantation capacities, and thus there is an intense drive to develop new pharmacological therapies that stall or reverse liver scarring. Recent late-stage failures of lead compounds have highlighted the challenges of resolving fibrosis, which has developed and stabilized over many years and varies in nature and composition from individual to individual. Hence, preclinical tools are being developed in both the hepatology and tissue engineering communities to elucidate the nature, composition, and cellular interactions of the hepatic extracellular niche in health and disease. In this protocol, we describe strategies for decellularizing cirrhotic and healthy human liver specimens and show how these can be used in simple functional assays to detect the impact on stellate cell function. Our simple, small-scale approach is translatable to diverse lab settings and generates cell-free materials which could be used for a variety of in vitro analyses as well as a scaffold for repopulating with key hepatic cell populations.

Key words Cirrhosis, Human, Liver, ECM, Decellularization, Cell adhesion, Migration

1 Introduction

Hepatic fibrosis remains an enormous and increasing clinical challenge with patients who present with advanced disease having limited, non-transplant therapeutic options [3]. Accumulation and breakdown of matrix within the liver is an incremental process, with development of significant fibrosis and cirrhosis typically taking many years. Responses vary among individuals [2], but the deposited matrix provides cues for cell orientation, migration, and differentiated function [12]. Hepatic stellate cells (HSC) are highly responsive to their external environment, which tunes their behavior to drive fibrogenesis or extracellular matrix turnover as

Ralf Weiskirchen and Scott L. Friedman (eds.), *Hepatic Stellate Cells: Methods and Protocols*, Methods in Molecular Biology, vol. 2669, https://doi.org/10.1007/978-1-0716-3207-9_14,

appropriate. While many studies describe the impact of ECM on functions of hepatocytes [10] and cholangiocytes [15], studies on HSC biology are limited. Comparative studies of pancreatic stellate cells confirm that differences in matrix composition that accompany fibrosis have a profound impact on function and gene expression [1]. Few studies have compared the impact of a complex and three dimensional, cross-linked fibrotic liver ECM, of the type that would be present in a human with established fibrosis on hepatic stellate cell biology. This unmet need is stalling development of new therapies, with many compounds identified in rodent models or preclinical in vitro systems subsequently performing poorly in large scale human trials. The tissue engineering field has provided solutions using traditional matrices and hydrogels [9], but often the models are either too complex or undefined or lack the minimal matrix constituents to recreate advanced disease. Decellularized extracellular matrix (dECM) from native tissue retains structural and biochemical characteristics and is being widely explored for tissue engineering applications [15], albeit with the challenge that large amounts of tissue are required to generate sufficient cells and matrix for repopulation studies. Published procedures are available describing cannulation and perfusion approaches with entire human liver lobes or intact rodent or pig livers. This is clearly not amenable to all lab settings/surgical samples; therefore, we have generated a protocol that works with smaller samples. Generation of small amounts of donor-specific material from different diseased livers provides opportunities to study matrix remodeling, interactions between HSC and the matrix that may influence metabolism, proliferation or senescence/apoptosis, and the impact of the environment on cell function. It also opens the potential for proteomic/matrisomal analysis to identify factors which may limit HSC proliferation or drive tissue regeneration or to determine the impact of candidate therapeutic models in a preclinical setting. In this chapter, we describe our approach for generation of dECM from donor and cirrhotic human tissue specimens and demonstrate how this can be used to test the impact of potential inhibitory molecules on cell function.

2 Materials

Below we describe the materials and protocols we use for generating decellularized ECM from human tissue and its use in functional assays. It is important to consider the relevant local regulation around using human biomaterials and data for such studies. In our program, samples are sourced from patients attending the Liver Unit at the Queen Elizabeth Hospital in Birmingham, UK, and all patients provide written informed consent for use of their anonymized data and samples. We also have local ethical review board permission for all studies and adhere to local policies on data and tissue storage and use. All procedures involving intact human tissue and cells are performed in a Class 2 Safety cabinet using sterile plastics, glassware, and instruments. All liquid waste that had been in contact with human biomaterials was decontaminated using Virkon before disposal.

2.1 Decellularization Reagents and Equipment

1. Benchtop orbital shaker (e.g., Cole-Parmer slow speed orbital shaker).
2. Sterile 500 mL glass beakers; autoclavable metal mesh cell strainer, glass, or stirring rod.
3. Sterile distilled water (dH_2O).
4. Decellularization reagent media: 3% sodium deoxycholate, 1% sodium dodecyl sulfate, 0.6% Triton X-100, 4.3% sodium chloride, and 0.005% trypsin-EDTA.
5. Hypertonic saline (9% solution sodium chloride).
6. 1% sodium dodecyl sulfate (1% SDS) in sterile PBS.
7. Detergent mix (0.1% (w/v) Triton X-100).
8. (Optional) Fat digestion solution (99% isopropanol).
9. (Optional) Tissue bleaching and sterilization solution sterile PBS containing 0.1% (v/v) peracetic acid and 4% (v/v) ethanol.
10. (Optional) Antibiotic storage solution: Premix sterile PBS containing 2% penicillin-streptomycin, 10 μg/mL gentamicin, and 2.5 μg/mL amphotericin B.

2.2 Lyophilization Reagents

1. Dounce homogenizer.
2. Auto freeze dryer (e.g., ScanVac CoolSafe).
3. Acetic acid solution 0.5 M acetic acid in PBS (supplemented with 3 mg/mL pepsin [5, 6]).
4. pH adjustment solution (pH of the liver dECM solution was adjusted to 7.4 with 10 M sodium hydroxide).
5. Plastic petri dishes; 15 cm sterile dishes.
6. Sterile syringe filters.

2.3 Cell Culture Reagents

1. Collagenase solution (type-1A collagenase) (0.4 μg/mL final concentration).
2. Sterile cell sieves.
3. Percoll gradients. Percoll made up in PBS to final concentrations of 33% and 77% (v/v).
4. Sterile PBS.
5. Immunomagnetic selection reagents (the antibody target varies depending on nature of cell we wish to select. We use mouse anti human-Epcam-1 antibody (HEA125, 10 μg/mL stock)

and Pan mouse IgG Dynabeads or anti-human CD31 antibody-coated dynabeads).

6. Magnet for immunomagnetic selection (we use DynaMag).
7. Complete media (medium (DMEM), high glucose supplemented with 2% fetal bovine serum (FBS), L-glutamine, and penicillin/streptomycin).
8. Cell dissociation reagent – TryPLE trypsin.
9. Wash solution PBS/0.1% BSA.
10. Gelatin solution (2% bovine skin gelatin made up in PBS).

2.4 Adhesion Assay Reagents

1. 96 well plates.
2. High content imaging platform – we use CellInsight™ CX5 High Content Screening (HCS) Platform and integrated HCS Studio™ Cell Analysis Software.
3. Cell label for quantitation (e.g., crystal violet or CellTracker dyes).

3 Methods

3.1 Decellularization of Human Liver Tissue Pieces

1. Set up incubation vessel in a cold room or on ice. This involves suspending a sterile cell strainer below a glass stirring rod over a beaker which can contain the incubation reagents (see Fig. 1b). This should be placed on the plate of an orbital shaker.
2. Place fresh tissue sample onto a sterile petri dish and cut into cube-shaped chunks approximately 0.5 cm^3 (Fig. 1b; *see* **Note 1**).
3. Weigh the tissue and place each sample into an Eppendorf tube (*see* **Notes 2** and **3**).
4. Add 1 mL dH_2O and freeze sample in a −80 °C freezer for at least an hour to rupture cells.
5. Remove the samples from the freezer and thaw, place the tissue samples inside the cell strainer, and immerse the sieve in fresh dH_2O (500 mL) in the beaker. Mix the samples on the platform mixer for 10 min (Fig. 1).
6. Repeat the mixing with fresh dH_2O twice more.
7. Replace the water with the decellularization reagent media, and mix again for a further 30 min.
8. Discard the buffer and replace with hypertonic saline solution. Mix for 10 min, discard, and repeat with fresh buffer for a further 10 min.
9. Replace the buffer with fresh dH_2O and mix for 10 min.

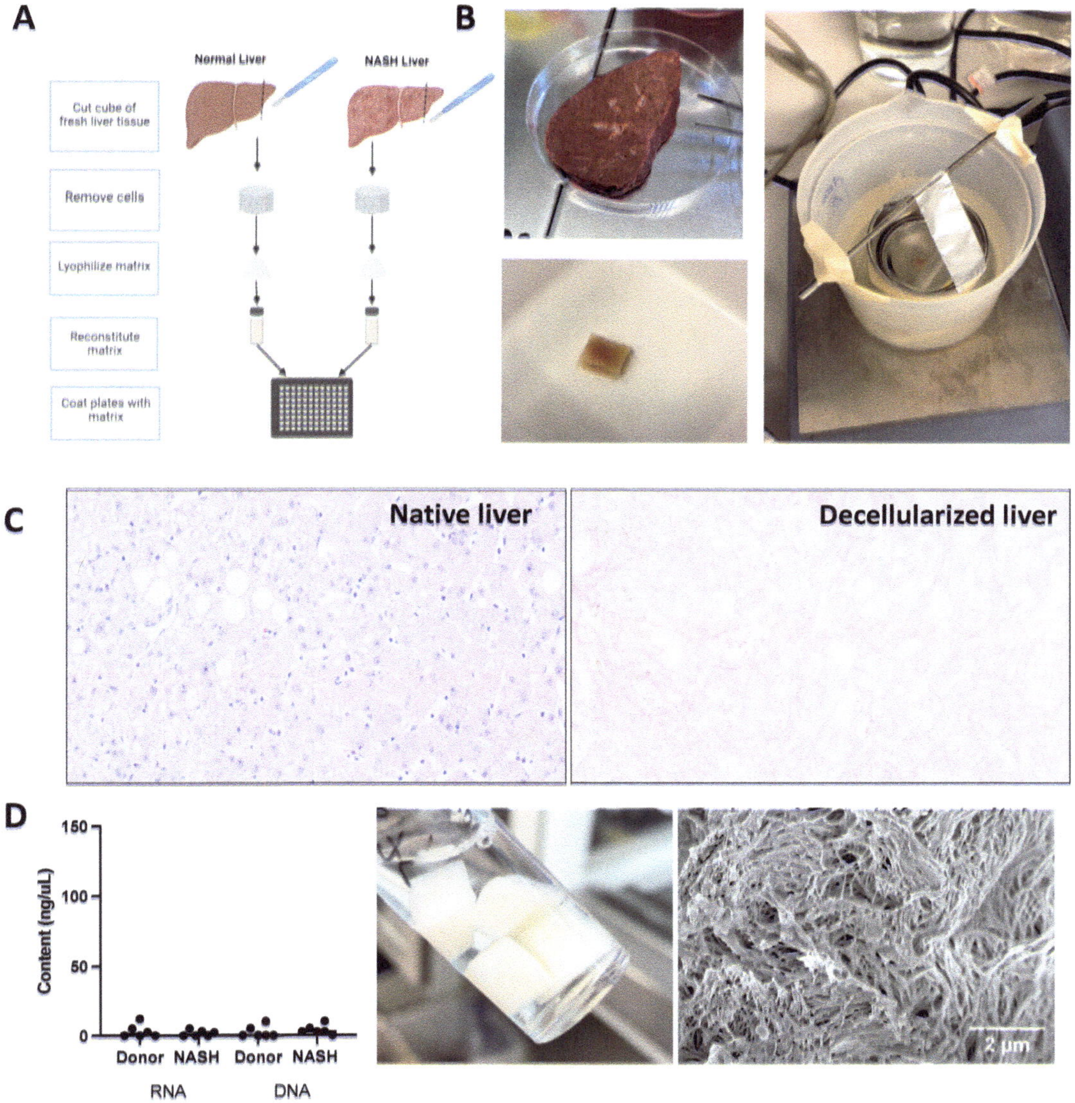

Fig. 1 Description and validation of our dECM protocols. Schematic of our experimental workflow (**a**) which utilizes chunks of fresh tissue (approx. 0.5 cm^3); panel **b**) cut from slices of explanted human liver tissue (top image, panel **b**). These are immersed in sequential detergents, cell lysis solutions, and wash reagents by suspension on a cell strainer over a media reservoir (panel **b**, right image) on a mixing platform. Representative hematoxylin and eosin staining of a liver sample at the start of the procedure (**c**, left), and Alcian blue stain of dECM sample (**c**, right panel) confirms removal of cells with preservation of ECM architecture. RNA and DNA quantitation of six samples of healthy donor and NASH cirrhotic tissue confirms removal of cell material (**d**, left graph) from the dECM chunks (see examples in center image, panel **d**). Scanning electron microscopy (**d**, right panel) also shows preservation of mature collagen fibrils and absence of cells in final samples

10. Replace the water with the decellularization reagent media and mix again overnight at 4 °C.

11. Replace the buffer with fresh dH_2O and mix for 10 min. Repeat this step twice more.
12. Replace the water with 1% SDS and mix overnight at 4 °C.
13. Wash as per **step 11** (3× dH_20) and then replace solution with detergent mix. Agitate overnight as before at 4 °C.
14. Finally, wash thoroughly in three changes of PBS for 10 min each.
15. dECM can then be processed or stored for downstream applications (*see* **Note 4**).

3.2 Preparation of dECM Product for Cell Culture

1. Sterilize dECM chunks in peracetic acid by immersion for 2 h at 4 °C in an Eppendorf tube (*see* **Note 5** [11]).
2. Snap freeze dECM cube either by immersion in liquid nitrogen or with a cryospray.
3. Lyophilize the sample by freeze drying under vacuum.
4. Grind the sample into a fine powder using a Dounce homogenizer.
5. Solubilize the powder in 0.1–0.5 M acetic acid and sterile filter (*see* **Note 6**).
6. Centrifuge at 3000–3500 g for 10 min.
7. Neutralize pH if desired using 10 M sodium hydroxide.
8. Protein concentration can be determined at this point using your preferred method.
9. Aliquot and store at −80 °C.

3.3 Culture of Primary Human Liver Myofibroblasts

We are fortunate to have access to human liver tissue samples which we use to isolate hepatic liver myofibroblasts (activated stellate cells) using our published methodologies which are summarized below [4]. We have also used immortalized cells lines such as LX-2 (*see* **Note** 7). Either can then be used in the adhesion assay protocols.

1. To isolate hepatic myofibroblasts, a 30–50 g slice of liver (see typical image in Fig. 1b) was diced to a fine consistency in a sterile petri dish using sterile scalpels.
2. The tissue pieces were placed into a sterile glass beaker, and 20 ml of sterile PBS is added along with one aliquot of collagenase. This is then incubated with constant agitation at 37 °C for up to 40 min (*see* **Note 8**).
3. The tissue digest is then strained through a cell sieve into a sterile beaker to remove large debris/connective tissue. The sample is flushed through with an excess of sterile PBS (typically 100 mL) to ensure all cells are collected.

4. The sieved sample should be adjusted to 200 mL volume and distributed into eight sterile universal tubes. These should be spun at 800 g for 5 min.
5. The supernatant should be carefully decanted to waste and the cell pellets combined and reconstituted into 100 mL PBS. This should again be centrifuged at 800 g for 5 min.
6. **Step 5** should then be repeated, and the cell pellets should be made up to 24 mL in PBS.
7. During the final wash step above, Percoll gradients should be prepared by layering 3 mL of 33% solution above 3 mL of 77% Percoll in 8 × 15 mL centrifuge tubes.
8. Carefully layer 3 mL of cell suspension on top of each gradient, and transfer to a centrifuge without agitating the layers.
9. Set the brake on the centrifuge to zero and spin the samples at 800 g for 25 min.
10. Collect the non-parenchymal cells at the interface of the gradient solutions, and resuspend in 100 mL of PBS. Centrifuge at 800 g for 5 min to pellet cells. This wash step can be repeated twice to ensure debris and residual Percoll is removed.
11. Finally combine the cell pellets in a volume of 500 μL of ice-cold PBS to facilitate positive selection of cells for removal from the myofibroblasts.
12. To deplete cholangiocytes, add 50 μL of EpCAM antibody to the cells in a 15 mL centrifuge tube, and incubate at 37 °C for 30 min. It is helpful to shake the tube occasionally during this incubation.
13. Top the tube up to 15 mL with sterile PBS and centrifuge (800 × g/5 min).
14. Resuspend the pellet in 500 μL ice-cold PBS; add 10 μL washed pan mouse IgG Dynabeads. Incubate at 4 °C for 30 min with constant agitation.
15. Add 5 mL of ice-cold PBS, mix well, and place the tube in a magnet for 2 min. With the tube still in the magnet, carefully pipette off the supernatant containing unlabeled cells and transfer to fresh tube (*see* **Note 9**).
16. Centrifuge the unlabeled cells as before to generate a cell pellet.
17. Resuspend the pellet in 500 μL ice-cold PBS; add 10 μL washed anti-CD31 coated Dynabeads. Incubate at 4 °C for 30 min with constant agitation.
18. Repeat **step 15** to use the magnet to select the labeled CD31 positive endothelial cells. Keep the unlabeled flow through containing the myofibroblasts.
19. Wash the unlabeled cells in PBS as before and resuspend the cell pellet in complete media.
20. These cells can now be seeded into gelatin coated T25 tissue culture flasks and maintained in complete media prior to use

for functional assays. We typically use our primary cells up to passage 5 in culture.

3.4 High Content Cell Adhesion Assay

1. Adhesion of LX-2 cells or primary human hepatic myofibroblasts to matrix was assessed using 96 well plates coated with reconstituted decellularized matrix protein mix. This was prepared from either donor liver or cirrhotic liver (*see* **Note 10**).
2. Dilute dECM solution to 0.2 mg/mL in appropriate buffer (*see* **Note 11**), and incubate in 96 well plates for 2 h at 37 °C to allow coating.
3. Aspirate matrix proteins and wash wells twice with PBS/0.1% BSA prior to use.
4. Trypsinize cells from your culture plates, wash and count the cells, and resuspend to 10,000 cells per 100 μL in DMEM
5. Add 100 μL cells per well to coated plates, and leave to settle at room temperature for 20 min before transferring to 37 °C incubator for 2 h.
6. At this stage, any inhibitory molecules can be titrated into test wells at desired final concentration (see example in Fig. 2).
7. Aspirate off non adherent cells and wash wells gently with PBS twice.
8. Adherent cells were fixed by addition of 100 μL methanol and stained with desired contrast/detection agent (see Fig. 2).
9. Adherent cell counts can then be performed using a High Content Screening (HCS) Platform. In our case, a CellInsight™ CX5 was utilized to quantify cell number in an automated, randomized, and high-throughput manner.
10. We typically acquire images from nine visual fields per well, with triplicate wells per experimental condition. A Cell Health Profiling algorithm was applied whereby individual cells were detected, in background corrected images, based on thresholding of label positivity followed by object smoothing and segmentation. Cellular debris was excluded from the analysis based on object size, as were objects at the edges of each visual field to avoid duplication of cell counts. The mean object count across all nine visual fields was used to give an average cell count per well (see Fig. 2).

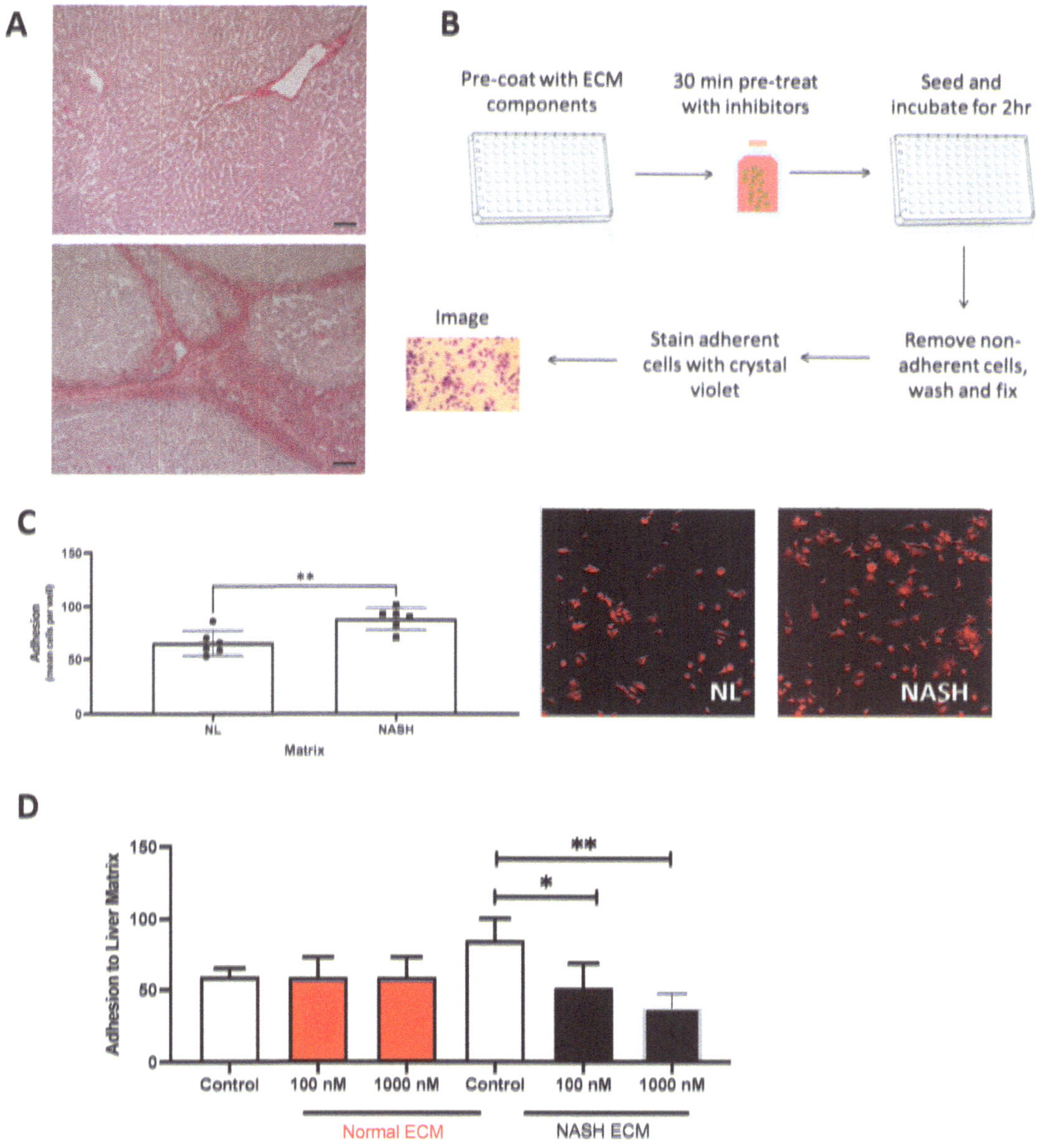

Fig. 2 dECM can be used as an adhesive substrate to test the impact of preclinical molecules of interest. dECM was generated from normal and cirrhotic livers (see example histological appearance in **a**) and used to coat 96 well plates as for adhesion assays as illustrated in schematic (**b**). Representative data shows that primary liver myofibroblasts bind in increased numbers to cirrhotic matrix (**c**, symbols represent individual donors, $p < 0.01$ paired t test). (**d**) Use of an $\alpha v \beta 1$ integrin inhibitor (C8, 100 uM) confirmed that adhesion of primary myofibroblasts but not LX-2 cells to NASH dECM was significantly reduced in the presence of inhibitor (Change versus DMSO vehicle control was assessed using ordinary one-way ANOVA with Dunnett's multiple comparisons test. $^{*}p < 0.05$, $^{**}p < 0.01$)

4 Notes

1. We typically begin our preparation with fresh tissue collected as soon as possible after surgery. We have successfully used both non-diseased donor tissue declined for transplantation and cirrhotic material collected as an explant during transplantation. However, the procedure also works perfectly well with tissue we have harvested and then immediately snap frozen in liquid nitrogen prior to long term storage at −80 °C. This has the advantage of allowing batch processing of material from multiple donors.
2. We find that approximately 125 mg of tissue works well using our decellularization protocols. However, if entire rodent livers are to be used, or decellularization reagents are to be perfused into the tissue via vascular access, then larger samples can be processed. We have noted that it is better to do multiple incubations with the reagent solution compared to increasing the incubation times without changing the solution (e.g., to do 3 × 15-min incubations rather than a single 45-min incubation), hence the multiple repeats in our final protocol.
3. Some authors [8, 13] suggest that isopropanol can be used as a means to enhance lipid extraction from fatty samples such as adipose tissue to improve the extent and efficacy of the decellularization procedure. We tested this approach by immersing out tissue in 5 ml isopropanol and shaking for 30 min, prior to centrifugation at 1800 g × 10 min. This did yield a considerable fatty fraction which could be aspirated from the top of the tube. However, this did not completely remove all lipids, and it also "fixed" the tissue, which extended the remaining period required in decellularization buffer significantly. Thus, we have not persisted with this approach, but it may be useful for rodent samples or tissues other than the liver.
4. We vary our subsequent processing dependent upon our downstream application. We tend to snap freeze chunks of dECM in liquid nitrogen or using cryospray for long term storage. We also keep a sample in RLT buffer for subsequent RNA/DNA quantitation to confirm efficacy of decellularization. Similarly, we place some into RIPA buffer for blotting. However, if the ambition is to reseed the intact matrix with cells, then it can be stored in PBS containing antibiotics at 4 °C prior to use and flushed with PBS just before cells are added.
5. Authors using rodent dECM note that the material can be bleached with peracetic acid [11] to normalize the color after decellularization and to sterilize the material. This did not work

very well in our experience, as adult human livers in that tissue remained opaque in many cases. However, the suggestion to sterilize material for long term storage is sensible.

6. We find that it is necessary to mix the sample thoroughly to facilitate dissolution. The concentration of acetic acid used will vary depending on the size of your sample and ultimate ECM concentration. We find a concentration of 0.1 M works well for protein in the range of 2 mg/mL.
7. It is possible to use the immortalized stellate cell line LX-2 for these assays. Our LX-2 cells (a kind gift of Scott Friedman [16]) were cultured in Dulbecco's Modified Eagle Medium (DMEM), high glucose supplemented with 2% fetal bovine serum (FBS), L-glutamine, and penicillin/streptomycin as previously described [14].
8. The incubation period required to achieve good digestion of the human liver varies with the nature of the tissue. Typically, a healthy donor liver takes less time than a very cirrhotic donor. We tend to judge by eye to maximize tissue breakdown while aiming to minimize damage to individual cells. Typically, we incubate for between 15 and 30 min.
9. Cells that are positively selected using this protocol (i.e., cholangiocytes and liver sinusoidal endothelial cells) can be collected and plated into collagen-coated flasks for other experiments as desired [7].
10. We have used different varieties of cirrhotic tissue explanted during transplantation for the purpose of generating dECM. In the current protocols, we have shown examples from patients with nonalcoholic steatohepatitis (NASH), but the methods work regardless of etiology.
11. The diluent used for generating tissue culture coating and the final concentration of matrix used can vary according to your application, with the literature suggesting use of 30% ethanol, up to 2% acetic acid, and sterile water, for example. Since we solubilized our original protein into acetic acid, we reconstituted in sterile water to coat our plates.

Financial Support

This study includes independent research supported by the Birmingham National Institute for Health Research (NIHR), Birmingham Biomedical Research Centre, based at the University of Birmingham. The views expressed are those of the authors and not necessarily those of the NHS, the National Institute of Health Research, or the Department of Health and Social Care.

Acknowledgments

We are grateful to the physicians and patients at the Queen Elizabeth Hospital in Birmingham who donated tissues for our investigations.

References

1. Friedman SL, Pinzani M (2022) Hepatic fibrosis 2022: unmet needs and a blueprint for the future. Hepatology 75:473–488
2. Decaris ML, Emson CL, Li K et al (2015) Turnover rates of hepatic collagen and circulating collagen-associated proteins in humans with chronic liver disease. PLoS One 10: e0123311
3. Uygun BE, Soto-Gutierrez A, Yagi H et al (2010) Organ reengineering through development of a transplantable recellularized liver graft using decellularized liver matrix. Nat Med 16:814–820
4. Thanapirom K, Caon E, Papatheodoridi M et al (2021) Optimization and validation of a novel three-dimensional co-culture system in decellularized human liver scaffold for the study of liver fibrosis and cancer. Cancers (Basel) 13(19):4936
5. Willemse J, Van Tienderen G, Van Hengel E et al (2022) Hydrogels derived from decellularized liver tissue support the growth and differentiation of cholangiocyte organoids. Biomaterials 284:121473
6. Apte MV, Yang L, Phillips PA et al (2013) Extracellular matrix composition significantly influences pancreatic stellate cell gene expression pattern: role of transgelin in PSC function. Am J Physiol Gastrointest Liver Physiol 305: G408–G417
7. Sorrentino G, Rezakhani S, Yildiz E et al (2020) Mechano-modulatory synthetic niches for liver organoid derivation. Nat Commun 11: 3416
8. Lee H, Han W, Kim H et al (2017) Development of liver decellularized extracellular matrix bioink for three-dimensional cell printing-based liver tissue engineering. Biomacromolecules 18:1229–1237
9. Saldin LT, Cramer MC, Velankar SS et al (2017) Extracellular matrix hydrogels from decellularized tissues: structure and function. Acta Biomater 49:1–15
10. Uygun BE, Price G, Saedi N et al (2011) Decellularization and recellularization of whole livers. J Vis Exp 48:2394
11. Holt AP, Haughton EL, Lalor PF et al (2009) Liver myofibroblasts regulate infiltration and positioning of lymphocytes in human liver. Gastroenterology 136:705–714
12. Wang L, Johnson JA, Zhang Q et al (2013) Combining decellularized human adipose tissue extracellular matrix and adipose-derived stem cells for adipose tissue engineering. Acta Biomater 9:8921–8931
13. Song M, Liu Y, Hui L (2018) Preparation and characterization of acellular adipose tissue matrix using a combination of physical and chemical treatments. Mol Med Rep 17:138–146
14. Xu J, Lee G, Wang H et al (2004) Limited role for CXC chemokines in the pathogenesis of alpha-naphthylisothiocyanate-induced liver injury. Am J Physiol Gastrointest Liver Physiol 287:G734–G741
15. Weston CJ, Shepherd EL, Claridge LC et al (2015) Vascular adhesion protein-1 promotes liver inflammation and drives hepatic fibrosis. J Clin Invest 125:501–520
16. Shepherd EL, Saborano R, Northall E et al (2021) Ketohexokinase inhibition improves NASH by reducing fructose-induced steatosis and fibrogenesis. JHEP Rep 3:100217

Chapter 15

Multiplex Immunostaining to Spatially Resolve the Cellular Landscape in Human and Mouse Livers

Adrien Guillot, Marlene Sophia Kohlhepp, and Frank Tacke

Abstract

Histological techniques based on tissue colorations (e.g., hematoxylin-eosin, Sirius red) and immunostaining remain gold standard methodologies for diagnostic or phenotyping purposes in liver disease research and clinical hepatology. With the development of -omics technologies, greater information can be extracted from tissue sections. We describe a sequential immunostaining protocol consisting of repetitive cycles of immunostaining and chemically induced antibody stripping that can be readily applied to various formalin-fixed tissues (liver or other organs, mouse or human) and does not require specific equipment or commercial kits. Importantly, the combination of antibodies can be adapted according to specific clinical or scientific needs.

Key words Multiplex immunostaining, Antibodies, Tissue section, Immune microenvironment, Imaging cytometry, Liver fibrosis, Antibody stripping

1 Introduction

With the emergence of large scale, unbiased, and multidimensional -omics analytical methods applied to biomedical research in recent years, we have considerably expanded our understanding of cellular functions and tissue organization. These studies establish that the cellular microenvironment plays a critical role in liver homeostasis and at all stages of liver disease. The liver parenchymal cells (hepatocytes, cholangiocytes) are spatially organized to allow communication with non-parenchymal cells – hepatic stellate cells, Kupffer cells (liver resident macrophages), sinusoidal endothelial cells – as well as with bypassing or immune cells being recruited to the liver [1]. Thus, it is critical to characterize each of the different cell populations – e.g., a quiescent or activated hepatic stellate cell – in their spatial context within the periportal or perivenular regions or in a fibrotic scar [2].

Ralf Weiskirchen and Scott L. Friedman (eds.), *Hepatic Stellate Cells: Methods and Protocols*, Methods in Molecular Biology, vol. 2669, https://doi.org/10.1007/978-1-0716-3207-9_15,

A gold standard method for decades in pathology relies on tissue fixation, sectioning, and coloration. As such, formalin-fixed paraffin embedded (FFPE) tissue samples represent a very widespread source of biomaterials for research or diagnostic purposes. FFPE tissues can be generated at a relatively low cost, easily stored long term, or shipped between different institutes at room temperature. This processing method is the best currently available for the preservation of the tissue organization while also preserving protein antigenicity, which is the basis for antibody-based staining (immunostaining). There are a number of platforms that enable multiplex immunostaining (e.g., imaging mass cytometry, barcode-conjugated antibodies), but these approaches require specific training and specialized and costly instrumentation and generally also rely on expensive commercial kits [3]. Those limitations prevent most laboratories from performing multiplex (i.e., >6 targets of interest) immunostaining on a large scale or for exploratory purposes.

Alternative methods have been developed to allow for cost-effective and rapid, yet comprehensive, analysis of cellular cross talk. Here, we describe a protocol for multiplex immunostaining (10–20 markers) we have recently implemented in our routine laboratory techniques, which is optimized for the study of liver tissue sections. This protocol consists of sequential cycles of immunostaining, imaging, and antibody stripping followed by image processing (Fig. 1). Importantly, this staining method relies on reagents widely available in laboratories performing classical immunostaining (Fig. 2), does not need any specific equipment, and requires very limited personnel training. The principle of the antibody stripping method used here was originally described elsewhere [4–6]. We optimized the immunostaining protocol, antibody panels, and bioinformatics for the study of the liver microenvironment and applied it to other organs as well. This is rendered possible by the tremendous flexibility of this protocol, since research question-specific antibodies may be added to a standard antibody panel (e.g., allowing for the visualization of the major liver cell types) at any point. As a result, this approach helped us to address a variety of research questions in human and mouse liver [7–11] and in mouse lung [12, 13], for example.

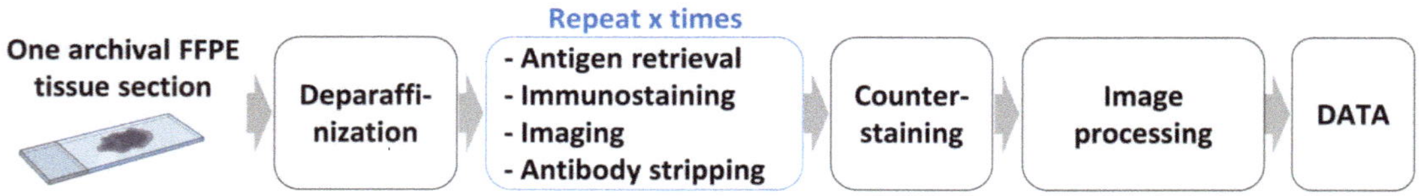

Fig. 1 Multiplex immunostaining workflow. This figure presents the successive steps of a multiplex immunostaining experiment

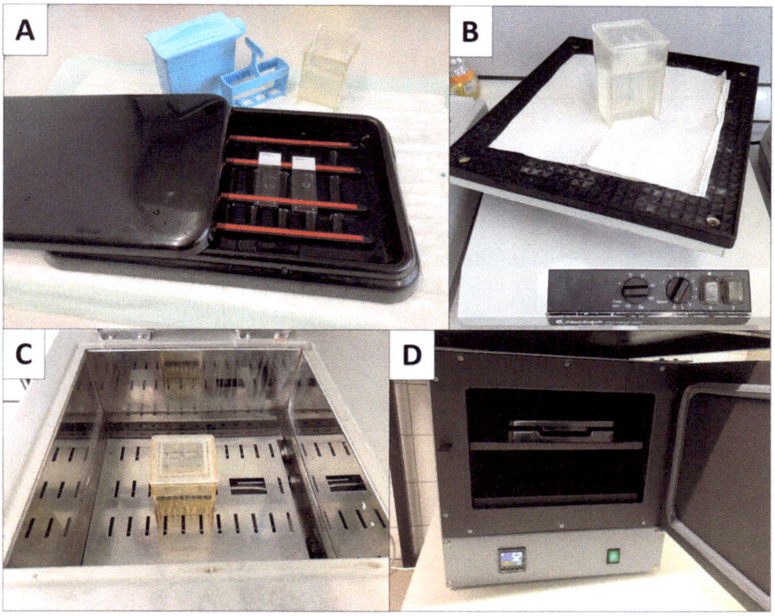

Fig. 2 Equipment needed for performing multiplex immunostaining. This protocol requires common laboratory equipment for immunohistochemistry such as (**a**) an opaque slide incubation chamber and plastic jars with slide holders, (**b**) an orbital shaker, (**c**) a water bath, and (**d**) an incubator

2 Materials

2.1 Equipment

1. Positively charged SuperFrost Plus Adhesion slides.
2. Cover glasses.
3. Opaque staining chamber.
4. Plastic slide jars (*see* **Note 1**).
5. Water bath (set at +98 °C).
6. Orbital shaker (*see* **Note 2**).
7. Slide incubator (set at +56 °C).

2.2 Reagents

1. De-paraffinization reagents: xylene, ethanol gradients (96%, 80%, 70%, 50%).
2. Antigen retrieval solution (*see* **Note 3**): Monosan HIER Citrate Buffer pH 6.0 or Tris-EDTA buffer pH 9.0. Dilute to 1× in deionized water prior to starting a new immunostaining series (*see* **Note 4**).
3. Image-iT FX Signal Enhancer (Cell Signaling).
4. Blocking solution: 1×-PBS with 2% normal goat serum (*see* **Note 5**).

Table 1
Primary antibodies used in Figs. 3 and 4 for mouse liver

Antigen	Cell type	Host species	Manufacturer
α-SMA	Myofibroblasts	Mouse	Agilent
CCR2	Recruited myeloid cells	Rabbit	Abcam
CD3	T lymphocytes	Rabbit	Abcam
CD11b	Myeloid cells	Rabbit	Abcam
CD45R	B lymphocytes	Rat	BioLegend
CK19	Ductular cells	Rat	Developmental Studies Hybridoma Bank. TROMA-III was deposited to the DSHB by Kemler, R. (DSHB Hybridoma Product TROMA-III)
CLEC4F	Kupffer cells	Rat	R&D Systems
HepPar1	Hepatocytes	Mouse	Dako
HNF1β	Ductular cells	Rabbit	Abcam
IBA1	Macrophages	Rabbit	VWR
MPO	Granulocyte neutrophils	Rabbit	Abcam
NA/K ATPase	Cell membranes	Mouse	Abcam
PCNA	Proliferation	Mouse	Abcam
PDGF-Rβ	Fibroblasts	Rabbit	Abcam

Table 2
Secondary antibodies used in Figs. 3 and 4

Target species	Fluorophore (Alexa Fluor)	Manufacturer
Mouse	488 / 555 / 647	Cell Signaling
Rat	647	Cell Signaling
Rabbit	647	Cell Signaling
Rabbit	750	ThermoFisher

5. Antibody dilution buffer: 1×-PBS supplemented with 1% bovine serum albumin.
6. Primary antibodies (*see* **Note 6**) prepared in the antibody dilution buffer (Table 1).
7. Secondary antibodies (*see* **Note 7**) prepared in the antibody dilution buffer (Table 2).

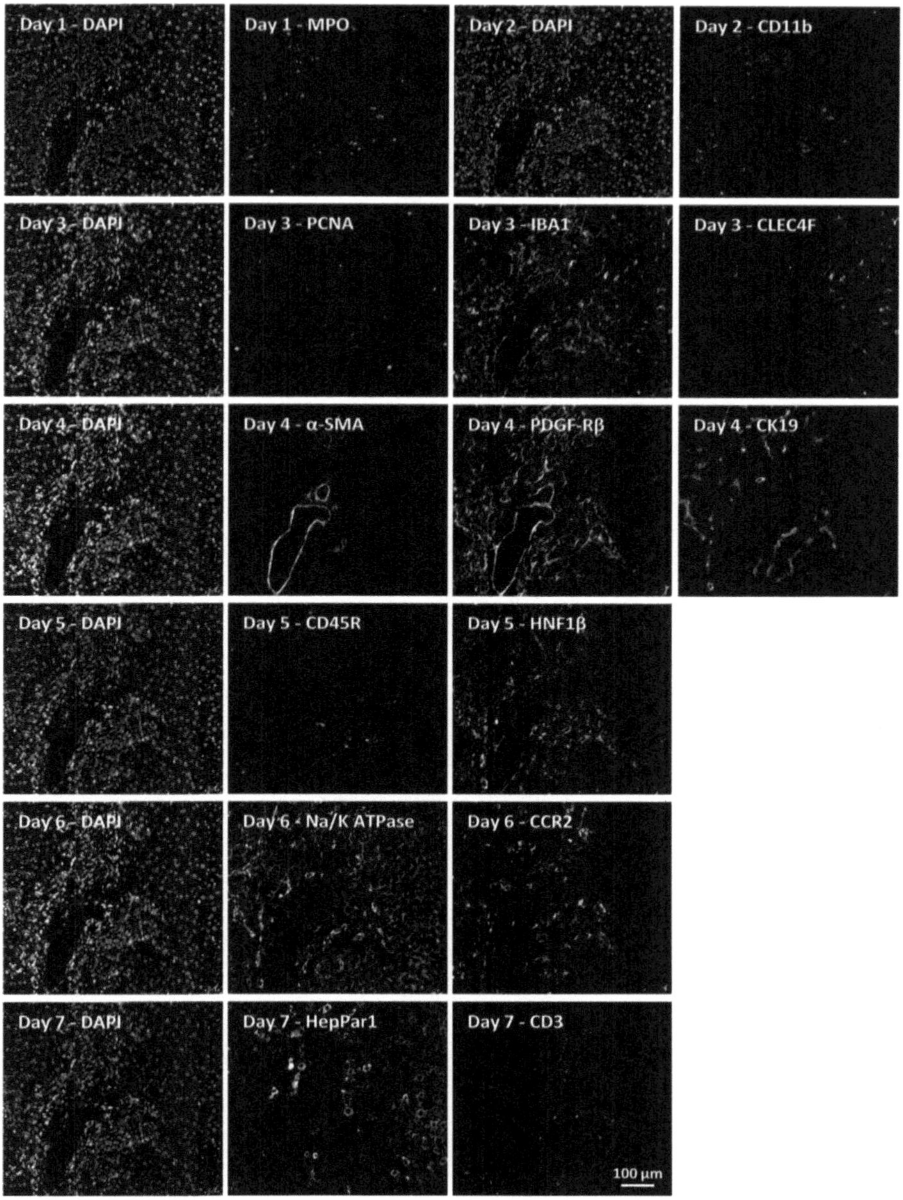

Fig. 3 Typical results after successive days of immunostaining. A formalin-fixed paraffin embedded (FFPE) mouse liver tissue section was subjected to multiplex immunostaining. Each panel displays the acquired image of the indicated day, from the same field of view, after registration and background subtraction. The tissue section was imaged for 7 consecutive days with a total of 14 primary antibodies. The images reveal intense ductular reaction with myeloid cell-driven inflammation and portal fibrosis, in the liver of a 24-week-old *Mdr2*-deficient female mouse

8. PBS-T: 1×-PBS + 0.1% (v/v) Tween-20.
9. 430 ng/mL 4′,6-diamidino-2-phenylindole (DAPI) working solution prepared in 1×-PBS.

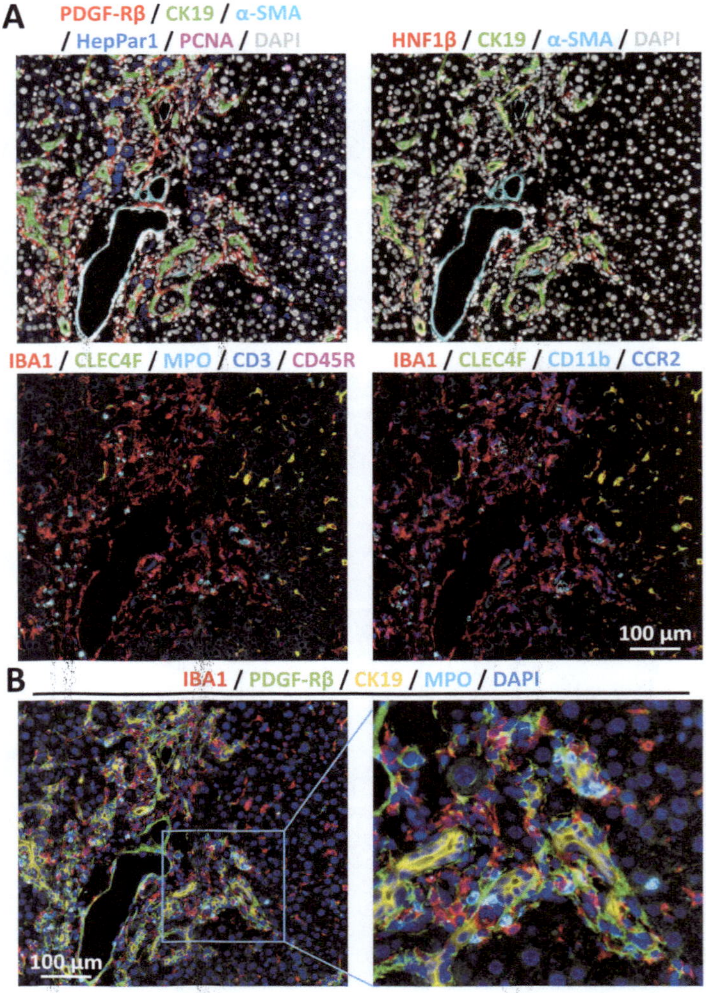

Fig. 4 Typical multiplex immunostaining images reveal complex cellular landscapes. (**a, b**) Singular channel images from Fig. 3 were colored and merged. The enlarged view on panel B demonstrates close localization of macrophages (IBA1$^+$ cells), fibroblasts (PDGF-Rβ^+), ductular cells (CK19$^+$), and neutrophils (MPO$^+$), suggestive of intense cellular interactions between those cell populations

10. Mounting medium (*see* **Note 8**): VectaMount AQ Aqueous Mounting Medium (Vector Laboratories).
11. Antibody stripping buffer: 675 μL distilled water +125 μL 0.5 M Tris-HCL pH 6.8 + 200 μL 10% (w/v) sodium dodecyl sulfate.
12. Counterstaining: Trichrome Stain Kit (Connective Tissue Stain).

3 Methods

3.1 Tissue Section Preparation

1. Tissue fixation and embedding: Tissues should be fixed in 4% formalin and embedded in paraffin according to standard protocols (*see* **Note 9**). Tissue blocks and tissue sections should be stored at +4 °C, to preserve tissue and antigen properties.
2. Tissue sections are cut at a thickness of 2–4 μm and used within a few months for immunostaining. Positively charged microscope slides are used to electrostatically fix the tissue sections to the slide.

3.2 First Cycle

1. Collect the tissue sections to be processed through the protocol, and let them dry overnight on the bench or >1 h under airflow at room temperature (*see* **Note 10**).
2. Pre-warm the antigen retrieval solution in a plastic jar placed in a water bath, set at +98 °C (*see* **Note 11**).
3. Deparaffinize the tissue sections by immersing into two consecutive baths of xylene for at least 5 min each.
4. Rehydrate the tissues by successive immersions for 5 min in 96% ethanol, 80% ethanol, 70% ethanol, and 50% ethanol. Finally, immerse the slides in deionized water, and rinse well to remove all residual traces of ethanol before going to the next step.
5. Transfer the slides to the pre-warmed antigen retrieval solution. Incubate for 20 min in the water bath at +98 °C, followed by a 30-min cooldown on the bench.
6. Rinse the tissue sections in 1×-PBS, 3× 2 min.
7. Draw a hydrophobic barrier around the tissue sections.
8. Apply one to two drops of Image-iT FX Signal Enhancer on each tissue section, and incubate for 30 min at room temperature.
9. Rinse the tissue sections in 1×-PBS, 3× 2 min.
10. Apply the blocking solution to the tissue sections, and incubate for 1 h at room temperature.
11. Rinse the tissue sections in 1×-PBS, 3× 2 min.
12. Apply the primary antibody solution and incubate overnight at +4 °C.
13. Rinse the tissue sections once in 1×-PBS-T for 5 min, followed by 1×-PBS, 3× 2 min.
14. Apply the secondary antibody solution and incubate for 1 h at room temperature. From this point, the tissue sections should not be exposed to direct light to minimize photobleaching.

15. Rinse the tissue sections once in 1×-PBS-T for 5 min, followed by 1×-PBS, 3× 2 min.
16. Incubate in DAPI solution for 3 min at room temperature.
17. Rinse the tissue sections in deionized water, 3× 2 min.
18. Bring the slides immersed in deionized water to the microscopy room. Three minutes prior to imaging, apply mounting medium and mount the slides with a cover glass.
19. Perform imaging (*see* **Note 12**). Representative results are shown in Fig. 3.
20. Immediately after imaging, immerse the slide in a jar containing deionized water. After 5–10 min, the cover glass should slip from the tissues without any damage to the sample.
21. If additional immunostaining cycles are to be performed, go to Subheading 3.3. If this was the last immunostaining cycle, go to Subheading 3.4. Alternatively, slides can be safely stored overnight or up to 2 weeks in deionized water at +4 °C until further processing.

3.3 Cycles 2 – n

1. Perform antibody stripping by incubating the slides in the antibody stripping buffer for 1 h at +56 °C. Afterwards, rinse the slides 3 × 20 min in 1 ×-PBT-T, then in deionized water.
2. Antigen retrieval: Proceed as described in Subheading 3.2, **steps 2** and **5**.
3. Repeat Subheading 3.2, **steps 11–21**, with the adequate antibody solutions.

3.4 Counterstaining

1. Follow the standard protocol of choice for counterstaining, and finish by mounting the slides with a permanent mounting medium (*see* **Note 13**).
2. Image the tissue with a bright field microscope.

3.5 Image Processing and Data Generation

1. As appropriate, apply a shading correction to all images. If tile scanning was performed, tile stitching is to be performed at this step. As appropriate, perform background subtraction (*see* **Note 14**).
2. Sequential image registration, meaning the alignment of consecutive days of imaging, is performed using the DAPI signal as a reference. For this purpose, export daily acquisitions as a stack of TIF images, and carry out hyperstack registration in FIJI, as previously described [10]. Successful alignment is verified by merging DAPI images of consecutive days. At this stage, single channel pictures may be colored and merged to visualize complex cellular landscapes (Fig. 4).

3. Continue with image segmentation and analysis with the preferred software solutions, depending on specific project needs. In our experience, there is no generic way of analyzing imaging data. Thus, specific tools and algorithms must be tailored to each research question, types of tissues, and antibody panels. In our laboratory, we usually perform image segmentation with the machine-learning software ilastik [14] and imaging data extraction with CellProfiler [15].

4 Notes

1. Plastic jars are preferable, especially in the heated water bath.
2. It is recommended to perform all washing steps on an orbital shaker for the indicated times.
3. Each antibody has to be tested for identifying the optimal antigen unmasking reagent. In our hands, EDTA based (pH 9.0) and citrate based (pH 6.0) buffers give the best results. We did not observe advantages in switching from one solution to another between different cycles.
4. The 1× antigen retrieval solution may be reused for up to five consecutive cycles. It is strongly advised to verify the absence of any particles prior to reuse.
5. If primary antibodies generated in goat are to be used, it may be necessary to substitute normal goat serum with horse serum or bovine serum albumin. Additionally, some primary antibodies may require specific blocking reagents.
6. No specific formulation is required (e.g., the antibodies do not need to be carrier-free). In the current protocol, we recommend using unconjugated primary antibodies. Despite allowing for greater flexibility in antibody panel design and serving for signal amplification, this approach has the limitation that only one antibody of each host species can be used in each cycle. This can be circumvented by using directly conjugated primary antibodies or secondary antibodies directed against specific IgG subtypes.
7. It is important to choose fluorophores that will result in a high signal to background ratio. No specific formulation is required, but secondary antibodies should be cross-adsorbed against the target sample species.
8. It is crucial to use an aqueous mounting medium, to ease cover glass removal and prevent tissue damage.
9. This protocol is not suitable for cryosections.
10. Some immunostaining protocols recommend incubating FFPE tissue sections for 1 h at +60 °C prior to starting the

immunostaining to melt paraffin. In our experience, this step is optional.

11. For heating, we use a water bath because this approach appeared more gentle and more reproducible than alternative methods (e.g., microwave). Alternative approaches may be necessary for specific primary antibodies.
12. Imaging should be performed in the optimal conditions to be determined for each project. It is important, though, that a similar image is acquired at every cycle through the whole series. In our case, DAPI serves as a reference channel for image alignment during image processing.
13. We usually perform Masson's trichrome staining at the end of a multiplex immunostaining series. Nuclei may appear dimmer than when performed on fresh tissue sections. This is presumably due to the repetitive immunostaining cycles and tissue treatments. Noteworthy, DAPI images may be superposed with the Masson's trichrome image to visualize nuclear morphology, when necessary. We also have experience with performing alternative counterstaining, such as hematoxylin and eosin.
14. Rolling ball background subtraction may be considered as image alteration, as it changes signal to background ratio locally and not on the whole image. If background subtraction is performed, it must be clearly stated when the findings are published. Moreover, background subtraction may not be performed prior to staining intensity measurements, as this would influence the outcome.

Acknowledgments

We are sincerely grateful to other members of our groups and collaborating scientists for their precious input and continuous support.

References

1. Wallace SJ, Tacke F, Schwabe RF, Henderson NC (2022) Understanding the cellular interactome of non-alcoholic fatty liver disease. JHEP Rep 4(8):100524. https://doi.org/10.1016/j.jhepr.2022.100524
2. Schwabe RF, Tabas I, Pajvani UB (2020) Mechanisms of fibrosis development in nonalcoholic steatohepatitis. Gastroenterology 158(7):1913–1928. https://doi.org/10.1053/j.gastro.2019.11.311
3. van Dam S, Baars MJD, Vercoulen Y (2022) Multiplex tissue imaging: spatial revelations in the tumor microenvironment. Cancers (Basel) 14(13). https://doi.org/10.3390/cancers14133170
4. Gendusa R, Scalia CR, Buscone S, Cattoretti G (2014) Elution of high-affinity (>10-9 KD) antibodies from tissue sections: clues to the molecular mechanism and use in sequential immunostaining. J Histochem Cytochem

62(7):519–531. https://doi.org/10.1369/0022155414536732
5. Bolognesi MM, Manzoni M, Scalia CR, Zannella S, Bosisio FM, Faretta M et al (2017) Multiplex staining by sequential immunostaining and antibody removal on routine tissue sections. J Histochem Cytochem 65(8):431–444. https://doi.org/10.1369/0022155417719419
6. Manzoni M, Bolognesi MM, Antoranz A, Mancari R, Carinelli S, Faretta M et al (2020) The adaptive and innate immune cell landscape of uterine leiomyosarcomas. Sci Rep 10(1):702. https://doi.org/10.1038/s41598-020-57627-1
7. Ma J, Guillot A, Yang Z, Mackowiak B, Hwang S, Park O et al (2022) Distinct histopathological phenotypes of severe alcoholic hepatitis suggest different mechanisms driving liver injury and failure. J Clin Invest 132(14). https://doi.org/10.1172/JCI157780
8. Dudek M, Pfister D, Donakonda S, Filpe P, Schneider A, Laschinger M et al (2021) Auto-aggressive CXCR6(+) CD8 T cells cause liver immune pathology in NASH. Nature 592(7854):444–449. https://doi.org/10.1038/s41586-021-03233-8
9. Guillot A, Guerri L, Feng D, Kim SJ, Ahmed YA, Paloczi J et al (2021) Bile acid-activated macrophages promote biliary epithelial cell proliferation through integrin alphavbeta6 upregulation following liver injury. J Clin Invest 131(9). https://doi.org/10.1172/JCI132305
10. Guillot A, Kohlhepp MS, Bruneau A, Heymann F, Tacke F (2020) Deciphering the immune microenvironment on a single archival formalin-fixed paraffin-embedded tissue section by an immediately implementable multiplex fluorescence immunostaining protocol. Cancers (Basel) 12(9). https://doi.org/10.3390/cancers12092449
11. Lefere S, Puengel T, Hundertmark J, Penners C, Frank AK, Guillot A et al (2020) Differential effects of selective- and pan-PPAR agonists on experimental steatohepatitis and hepatic macrophages(). J Hepatol 73(4):757–770. https://doi.org/10.1016/j.jhep.2020.04.025
12. Gunes Gunsel G, Conlon TM, Jeridi A, Kim R, Ertuz Z, Lang NJ et al (2022) The arginine methyltransferase PRMT7 promotes extravasation of monocytes resulting in tissue injury in COPD. Nat Commun 13(1):1303. https://doi.org/10.1038/s41467-022-28809-4
13. Conlon TM, John-Schuster G, Heide D, Pfister D, Lehmann M, Hu Y et al (2020) Inhibition of LTbetaR signalling activates WNT-induced regeneration in lung. Nature 588(7836):151–156. https://doi.org/10.1038/s41586-020-2882-8
14. Berg S, Kutra D, Kroeger T, Straehle CN, Kausler BX, Haubold C et al (2019) Ilastik: interactive machine learning for (bio)image analysis. Nat Methods 16(12):1226–1232. https://doi.org/10.1038/s41592-019-0582-9
15. Stirling DR, Swain-Bowden MJ, Lucas AM, Carpenter AE, Cimini BA, Goodman A (2021) CellProfiler 4: improvements in speed, utility and usability. BMC Bioinf 22(1):433. https://doi.org/10.1186/s12859-021-04344-9

Chapter 16

Single Cell Secretome Analyses of Hepatic Stellate Cells: Aiming for Single Cell Phenomics

Richell Booijink, Leon Terstappen, and Ruchi Bansal

Abstract

Activated hepatic stellate cells (HSCs) that secrete large amounts of extracellular matrix (ECM) proteins, primarily collagens, are recognized as the key pathogenic cells in liver diseases. Excessive ECM accumulation results in tissue scarring, referred to as liver fibrosis, that progresses to liver cirrhosis (liver dysfunction) and hepatocellular carcinoma. Recent studies using single cell RNA sequencing have discovered various subpopulations of HSCs with high degree of heterogeneity in quiescent, activated, as well as inactive (identified during disease regression) HSCs. However, little is known about the role of these subpopulations in ECM secretion and cell-cell communication or if they respond differently to different exogenous and endogenous factors. Moreover, how the heterogenous single cell transcriptome translates into the single cell secretome and "communicatome" (cell-cell communication) remains largely underexplored. In this chapter, we describe the method (modified enzyme-linked immunosorbent spot, ELISpot) for analyzing collagen type 1 secretion of HSCs at the single cell level, enabling a deeper understanding into the HSC secretome. In the near future, we aim to develop an integrated platform with which we can study secretome of individual cells identified by immunostaining-based fluorescence-activated cell sorting derived from healthy and diseased liver. Through the use of the VyCAP 6400-microwell chip in combination with their puncher device, we aim to perform single cell phenomics by analyzing and correlating phenotype, secretome, transcriptome, and genome of the single cells.

Key words Hepatic stellate cells, Collagen type 1, Single cell secretome, Image analysis, ELISpot

1 Introduction

Liver diseases are a growing health concern with high mortality and morbidity worldwide [1]. Due to the expanding pandemic of obesity and diabetes, metabolic-associated liver diseases are increasing exponentially and are common indications for liver transplantation [2–4]. Activated hepatic stellate cells (HSCs) are the main drivers of liver fibrosis, due to their differentiation into myofibroblasts and their secretion of high amount of extracellular matrix (ECM) pro-

Ralf Weiskirchen and Scott L. Friedman (eds.), *Hepatic Stellate Cells: Methods and Protocols*, Methods in Molecular Biology, vol. 2669, https://doi.org/10.1007/978-1-0716-3207-9_16,

teins, mainly collagens, resulting in the formation of fibrous scar tissue [5]. Therefore, HSCs have been the focus of numerous studies, with varying outcomes [6–9].

Recent single cell analysis of liver cells, and HSCs specifically, has reported multiple subpopulations in quiescent, activated, or inactivated HSCs, both in mouse models and human [10, 11]. Besides heterogeneity in activated HSCs, Rosenthal et al. [12] showed distinct subclusters in quiescent HSCs. These recent findings underline the need for a method that can confirm whether the observed genetic/transcriptomic heterogeneity translates into cell behavior. In other words, do these HSC subpopulations demonstrate heterogeneity in ECM secretion? Does this correlate with genetic/transcriptomic heterogeneity? Moreover, how do these secretory profiles translate into cellular function and cell-cell communication, and how can they be used to predict disease progression/regression?

Collagen type 1, alpha 1 (COL1A1), secreted by HSCs, is one of the most abundant proteins that constitute the fibrillar ECM during liver fibrosis. Collagen expression is normally evaluated using immunostainings (intracellular and extracellular), quantitative PCR (gene expression), Western blots (intracellular in the cell lysates and extracellular in the culture supernatant), hydroxyproline assays, or enzyme-linked immunosorbent assays (ELISA).

This chapter describes a modified ELISpot assay that visualizes single HSC collagen type 1 secretion (Fig. 1). With this assay, the effect of different stimuli, inhibitors, conditioned medium derived from other cells (paracrine/endocrine communication), and direct cell-cell communication (coculture) on the ECM production of single HSCs can be evaluated (Fig. 2). It can therefore be used to evaluate novel therapeutics targeting/inhibiting activated HSCs, as well as assist in fundamental research toward HSC subpopulations or into the mechanisms that drive ECM accumulation. With this method, we are able to capture and quantify collagen 1 secretion of single HSC and evaluate the effect of transforming growth factor beta 1 (TGF-β1) activation on the single cell collagen 1 production.

Moreover, we are currently integrating this method with the VyCAP 6400-microwell and puncher device. This allows us to culture and visualize one single cell per microwell (phenotypic analysis), capture the proteins secreted by this cell, and correlate the secretion with the cell in the well (secretome analysis). Thereafter, we can use the puncher system to punch this single cell out of the well for further downstream genome-transcriptome analysis (Fig. 3) [13]. With this integrated phenomics (combining phenotype analysis with genome-transcriptome-secretome analysis) platform, we aim to map the complete "communicatome" of the HSCs in the liver disease. Currently, this method is being used to detect

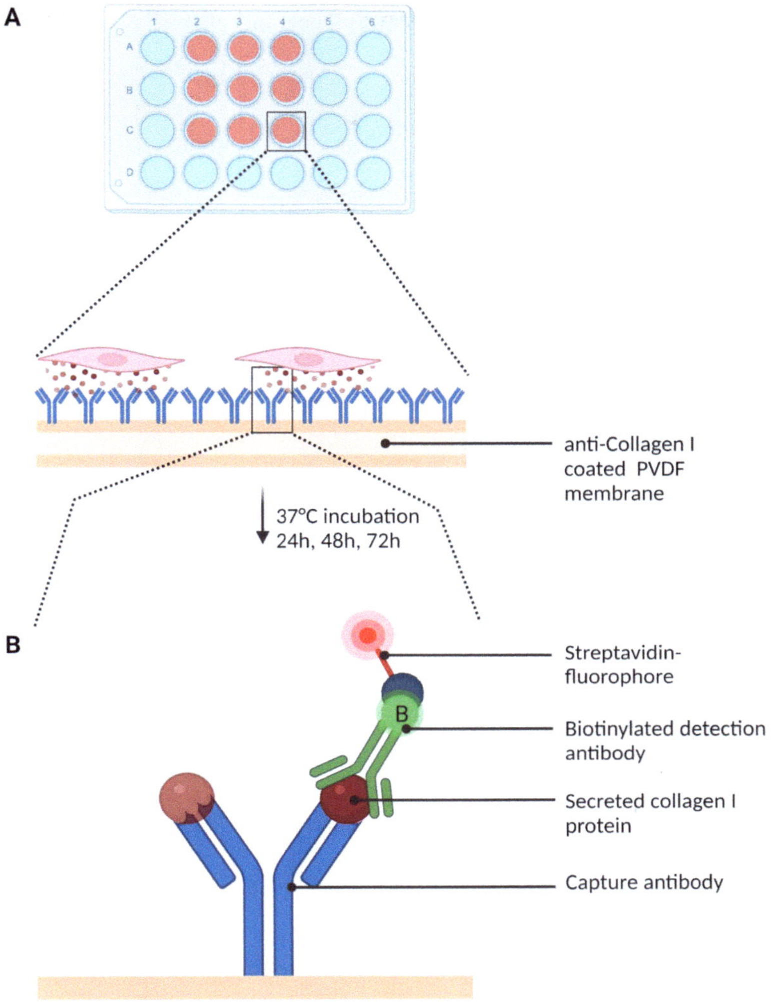

Fig. 1 Schematic representation of measuring single cell collagen 1 secretion. (**a**) Collagen I, secreted by LX2 cells cultured in 24-well plate, is captured by the anti-collagen 1 coated PVDF membrane. (**b**) After incubation at 37 °C, cells are washed off, and the captured collagen 1 is visualized using biotinylated anti-collagen 1 polyclonal antibody and fluorophore conjugated streptavidin

and quantify antibodies and prostate-specific antigen (PSA) secreted by single hybridoma and prostate cancer cells, respectively [14, 15], and is currently being optimized to measure collagen 1 secretion by HSCs.

The method reported here can be easily used for the analysis of collagen 1 production from single HSCs and can be adapted to analyze any secretory protein analysis following optimization.

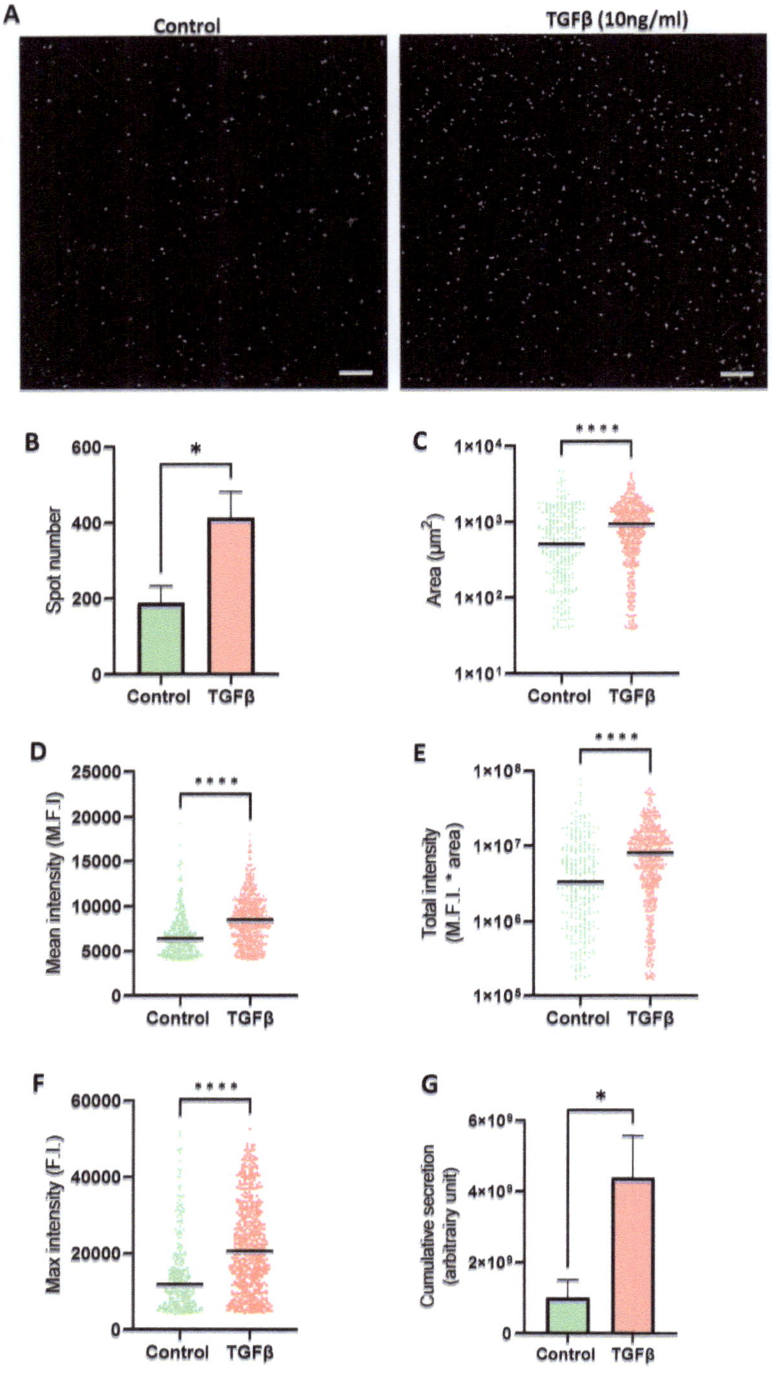

Fig. 2 Example of single cell collagen 1 secretion by LX-2 cells, imaged and analyzed. (**a**) Collagen type 1 secreted by LX-2 cells (immortalized human HSCs) with or without TGFβ stimulation. Each spot represents the amount of collagen 1 secreted by one single cell after 24 h of incubation. Heterogeneity is observed in terms of spot size and intensity. Scale bar is 500 μm. (**b–g**) Quantification of collagen 1 secretion, done as described in 3.7, with (**b**) spot number, $n = 3$, (**c**) area per spot, in μm^2, (**d**) mean fluorescent intensity (MFI) per spot, (**e**) total intensity per spot (calculated through area * MFI), (**f**) maximum intensity of each spot, and (**d**) cumulative collagen 1 secretion, taken by the sum of all total intensities ($n = 3$). Statistical analysis was performed by ratio paired (**b&g**) or unpaired Student's t-test (**c–f**)

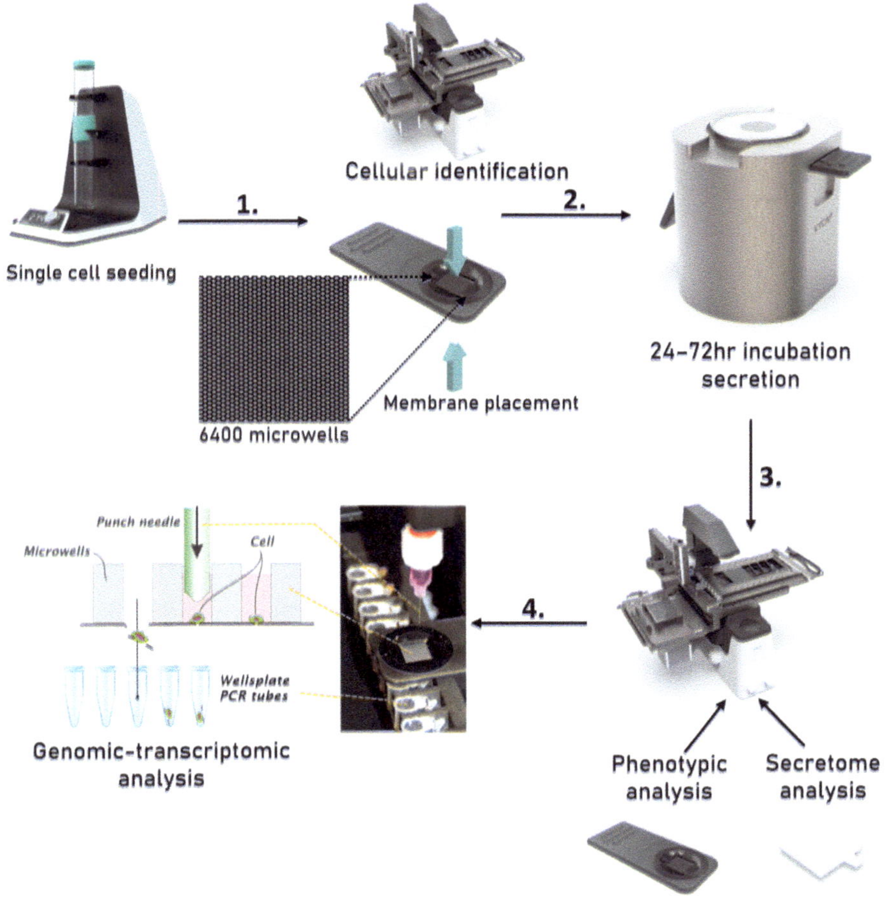

Fig. 3 Single cell phenomics with the VyCAP microwell and puncher system. Step 1. Flow is applied through the 6400-microwell chip and the 5 μm pore at the bottom of each well, which results in an individual distribution of cells in the microwells. An activated membrane is placed underneath the chip to capture the proteins secreted by a single cell. Step 2. Chip with membrane is incubated in the clamp unit and placed in the incubator set at 37 °C/5% CO_2. Step 3. Cells in the chip can be visualized for phenotypic analysis, and the membrane is removed from the chip, developed, and imaged, and the location of each spot can be correlated to the location of the cell inside the microwell chip using a spot software integrated in the device. Step 4. Interesting cells – e.g., high or low producing cells – can be punched out from the microwell chip for downstream analysis, such as genomic (DNA)/transcriptomic (RNA) sequencing, and/or for clonal expansion or cultured for further characterization. (Images are adapted from VyCAP website)

2 Materials

All the materials and reagents listed in Subheadings 2.1 and 2.2 are sterilized before use in a laminar flow cabinet, unless indicated otherwise. Cells are cultured under standard conditions in an incubator set to 37 °C, 5% CO_2. Materials and reagents are stored at room temperature, unless specified otherwise.

2.1 Coating

1. Clean, sharp scissors (*see* **Note 1**).
2. Fine, stainless steel bent tweezers.
3. 24-well plate mold.
4. Sterile petri dish, 9 cm diameter (*see* **Note 2**).
5. Sterile, 12-well tissue culture plate, tissue culture treated.
6. Low fluorescent 0.45 μm polyvinylidene fluoride (PVDF) membranes.
7. Analytical grade 100% methanol (*see* **Note 3**).
8. Sterile, filtered phosphate buffered saline (PBS), pH 7.4. Store at 4 °C.
9. Sterile, human anti-collagen 1 alpha 1 monoclonal antibody. Store at −20 °C (*see* **Note 4**).

2.2 Blocking and Cell Culture

1. Blocking buffer: 3% bovine serum albumin (BSA) in PBS (*see* **Note 5**).
2. 0.2 μm non-pyrogenic filter unit and syringe.
3. Human hepatic stellate cell line LX-2, 85–90% confluent, p27–p60, cultured under standard culturing conditions at 37 °C, 5% CO_2 (*see* **Note 6**).
4. Culture medium: Dulbecco's Modified Eagle Medium (DMEM), high glucose with GlutaMAX supplement, with 1% penicillin/streptomycin and 10% fetal bovine serum (FBS). Store at 4 °C.
5. Sterile PBS. Store at 4 °C.
6. Sterile 0.05% trypsin-ethylenediaminetetraacetic acid (EDTA), phenol red. Store at 4 °C.
7. Starvation medium: DMEM, high glucose with GlutaMAX supplement, with 1% penicillin/streptomycin (no FBS). Store at 4 °C.
8. Recombinant CHO-cell derived human TGF-β1. Store at −20 °C (*see* **Note 7**).

2.3 Spot Development

1. Washing solution: PBS with 0.05% Tween 20.
2. Diluent: filtered 1% BSA in PBS.
3. Plate shaker.
4. Primary human anti-collagen 1alpha (α)1 monoclonal antibody, biotin conjugated.
5. Fluorescently labeled secondary antibody, streptavidin conjugated.
6. Ultrapure water.

2.4 Imaging and Quantification

1. Microscope glass slides.
2. Cover glass 24 × 50 mm.
3. Scanning fluorescent microscope with the ability to scan a complete slide.
4. ImageJ, image processing and analysis software, version Java 1.8.0_172.

3 Methods

Perform all procedures at room temperature unless specified otherwise.

3.1 Coating

1. Using clean, sharp scissors, cut out 15 mm-diameter rounds from the PVDF membranes, using the bottom of a 24-well plate as a mold (*see* **Note 8**). Cut out one PVDF round per sample, including a negative and positive control, and bring the PVDF rounds inside a laminar flow hood.
2. Inside the laminar flow hood, remove the protective paper from the PVDF membrane (*see* **Note 9**), place in a 9 cm petri dish, and activate each membrane with 250 μL 100% methanol for 1 min (*see* **Note 10**).
3. With tweezers, pick up the membrane at the edge (*see* **Note 11**), and move the membranes to a 12-well plate, one membrane per well. Wash for 5 min by adding 500 μL sterile PBS on top of each membrane (*see* **Note 12**).
4. Remove PBS and coat the membranes with 200 μL of 4 μg/mL anti-collagen 1α1 monoclonal antibody overnight at 4 °C (*see* **Note 13**).

3.2 Cell Starvation

1. Preheat the cell starvation medium to 37 °C.
2. Take out the LX-2 cells from the incubator, aspirate cell culture medium, add fresh starvation medium to the cells, and place them back into the incubator (*see* **Note 14**).

3.3 Blocking

1. The next day, filter the 3% BSA blocking buffer in the laminar flow cabinet by passing it through a syringe with a 0.2 μm filter.
2. Remove coating antibody from membranes, wash with sterile PBS for 5 min, and incubate with 500 μL blocking buffer for at least 1.5 h at room temperature.
3. Remove the blocking buffer, and incubate membranes with starvation medium at room temperature until further use (*see* **Notes 15** and **16**).

3.4 Culture

1. Preheat cell culture medium, cell starvation medium, sterile PBS, and 0.05% trypsin-EDTA to 37 °C.
2. Remove the cells from the incubator, and wash twice with PBS, followed by trypsinization for 5 min and resuspension of the cell suspension in culture medium *with FBS* (*see* **Note 17**). Count the cells and create a cell suspension of 20,000 cells per mL in starvation medium *without FBS*.
3. Prepare a working concentration of TGF-β1 of 20 ng/mL in starvation medium.
4. With tweezers, add the round PVDF membrane on the bottom of a 24-well plate, and add 250 μL cell suspension (5,000 cells) on top of the membrane. Thereafter, add 250 μL starvation medium with or without stimuli/TGF-β1 (final concentration, 10 ng/mL) to the wells. Incubate the membranes with cells for either 24, 48, or 72 h at 37 °C with 5% CO_2 (*see* **Note 18**).

3.5 Spot Development

1. Remove the membranes from the incubator, remove medium (*see* **Note 19**), and add 500 μL 0.05% Tween20 in PBS to each membrane. Aspirate this, add 500 μL fresh 0.05% Tween20 in PBS, and incubate this for 20 min (*see* **Note 20**) on a plate shaker shaking at 300 rpm.
2. Wash three times with 1% BSA, 5 min on the plate shaker (300 rpm) each time.
3. Add 250 μL biotin conjugated polyclonal primary human anti-collagen 1α1 antibody (*see* **Note 21**) (100 ng/mL, diluted in diluent), and incubate for 2 h (*see* **Note 22**) on the plate shaker (300 rpm).
4. Wash three times with 1% BSA, 5 min on the plate shaker each time (300 rpm).
5. Add 250 μL fluorescently labeled streptavidin (1:100, diluted in diluent), and incubate for 1 h (*see* **Note 22**) in the dark.
6. Wash two times with 1% BSA, 5 min on the plate shaker each time (300 rpm), while keeping the membranes in the dark by covering the plate with aluminum foil.
7. Wash two times with ultrapure MilliQ water.
8. Using fine bent tweezers, gently remove the membrane from the well (*see* **Note 23**), dip the side of the membrane dry onto disposable professional tissue wipes, and add the membrane to a clean and dry plate. Leave the membranes to dry overnight.

3.6 Imaging

1. Mount the membrane between a microscope slide and a cover-slip (*see* **Note 24**).
2. Use a scanning fluorescent microscope to image and scan the complete membrane with a magnification of at least ten times (*see* **Note 25**).

3.7 Quantification

1. Open the scan of the membrane in ImageJ (*see* **Note 26**). Crop the image to a predetermined size and duplicate the image.
2. For one image, set the threshold to highlight only the structures that you want to analyze (*see* **Note 27**), creating a binary version of the image.
3. Using the Analyze, set measurements tab, select the properties that are of interest in your analysis, and redirect to the original image.
4. In the Analyze tab, select analyze particles, and choose display results, summarize, and in situ show (*see* **Note 28**).
5. Copy data to an analysis software program for further processing (*see* **Note 29**).

4 Notes

1. Instead of using scissors, you can also use a punching device that cuts out 15 mm rounds. This eliminates size differences between the paper rounds, but it is not guaranteed that the punching device is sharp enough to cut through the membrane and both layers of the protective paper.
2. Other material that is sterile, has a clean surface, and can be covered with a lid is also suitable.
3. Methanol should be discarded in appropriate hazardous waste containers.
4. To prevent frequent freeze-thaw cycles after reconstitution of the antibody (in PBS), make aliquots. Store the main stock at −20 °C and the aliquot in use at 4 °C.
5. Best results are obtained when the blocking solution is prepared fresh before each experiment.
6. Similar results are obtained using primary hepatic stellate cells (human liver myofibroblasts), indicating that the method could be used for other cells as well.
7. Instead of, or in combination with, TGF-β1, other stimuli, inhibitors, or substances could be added to examine the effect of these on the single cell collagen 1α1 secretion.
8. The PVDF membrane should stay clean and unaffected. Therefore, do not touch the membrane with your hands, wear gloves, and make sure the scissors are clean and sharp. We recommend using tweezers while cutting the membrane. Keep the protective paper around the membrane during and after cutting.
9. Do not touch the PVDF while removing the protective layer. You can take up the PVDF with tweezers at the very edge, then hold the PVDF between your thumb and index finger, and rub

the thumb against the fingertip to gently separate the protective paper from the PVDF membrane.

10. Pure methanol is a volatile, flammable liquid and is normally used in a fume hood. Therefore, incubate the membranes in methanol in a closed environment, e.g., in a petri dish with the lid on top.
11. Try to remove as much methanol as possible by wiping it off from the membrane at the edge of the petri dish.
12. Once activated, it is important that the membrane does not dry. Therefore, it is advised to move a maximum of three membranes at a time, add PBS as fast as possible, and then continue with another set.
13. The antibody solution should be added directly on top of the membrane, forming a droplet. It is advised to use the well plate dimensions and volumes specified, as too much volume or a smaller well diameter might result in the membrane "floating" on top of the antibody solution, resulting in improper coating.
14. To synchronize cell cycle stage, LX-2 cells should be starved 24 h before incubating them on the membranes.
15. Incubating the membranes with starvation medium prior to cell culture improves the viability of cells on the membrane.
16. Use the same medium that you will use for your experiment.
17. The cell suspension is created in culture medium with FBS to neutralize the trypsin. Trypsin neutralizing solution can also be used instead of FBS, when handling very sensitive cells.
18. Always include an empty membrane, without cells, in your experiment. This can be used as negative control and reference with respect to background staining and unspecific antibody binding.
19. The cell culture medium can be stored and used for other analytical purposes.
20. The washing time of 20 min is based on in vitro cell culture of LX-2 cells. If other cell types or derivatives are used, washing times might have to be optimized.
21. Although other conjugated primary collagen 1α1 antibody could be used for the detection of secreted collagen, the details of this method have been optimized using biotin-streptavidin binding. If other antibodies are used, optimal concentrations must be tested.
22. Optimal incubation times might vary depending on the specific antibody used.
23. Use closed bent tweezers to make a circular motion around the edges of the well; do this until the membrane lifts up from the bottom. At this point, use the tweezers to go underneath the membrane and lift it from the well.

24. Place the membrane on the microscope slide and the coverslip on the membrane. Use your thumb and index finger to press the membrane tightly between the coverslip and the glass slide, and use tape to attach the two. To avoid fingerprints on the glass, wear gloves.
25. Also scan at a wavelength that does not have the fluorophore, to correct for autofluorescence.
26. If necessary, one can choose to first subtract background of the image. This is done using the tool "subtract background." This tab replaces the intensity of pixels that are similar to the background with the mean background intensity value.
27. This threshold should be the same for each image, and in your negative control, no structures should be visible with this threshold.
28. In this tab, you can choose to show the outline, i.e., number and outline each spot. With this, the spot can be correlated to the data point, and one can validate by eye if the automated quantification is properly performed.
29. Spots with an area smaller than 40 μm^2 are excluded from further data processing. Experience has shown that this cutoff eliminates false positives – generated by automated particle analysis – from the data set.

References

1. Asrani SK, Devarbhavi H, Eaton J, Kamath PS (2019) Burden of liver diseases in the world. J Hepatol 70(1):151–171. https://doi.org/10.1016/j.jhep.2018.09.014
2. Burra P, Becchetti C, Germani G (2020) NAFLD and liver transplantation: disease burden, current management and future challenges. JHEP Rep 2(6):100192. https://doi.org/10.1016/j.jhepr.2020.100192
3. Friedman SL, Pinzani M (2022) Hepatic fibrosis 2022: unmet needs and a blueprint for the future. Hepatology 75(2):473–488. https://doi.org/10.1002/hep.32285
4. van Kleef LA, Ayada I, Alferink LJM, Pan Q, de Knegt RJ (2022) Metabolic dysfunction-associated fatty liver disease improves detection of high liver stiffness: the Rotterdam Study. Hepatology 75(2):419–429. https://doi.org/10.1002/hep.32131
5. Loomba R, Friedman SL, Shulman GI (2021) Mechanisms and disease consequences of nonalcoholic fatty liver disease. Cell 184(10): 2537–2564. https://doi.org/10.1016/j.cell.2021.04.015
6. Higashi T, Friedman SL, Hoshida Y (2017) Hepatic stellate cells as key target in liver fibrosis. Adv Drug Deliv Rev 121:27–42. https://doi.org/10.1016/j.addr.2017.05.007
7. Friedman SL (2008) Hepatic stellate cells: protean, multifunctional, and enigmatic cells of the liver. Physiol Rev 88(1):125–172. https://doi.org/10.1152/physrev.00013.2007
8. Wang S, Friedman SL (2020) Hepatic fibrosis: a convergent response to liver injury that is reversible. J Hepatol 73(1):210–211. https://doi.org/10.1016/j.jhep.2020.03.011
9. Tsuchida T, Friedman SL (2017) Mechanisms of hepatic stellate cell activation. Nat Rev Gastroenterol Hepatol 14(7):397–411. https://doi.org/10.1038/nrgastro.2017.38
10. Yang W, He H, Wang T, Su N, Zhang F, Jiang K et al (2021) Single-cell transcriptomic analysis reveals a hepatic stellate cell-activation roadmap and myofibroblast origin during liver fibrosis in mice. Hepatology 74(5): 2774–2790. https://doi.org/10.1002/hep.31987

11. Zhang W, Conway SJ, Liu Y, Snider P, Chen H, Gao H et al (2021) Heterogeneity of hepatic stellate cells in fibrogenesis of the liver: insights from single-cell transcriptomic analysis in liver injury. Cell 10(8). https://doi.org/10.3390/cells10082129
12. Rosenthal SB, Liu X, Ganguly S, Dhar D, Pasillas MP, Ricciardelli E et al (2021) Heterogeneity of HSCs in a mouse model of NASH. Hepatology 74(2):667–685. https://doi.org/10.1002/hep.31743
13. Stevens M, Oomens L, Broekmaat J, Weersink J, Abali F, Swennenhuis J et al (2018) VyCAP's puncher technology for single cell identification, isolation, and analysis. Cytometry A 93(12):1255–1259. https://doi.org/10.1002/cyto.a.23631
14. Abali F, Baghi N, Mout L, Broekmaat JJ, Tibbe AGJ, Terstappen L (2021) Measurement of the drug sensitivity of single prostate cancer cells. Cancers (Basel) 13(23). https://doi.org/10.3390/cancers13236083
15. Abali F, Broekmaat J, Tibbe A, Schasfoort RBM, Zeune L, Terstappen L (2019) A microwell array platform to print and measure biomolecules produced by single cells. Lab Chip 19(10):1850–1859. https://doi.org/10.1039/c9lc00100j

Chapter 17

Hepatic Stellate Cell Targeting Using Peptide-Modified Biologicals

Ruchi Bansal and Klaas Poelstra

Abstract

Liver diseases are a leading cause of death worldwide and are rising exponentially due to increasing prevalence of metabolic disorders. Hepatic stellate cells (HSCs) are recognized as a key therapeutic target in liver diseases as these cells, upon activation during liver damage and ongoing liver inflammation, secrete excessive amounts of extracellular matrix that leads to liver tissue scarring (fibrosis) responsible for liver dysfunction (end-stage liver disease) and desmoplasia in hepatocellular carcinoma. Targeting of HSCs to reverse fibrosis progression has been realized by several experts in the field, including us. We have developed strategies to target activated HSCs by utilizing the receptors overexpressed on the surface of activated HSCs. One well-known receptor is platelet derived growth factor receptor-beta (PDGFR-β). Using PDGFR-β recognizing peptides (cyclic PPB or bicyclic PPB), we can deliver biologicals, e.g., interferon gamma (IFNγ) or IFNγ activity domain (mimetic IFNγ), to the activated HSCs that can inhibit their activation and reverse liver fibrosis. In this chapter, we provide the detailed methods and the principles involved in the synthesis of these targeted (mimetic) IFNγ constructs. These methods can be adapted for synthesizing constructs for targeted/cell-specific delivery of peptides/proteins, drugs, and imaging agents useful for various applications including diagnosis and treatment of inflammatory and fibrotic diseases and cancer.

Key words Hepatic stellate cells, HSCs targeting strategies, Targeted biologicals, PDGFR-β, Receptor targeting peptides, Chemical conjugation

1 Introduction

Activated hepatic stellate cells (aHSCs) play a significant role in liver fibrosis [1, 2]. Upon liver injury, damaged hepatocytes instigate activation of resident Kupffer cells (KCs) that together with other resident hepatic and immune cells secrete chemotactic cytokines resulting in the infiltration of circulating monocytes and other immune cells. In particular, infiltrated monocytes and activated macrophages secrete many growth factors, chemokines, and cytokines that subsequently lead to the activation of HSCs [3]. These HSCs transdifferentiate into highly proliferative, contractile, and

Ralf Weiskirchen and Scott L. Friedman (eds.), *Hepatic Stellate Cells: Methods and Protocols*, Methods in Molecular Biology, vol. 2669, https://doi.org/10.1007/978-1-0716-3207-9_17,

fibrogenic myofibroblasts that produce excessive amounts of extracellular matrix (ECM), the hallmark of fibrotic diseases. Regardless of the etiology (alcohol/drug abuse, metabolic disorders, unhealthy diet-lifestyle, viral infections or genetic predisposition), aHSCs are responsible for liver tissue scarring (fibrosis), distorted liver architecture, end-stage liver disease (cirrhosis), and desmoplasia in hepatocellular carcinoma [1–4]. There are no treatments available against liver diseases, while the morbidity and mortality due to liver diseases are rising exponentially with more than two million deaths every year [5].

Due to the rich vasculature of the liver and unique metabolic capacity, uptake of drugs in the liver is generally high. However, most of the drugs are either metabolized by hepatocytes or removed by the reticuloendothelial system, making only low amounts of drug available for HSCs to generate a therapeutic effect. Moreover, most anti-fibrotic drugs fail in clinical trials due to low efficacy caused by uptake (and clearance) by nontarget organs or cells. Targeted delivery of drugs or biologicals to the key players in fibrogenesis (aHSCs) is therefore recognized as an important approach to ameliorate liver diseases at a clinically relevant stage of the disease [6, 7]. Knowing the central importance of platelet derived growth factor receptor-beta (PDGFR-β) in liver fibrosis, and its high expression on aHSCs [8, 9], we documented several targeting approaches using our PDGFR-β-recognizing peptide (PPB) for the delivery of biologicals including the anti-fibrotic cytokine interferon gamma (IFNγ) to PDGFR-β-expressing aHSCs in the fibrotic liver [9–12]. IFNγ is coupled to PPB either monocyclic PPB or bicyclic dimeric BiPPB via bifunctional polyethylene glycol (PEG) linkers or to human serum albumin (HSA). The latter can serve as a drug carrier for many types of therapeutics [9–12]. Monocyclic PPB can be applied as a homing moiety when multiple peptides are attached to a core molecule (such as proteins or nanoparticles where multiple coupling sites are available) [9, 12–18], whereas BiPPB can be used to target a small molecule (drug, peptide, or protein where only one coupling site is available) [10, 11, 19, 20]. This is because dimeric ligand (peptide)-receptor interaction is required for targeting dimeric PDGF-β-receptor [21, 22]. Targeted IFNγ showed HSC-specific uptake and improved therapeutic efficacy while reducing systemic side effects when compared to untargeted free or PEGylated IFNγ in carbon tetrachloride (CCl_4)-induced acute and/or chronic liver fibrosis mouse models [9–12]. Others have synthesized PPB-modified sterically stable liposomes (PPB-SSL) for HSC-specific delivery of IFNγ or IFNα or TNF-related apoptosis-inducing ligand (TRAIL) [15–18]. For the synthesis of PPB-SSL, PPB was first conjugated with Mal-PEG(3400)-DSPE and then purified and used for preparing liposomes using thin-film method where liposomes were rehydrated using PBS or PBS with biologicals. These liposomes evidenced potent therapeutic effects and reduced adverse effects [15–18]. Altogether, these results demonstrate the impact of cell-specific targeted therapies compared to nontargeted free biologicals. For clinical translation, targeted IFNγ was further miniaturized by synthesizing a chimeric molecule (mimγ-BiPPB) composed of mimetic IFNγ (the IFNγ signaling domain) and the BiPPB attached to each other either chemically using a PEG linker [11] or biotechnologically by recombinant expression in *E. coli* [10]. This mimetic IFNγ conjugate (mimIFNγ-BiPPB, also referred to as *Fibroferon* [19, 20]) substantially prevented the progression of fibrosis in CCl_4-induced mouse models of liver fibrosis [10, 11], offering novel therapeutic approaches not only for the treatment of liver fibrosis but also for curing other fibrotic diseases and cancer [13, 19, 23]. Besides as therapeutics, these constructs can also be used for monitoring the fibrogenic process in a theranostic (i.e., a combined therapeutic and diagnostic) approach [20, 21].

Besides PDGFR-β, other receptors including integrins such as αv-containing integrins (αvβ1, αvβ3, αvβ5, αvβ8, α5, and α11) [24–30], mannose-6-phosphate/insulin-like growth factor-II receptor (M6P/IGF-IIR) [31–35], fibroblast activated protein (FAP) [36], retinol binding protein (RBP) receptor [37], fibroblast growth factor (FGF)-inducible 14 [38], FGFR1 [39], relaxin receptor RXFP1 [40], etc. have been used in recent years to target drugs-biologicals to (a) HSCs to ameliorate liver fibrosis. Several specific receptor-recognizing peptides can be or have been designed for these receptors. Due to recent advances in single-cell RNA sequencing technologies, heterogenous population of HSCs have been identified [41]. With these data increasingly available, more HSC-specific targets can be identified, enabling the targeting of specific HSC populations in the future.

In this chapter, we present the materials and methods for synthesizing PDGF-β-receptor targeted (mimetic) IFNγ constructs (IFNγ-PEG-PPB, PPB-HSA-PEG-IFNγ, and mimetic IFN-γ-PEG-BiPPB). These methods are widely applicable for the conjugation of receptor-recognizing peptides to proteins or other drug carriers for the delivery of biologicals and small chemical entities to the designated target cells.

2 Materials

All the reactions were performed in low protein binding tubes (LoBind tubes, Eppendorf, Hamburg, Germany). The materials used for synthesis of PPB constructs are provided in Table 1. All solutions were prepared freshly before use unless indicated otherwise.

Table 1
Materials required for the synthesis and purification of PPB constructs

Chemicals	MW	Storage temperature	Cat. no./supplier
Cyclic ($C^aSRNLIDC^a$)-SATA (**PPB-SATA)**	1000	4 °C	Custom-designed, Ansynth B.V. (Roosendaal, The Netherlands)
$C^aSRNLIDC^aSRNLIDC^a$-ATA Bicyclic PPB (**BiPPB-ATA)**	2223	4 °C	Custom-designed, Genosphere (Paris, France)
Succinimidyl α-methylbutanoate **(mPEG-NHS, 2KDa PEG)**	2000	−20 °C	PT-02F-05/Nektar therapeutics, San Francisco, CA, US
Maleimide-PEG-succinimidyl carboxy methyl ester **(MAL-PEG-SCM, 2KDa PEG)**	1903	−20 °C	PHB-950/Creative PEGWorks, Hamburg, Germany
N-Succinimidyl-S-acetylthioacetate (**SATA**) (Cas. no. 76931-93-6)	231.23	20 °C	A-9043/Sigma-Aldrich, Zwijndrecht, The Netherlands
N-[γ-maleimidobutyryloxy] succinimide ester (**GMBS**) (Cas. no. 80307-12-6)	280.23	4 °C	63175/Sigma-Aldrich, Zwijndrecht, The Netherlands
Hydroxylamine (NH_2OH) (Cas. no. 5470-11-1)	69.49	Room temperature	159417/Sigma-Aldrich Zwijndrecht, The Netherlands
Ethylenediaminetetraacetic acid disodium salt dihydrate (**EDTA**) (Cas. no. 6381-92-6)	372.24	Room temperature	03677/Sigma-Aldrich Zwijndrecht, The Netherlands
N, N-dimethylformamide (**DMF**) (Cas. no. 68-12-2)	73.09	Room temperature	227056/Sigma-Aldrich Zwijndrecht, The Netherlands
L-Lysine (Cas. no. 56-87-1)	146.2	Room temperature	62840/Sigma-Aldrich Zwijndrecht, The Netherlands
Recombinant IFNγ (mouse)	15,600	70 °C	315-05/Peprotech, London, UK
AKFEVNNPQVQRQAFNELIRVVHQLLPESSLRKRKRSR)-SATA **Peptidomimetic IFNγ-SATA**	4689		Custom-designed, Ansynth B.V. (Roosendaal, The Netherlands)

PPB-modified HSA **(PPB-HSA)** (see the protocol)	75,000	−20 °C	Prepared in-house
Micro DispoDialyzers (10 KDa)		4 °C	74-0708/Harvard Apparatus, Massachusetts, USA
Micro DispoDialyzers (50 KDa)		4 °C	74-0712/Harvard Apparatus, Massachusetts, USA
Slide-A-Lyzer™ Dialysis Cassettes, 10K MWCO		Room temperature	66380/Thermo Scientific, Rockford, IL
7K Zeba spin desalting columns		4 °C	89882/Thermo Scientific, Rockford, IL

[a]Denotes the cyclization site via disulfide bridge

3 Methods

3.1 Synthesis of IFNγ-PEG-PPB (Table 2 and Fig. 1) [9]

1. Briefly centrifuge the tube containing IFNγ (100 μg) and dissolve the contents (IFNγ) in 1 mL of PBS, make 100 μL aliquots in low protein binding tubes, and store in −80 °C (*see* **Note 1**).
2. Prepare maleimide-PEG-succinimidyl carboxy methyl ester (Mal-PEG-SCM, 2 KDa) fresh by dissolving 1 mg of Mal-PEG-SCM in 100 μL of dimethylformamide (DMF) achieving a concentration of 10 mg/mL.

   Mal-PEG-SCM is a linear heterobifunctional PEG reagent with a maleimide and a succinimidyl NHS-ester group. Maleimide reacts with thiol, SH, sulfhydryl, or mercapto. Succinimidyl carboxyl methyl (SCM) ester reacts with primary amines (in this case, N-terminus amine and lysine residue side chains present in IFNγ; murine IFNγ contains 20 lysines per homodimer). Maleimide contains a reactive C=C double bond and is light or oxygen sensitive, and therefore either should be prepared fresh before use or should be stored appropriately (*see* **Note 2**).
3. IFNγ (0.256 nmol, 4 μg) is reacted with 25.6 nmol (51.2 μg) of Mal-PEG-SCM (*see* Table 2) for 2 h (30 min at room temperature followed by 90 min at 4 °C; *see* **Note 3**).
4. Dialyze against PBS using 10K MWCO Micro DispoDialyzers overnight.

   Micro DispoDialyzer is a disposable dialyzer that contains regenerated cellulose membranes. The dialyzer unit can be placed in a beaker on a magnetic stirrer with stir bar for constant agitation of the sample to reduce dialysis times. The unit floats directly in the dialysis buffer. Once dialysis is complete, invert the dialyzer into a new collection tube, and briefly

Table 2
Synthesis of IFNγ-PEG-PPB (final volume: 200 μL, 20 ng/μL, or 20 μg/mL)

Total reaction	nmol	μg	Stock conc.	Volume
(I) IFNγ (×1)	0.256	4.0	100 μg/mL (in PBS)	40 μL + 110 μL PBS
Mal-PEG-SCM (×100)	25.6	51.2	10 mg/mL (in DMF)	5.12 μL
React for 30 min at room temperature followed by 90 min at 4 °C *Dialyze against PBS using Micro DispoDialyzer 10K MWCO overnight*				
(II) + PPB-SATA (×100)	25.6	25.6	1 mg/mL (in DMF)	25.6 μL
+ NH_2OH (0.5 M) + EDTA (25 mM)				18 μL
React for 30 min at room temperature followed by overnight at 4 °C *Dialyze against PBS using DispoDialyzer 10K MWCO overnight*				

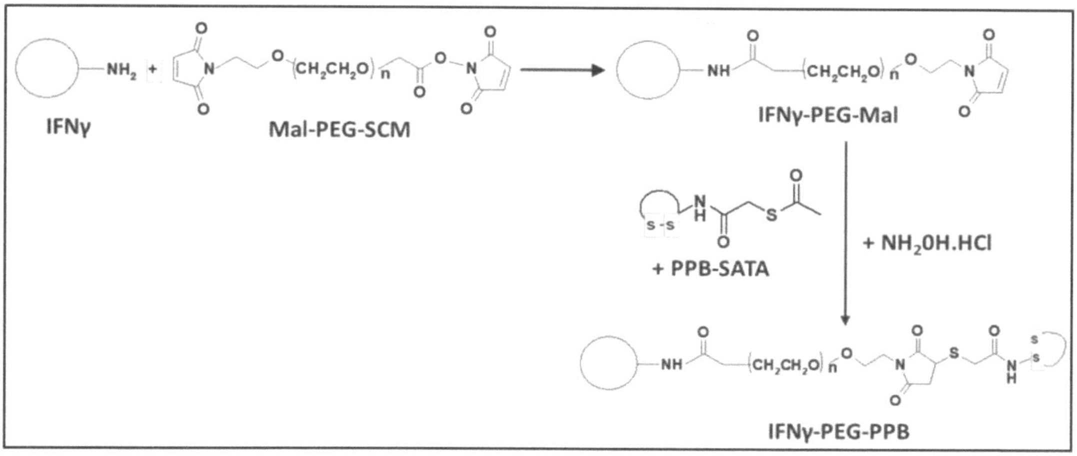

Fig. 1 Schematic showing stepwise the chemical reactions involved in the synthesis of IFNγ-PEG-PPB. Please see Subheading 3.1 for the details

centrifuge (500–2000 rpm for 1–2 s). The Micro DispoDialyzer comes with a foam float, cap, and two 1.5 mL collection tubes. It can handle sample sizes 5–100 μL, is easy-to-use and leak-proof, has low protein binding, and yields high sample recovery.

5. The purified product (IFNγ-PEG-Mal) is further reacted with 25.6 nmol (25.6 μg) of PPB-SATA (dissolved in DMF at the concentration 1 mg/mL) in the presence of deacetylating reagent (NH_2OH + EDTA). PPB-SATA was custom-prepared by modifying PPB with N-succinimidyl-S-acetylthioacetate (SATA). SATA is a short-chain (2.8 angstrom spacer arm) reagent used for covalent modification of primary amines to form stable amide bonds and adds a protected sulfhydryl group that can be deprotected by hydroxylamine. SATA also preserves protein activity with its mild, non-denaturing reaction conditions. In this case, deprotected sulfhydryl groups react with maleimide groups from IFNγ-PEG-Mal that results in the formation of a stable thioether linkage which is nonreversible.
6. The reaction is performed at room temperature for 30 min followed by 90 min at 4 °C (*see* **Note 3**). Finally, IFN-γ-PEG-PPB is extensively dialyzed overnight against PBS at 4 °C.
7. The IFNγ conjugate (IFNγ-PEG-PPB) is characterized using anti-IFNγ and anti-PPB antibodies by Western blotting as per standard procedures. Briefly, 150 ng of IFNγ conjugates are subjected to SDS-PAGE and blotted on a PVDF (polyvinylidene difluoride) membrane. The membranes are blocked with TBST (20 mM Tris-HCl, pH 7.6, 154 mM NaCl, 0.1% Tween 20) containing 5% nonfat dry milk for 1 h at room temperature

and further incubated with either rabbit polyclonal IFNγ antibody (ab9918, Abcam) or rabbit polyclonal PPB antibody (developed by Harlan, Zeist, The Netherlands) overnight at 4 °C. After washing with TBST, the blots are incubated with HRP-conjugated goat anti-rabbit antibody (DAKO, Glostrup, Denmark) for 1 h at room temperature. After washings with TBST, the blots are developed using Pierce ECL western blotting substrate as per manufacturer's instructions. The bands are analyzed using ImageJ software (NIH, USA) to determine the % band intensity, and molecular weights of the bands are estimated using molecular weight markers.

8. Besides analysis of successful conjugation, IFNγ activity is assessed using nitric oxide release assay. Briefly, RAW264.7 macrophage seeded in 96-well plates is incubated with either medium alone or different concentrations (5, 10, 20, and 50 ng/mL) of IFNγ conjugates and non-modified IFNγ (as a control; *see* **Note 4**). After 24 h, secreted nitrite is measured as absorbance at 550 nm using Griess reagent (1% sulfanilamide, 0.1% naphthylethylenediamine dihydrochloride, 3% H_3PO_4). RAW264.7 macrophages (until passage 20) are cultured in Dulbecco's Modified Eagle's Medium (DMEM, Invitrogen, Carlsbad, CA) supplemented with 10% FBS and antibiotics (50 U/mL penicillin and 50 ng/mL streptomycin).

3.2 Synthesis of PPB-HSA-PEG-IFNγ [12] (Table 3 and Fig. 2)

1. PPB-HSA is synthesized as described previously [21]. Briefly, HSA (1.5 μmol, dissolved in PBS) is reacted with γ-maleimidobutyryloxy-succinimide ester (GMBS, 30 μmol, dissolved in DMF) for 2 h at room temperature and extensively dialyzed against PBS using 10 kDa cutoff Slide-A-Lyzer™ dialysis cassette. PPB-ATA (34.5 μmol; dissolved in DMF) is added to the GMBS-modified HSA and allowed to react overnight, dialyzed against PBS, and then further dialyzed against ultrapure water using 10 kDa cutoff Slide-A-Lyzer™ dialysis cassette. The final product can be freeze-dried for long-term storage at −20 °C.

2. PPB-HSA-PEG-Mal conjugate: 133.3 nmol (100 μg) of PPB-HSA (dissolved in PBS at the concentration of 1 mg/mL) is reacted with 66.65 nmol (133.3 μg) of freshly prepared bifunctional Mal-PEG-SCM (*see* **Note 2**) (dissolved in DMF at the concentration of 20 mg/mL) overnight at 4 °C and purified extensively using 7K Zeba spin desalting columns (equilibrated with 1× PBS). Zeba spin desalting columns are polypropylene devices containing a proprietary high-performance size-exclusion chromatography resin that provides an excellent protein desalting and recovery in a centrifuge format. These columns (7K MWCO, 0.5 mL) are easy-to-use and provide fast and high protein recovery due to the low-binding resin.

Table 3
Synthesis of IFNγ-PEG-HSA-PPB (final volume: 200 μL, 25 ng/μL, or 25 μg/mL)

Total reaction	nmol	μg	Stock conc.	Volume
(I) PPB-HSA	133.3	100.0	1 mg/mL (in PBS)	100 μL
Mal-PEG-SCM (×50)	66.65	133.3	20 mg/mL (in DMF)	6.7 μL
React overnight at 4 °C *Dialyze against PBS using 7K Zeba spin desalting columns overnight*				
(II) IFNγ (×1)	0.3205	5.0	100 μg/mL (in PBS)	50 μL + 135 μL PBS
SATA (×25)	8.0125	1.85	1 mg/mL (in DMF)	1.85 μL
React overnight at 4 °C *Dialyze against PBS using 7K Zeba spin desalting columns overnight*				
(III) (I) (×1) + (II) (×3)	0.107 (I) + 0.3205 (II)	8.0 (I) + 5.0 (II)		8 μL (I) + (II)
+NH2OH (0.5 M) + EDTA (25 mM)				19 μL
React overnight at 4 °C *Dialyze against PBS using DispoDialyzer 10K MWCO overnight*				

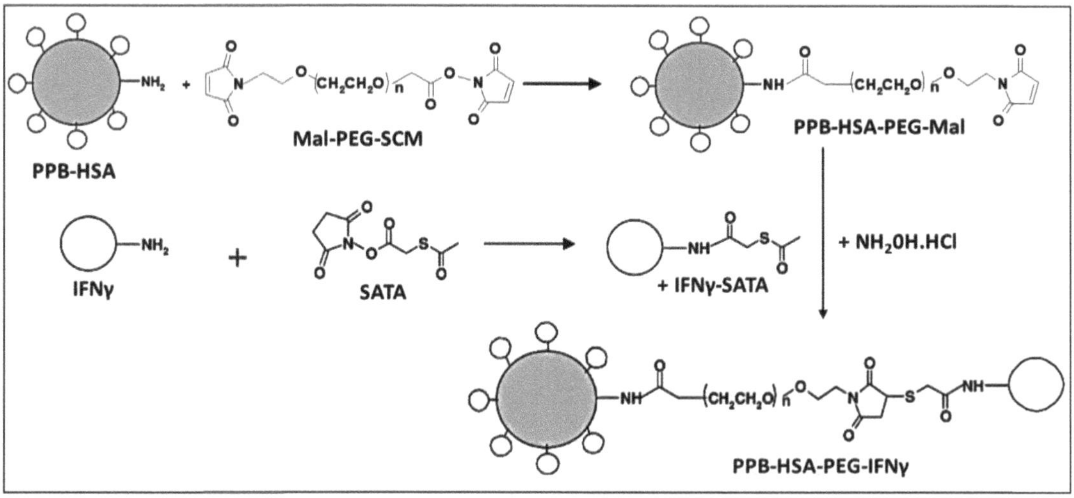

Fig. 2 Schematic showing stepwise the chemical reactions involved in the synthesis of PPB-HSA-PEG-IFNγ. Please see Subheading 3.2 for the details

3. IFNγ-SATA conjugate: 0.3205 nmol (5 μg) of IFNγ is reacted with 8.0125 nmol (1.853 μg) of freshly prepared N-succinimidyl-S-acetylthioacetate (SATA, dissolved in DMF at the concentration of 200 μg/mL) (*see* **Note 5**) overnight at 4 °C and purified by dialysis using 7K Zeba spin desalting columns (Thermo Scientific).
4. PPB-HSA-PEG-IFNγ conjugate: The purified products (PPB-HSA-PEG-Mal and IFNγ-SATA) are reacted at the ratio of 1:3 in the presence of deacetylating reagent (0.1 M hydroxylamine, 25 mM EDTA in PBS pH 7.2) overnight at 4 °C.
5. Finally, dialyze PPB-HSA-PEG-IFNγ extensively against PBS at 4 °C overnight using 50 kDa DispoDialyzers.
6. PPB-HSA-PEG-IFNγ conjugate is characterized by SDS-PAGE analysis followed by barium iodide PEG staining. PEG staining: Briefly, proteins (250 ng) are subjected to SDS-PAGE (10%) according to standard protocols. After running the gel, PEG staining was performed as follows: The gels are rinsed with water, followed by fixation in perchloric acid (0.1 M) for 15 min. The gels are washed again and treated with barium chloride (5%) for 10 min. Subsequently, the color is developed using Titrisol iodine solution (Sigma). The gels can be photographed using G-Box (Syngene, Cambridge, U.K.).
7. In addition to PEG staining, Western blotting was performed. Briefly, proteins (250 ng) are subjected to SDS-PAGE (10%) according to standard protocols. Separated proteins are transferred to polyvinylidene difluoride (PVDF) membranes (Roche, Mannheim, Germany). The membranes are blocked with TBST (20 mM Tris-HCl pH 7.6, 154 mM NaCl, 0.1% Tween 20) containing 5% skimmed milk and incubated with either rabbit polyclonal anti-IFNγ antibody (1:2000; Abcam, Cambridge, U.K.) or rabbit polyclonal anti-PPB antibody (1: 1000; custom-made, Harlan) overnight. After washings, the blots are incubated with HRP-conjugated goat anti-rabbit antibody (1:1000; DAKO, Glostrup, Denmark) for 1 h. Subsequently, the blots are washed and developed using Western Lightning-ECL reagent (Perkin Elmer, Boston, MA) according to the manufacturer's instructions.
8. Besides analysis of successful conjugation, the bioactivity of IFNγ and PPB-HSA-PEG-IFNγ is assessed by measuring accumulation of nitrite NO2, a stable nitric oxide (NO) metabolite produced by murine RAW macrophages. Briefly, cells (1×10^5cells/200 μL/well) seeded in 96-well plates are incubated either with medium alone or with different concentrations of PPB-HSA, IFNγ and IFNγ conjugate, PPB-HSA-PEG-IFNγ (5, 10, 20, and 50 ng/mL). After 24 h, the secreted nitrite is measured as described above.

3.3 Synthesis of mimIFNγ-PEG-BiPPB [11] (Table 4 and Fig. 3)

1. To synthesize targeted peptidomimetic IFNγ (mimIFN-γ-BiPPB) construct, 0.112 μmol BiPPB (0.25 mg, prepared in 1:1 DMF:PBS, *see* **Note 6**) is reacted with 0.337 μmol (0.67 mg) of Maleimide-PEG-succinimidyl carboxy methyl ester (Mal-PEG-SCM, freshly prepared in 1:1 DMF:PBS, *see* **Note 6**) for 3 h at room temperature.
2. Excess of Mal-PEG-SCM is blocked with lysine (0.337 μmol) for 1 h at room temperature (*see* **Note 7**).
3. Subsequently, the prepared BiPPB-PEG-Mal (0.112 μmol, 0.25 mg) is reacted overnight with mimIFNγ-SATA in the

Table 4
Synthesis of mimIFNγ-PEG-BiPPB

Total reaction	μmol	mg	Stock conc.	Volume
BiPPB (×1)	0.112	0.25	5 mg/mL (in PBS/DMF; 1:1)	50 μL
Mal-PEG-SCM (×3)	0.337	0.67	5 mg/mL (in PBS/DMF; 1:1)	134 μL
React for 3 h at room temperature				
BiPPB-PEG-Mal (×1)	0.112	0.25		184 μL
Lysine (×10)	1.12	0.163	5 mg/mL (in PBS)	32.6 μL
React for 1 h at room temperature				
BiPPB-PEG-mal (×1)	0.112	0.25		
mimIFNγ-SATA (×5)	0.56	2.63	20 mg/mL (in PBS)	
+NH2OH (0.5 M) + EDTA (25 mM)				
React overnight at 4 °C *Dialyzed against PBS using 7K Zeba spin columns*				

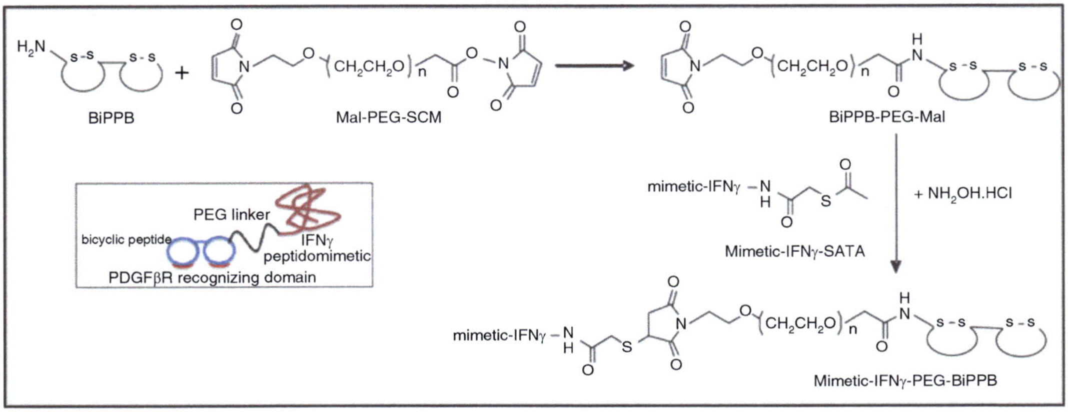

Fig. 3 Schematic showing stepwise the chemical reactions involved in the synthesis of mimIFNγ-PEG-BiPPB. Please see Subheading 3.3 for the details

presence of deacetylating reagent (0.1 M hydroxylamine, 25 mM EDTA in PBS, pH 7.2).

4. Finally, the prepared mimIFNγ-PEG-BiPPB (mimγ-BiPPB) conjugate (8.9 kDa) is extensively dialyzed against PBS using 7K Zeba spin desalting columns. The mimIFNγ-PEG-BiPPB construct is fully prepared under mild conditions to maintain the dimeric BiPPB ring structures for appropriate receptor interaction.

4 Notes

1. IFNγ (or other protein of interest) should not contain any preservatives or additives, for example BSA, which can influence the reactions. IFNγ (or other protein of interest) should be aliquoted in low protein binding tubes and all the reactions should be performed in low protein binding tubes to minimize any loss of protein due to binding to the Eppendorf tubes/microtubes. Minimize freeze-thaw cycles to avoid losing IFNγ (protein) function.
2. Storage Condition: PEG product should be stored in the original form as received in a freezer at −20 °C or lower for long-term storage. Stock solutions of PEG reagents that DO NOT contain oxygen or moisture sensitive functional groups (e.g., NHS, thiol etc.,) may be temporarily stored in a refrigerator for multiple days. Stock solution should avoid repeated freeze-and-thaw cycles. For moisture-sensitive PEG reagents (NHS-ester), anhydrous solvents are required. Also, light and oxygen sensitive PEG products including thiols and those with unsaturated carbon-carbon double bonds such as maleimide, DBCO, and acrylate, ideally shall be stored away from light (i.e., wrapped with aluminium foil) and in an air-free atmosphere. The best way to achieve inert atmosphere is to purge the vial with nitrogen or argon in an inert gas-filled glove box. To prepare stock solutions of oxygen sensitive reagents, degassing the solvent with nitrogen or argon is preferred. Light sensitive PEG products are bottled in amber glass vials or plastic vials. Handling: PEGs are highly hygroscopic and absorb moisture from air quickly. Follow these steps to aliquot: (1) Allow vials to thaw and equilibrate to room temperature before opening the vial; (2) Open vials and weigh the quantity you need quickly; (3) Flush vials with dry argon or nitrogen (if you have access to nitrogen or argon). Caution: adjust gas flow so that dry powders will not blow.

3. The temperatures are particularly important for the synthesis of the IFNγ constructs. Since the biological activity is influenced by conjugation steps and temperatures, mild conjugation reagents and mild temperatures should be used to maintain the activity of IFNγ. In our case, we conducted the conjugation for 30 min at room temperature and 90 min at 4 °C. All the purification steps are performed at 4 °C.
4. Controls are important. We recommend to use two different controls: (1) freshly diluted IFNγ (the native protein); (2) IFNγ that went to all the incubation and purification steps as the IFNγ constructs except the chemical coupling agents (sham-treated control).
5. SATA is moisture sensitive. Store desiccated at −20 °C. To avoid moisture condensation in the product, fully equilibrate the vial to room temperature before opening. Dissolve reagent immediately before use. The NHS-ester moiety readily hydrolyzes, making the reagent nonreactive; therefore, reagent solutions must not be stored as stocks. Discard any unused reagent solution. Avoid using buffers that contain amines (e.g., Tris and glycine) because amines directly compete with the reaction. Phosphate-buffered saline (PBS) or HEPES buffers at pH 7.2–8.0 are good options for applications involving proteins. Both the acylation reaction to primary amines and the hydrolysis (inactivation) of these NHS-ester reagents occur more rapidly at higher pH. For this reason, procedures for modification with SATA involve addition of a molar excess of reagent, and reactions proceed to completion (modification or hydrolysis) in min (pH 9) or hours (pH 7). As the target amines are more concentrated, the intended acylation reaction is more favored over hydrolysis.
6. Dissolve in DMF and then add equal volume of PBS to make 1: 1 DMF/PBS.
7. Lysine is used to block excess of Mal-PEG-SCM (added three times in excess). One molecule of Mal-PEG-SCM will be bound to BiPPB, and two free molecules of Mal-PEG-SCM will remain in solution. Normally, one would purify the conjugate, but to avoid loss of BiPPB during the purification steps, lysines were added to scavenge free molecules of Mal--PEG-SCM. This step is essential; otherwise, in the next steps where mimIFNγ-ATA is added, it will bind to free Mal-PEG-SCM and will result in impurity i.e., nontargeted mimIFNγ which cannot be purified from mimIFNγ-PEG-PPB due to the small difference in molecular weights.

References

1. Friedman SL (2008) Hepatic stellate cells: protean, multifunctional, and enigmatic cells of the liver. Physiol Rev 88(1):125–172. https://doi.org/10.1152/physrev.00013.2007
2. Yazdani S, Bansal R, Prakash J (2017) Drug targeting to myofibroblasts: implications for fibrosis and cancer. Adv Drug Deliv Rev 121: 101–116. https://doi.org/10.1016/j.addr.2017.07.010
3. Hernandez-Gea V, Friedman SL (2011) Pathogenesis of liver fibrosis. Annu Rev Pathol 6: 425–456. https://doi.org/10.1146/annurev-pathol-011110-130246
4. Bansal R, Nagorniewicz B, Prakash J (2016) Clinical advancements in the targeted therapies against liver fibrosis. Mediat Inflamm 2016: 7629724. https://doi.org/10.1155/2016/7629724
5. Asrani SK, Devarbhavi H, Eaton J, Kamath PS (2019) Burden of liver diseases in the world. J Hepatol 70(1):151–171. https://doi.org/10.1016/j.jhep.2018.09.014
6. Poelstra K, Schuppan D (2011) Targeted therapy of liver fibrosis/cirrhosis and its complications. J Hepatol 55(3):726–728. https://doi.org/10.1016/j.jhep.2011.04.008
7. Poelstra K, Prakash J, Beljaars L (2012) Drug targeting to the diseased liver. J Control Release 161(2):188–197. https://doi.org/10.1016/j.jconrel.2012.02.011
8. Borkham-Kamphorst E, Kovalenko E, van Roeyen CR, Gassler N, Bomble M, Ostendorf T et al (2008) Platelet-derived growth factor isoform expression in carbon tetrachloride-induced chronic liver injury. Lab Investig 88(10):1090–1100. https://doi.org/10.1038/labinvest.2008.71
9. Bansal R, Prakash J, Post E, Beljaars L, Schuppan D, Poelstra K (2011) Novel engineered targeted interferon-gamma blocks hepatic fibrogenesis in mice. Hepatology 54(2):586–596. https://doi.org/10.1002/hep.24395
10. Bansal R, Prakash J, De Ruiter M, Poelstra K (2014) Targeted recombinant fusion proteins of IFNgamma and mimetic IFNgamma with PDGFbetaR bicyclic peptide inhibits liver fibrogenesis in vivo. PLoS One 9(2):e89878. https://doi.org/10.1371/journal.pone.0089878
11. Bansal R, Prakash J, De Ruiter M, Poelstra K (2014) Interferon gamma peptidomimetic targeted to hepatic stellate cells ameliorates acute and chronic liver fibrosis in vivo. J Control Release 179:18–24. https://doi.org/10.1016/j.jconrel.2014.01.022
12. Bansal R, Prakash J, de Ruijter M, Beljaars L, Poelstra K (2011) Peptide-modified albumin carrier explored as a novel strategy for a cell-specific delivery of interferon gamma to treat liver fibrosis. Mol Pharm 8(5):1899–1909. https://doi.org/10.1021/mp200263q
13. Bansal R, Tomar T, Ostman A, Poelstra K, Prakash J (2012) Selective targeting of interferon gamma to stromal fibroblasts and pericytes as a novel therapeutic approach to inhibit angiogenesis and tumor growth. Mol Cancer Ther 11(11):2419–2428. https://doi.org/10.1158/1535-7163.MCT-11-0758
14. Jia Z, Gong Y, Pi Y, Liu X, Gao L, Kang L et al (2018) pPB peptide-mediated siRNA-loaded stable nucleic acid lipid nanoparticles on targeting therapy of hepatic fibrosis. Mol Pharm 15(1):53–62. https://doi.org/10.1021/acs.molpharmaceut.7b00709
15. Li F, Li QH, Wang JY, Zhan CY, Xie C, Lu WY (2012) Effects of interferon-gamma liposomes targeted to platelet-derived growth factor receptor-beta on hepatic fibrosis in rats. J Control Release 159(2):261–270. https://doi.org/10.1016/j.jconrel.2011.12.023
16. Li Q, Yan Z, Li F, Lu W, Wang J, Guo C (2012) The improving effects on hepatic fibrosis of interferon-gamma liposomes targeted to hepatic stellate cells. Nanotechnology 23(26): 265101. https://doi.org/10.1088/0957-4484/23/26/265101
17. Li Q, Yu Q, Ju J, You T, Yan Z, Nan X et al (2017) Long-circulating liposomal delivery system targeting at PDGFR-beta enhances the therapeutic effect of IFN-alpha on hepatic fibrosis. Curr Pharm Des 23(20):3034–3046. https://doi.org/10.2174/1381612822666161208144953
18. Li Q, Ding Y, Guo X, Luo S, Zhuang H, Zhou J et al (2019) Chemically modified liposomes carrying TRAIL target activated hepatic stellate cells and ameliorate hepatic fibrosis in vitro and in vivo. J Cell Mol Med 23(3):1951–1962. https://doi.org/10.1111/jcmm.14097
19. Poosti F, Bansal R, Yazdani S, Prakash J, Beljaars L, van den Born J et al (2016) Interferon gamma peptidomimetic targeted to interstitial myofibroblasts attenuates renal fibrosis after unilateral ureteral obstruction in mice. Oncotarget 7(34):54240–54252. https://doi.org/10.18632/oncotarget.11095
20. van Dijk F, Olinga P, Poelstra K, Beljaars L (2015) Targeted therapies in liver fibrosis:

combining the best parts of platelet-derived growth factor BB and interferon gamma. Front Med (Lausanne) 2:72. https://doi.org/10.3389/fmed.2015.00072

21. Beljaars L, Weert B, Geerts A, Meijer DK, Poelstra K (2003) The preferential homing of a platelet derived growth factor receptor-recognizing macromolecule to fibroblast-like cells in fibrotic tissue. Biochem Pharmacol 66(7):1307–1317. https://doi.org/10.1016/s0006-2952(03)00445-3
22. Rogers MA, Campana MB, Long R, Fantauzzo KA (2022) PDGFR dimer-specific activation, trafficking and downstream signaling dynamics. J Cell Sci 135(17). https://doi.org/10.1242/jcs.259686
23. Poosti F, Bansal R, Yazdani S, Prakash J, Post E, Klok P et al (2015) Selective delivery of IFN-gamma to renal interstitial myofibroblasts: a novel strategy for the treatment of renal fibrosis. FASEB J 29(3):1029–1042. https://doi.org/10.1096/fj.14-258459
24. Turaga RC, Yin L, Yang JJ, Lee H, Ivanov I, Yan C et al (2016) Rational design of a protein that binds integrin alphavbeta3 outside the ligand binding site. Nat Commun 7:11675. https://doi.org/10.1038/ncomms11675
25. Turaga RC, Satyanarayana G, Sharma M, Yang JJ, Wang S, Liu C et al (2021) Targeting integrin alphavbeta3 by a rationally designed protein for chronic liver disease treatment. Commun Biol 4(1):1087. https://doi.org/10.1038/s42003-021-02611-2
26. Schnittert J, Bansal R, Storm G, Prakash J (2018) Integrins in wound healing, fibrosis and tumor stroma: high potential targets for therapeutics and drug delivery. Adv Drug Deliv Rev 129:37–53. https://doi.org/10.1016/j.addr.2018.01.020
27. Patsenker E, Stickel F (2011) Role of integrins in fibrosing liver diseases. Am J Physiol Gastrointest Liver Physiol 301(3):G425–G434. https://doi.org/10.1152/ajpgi.00050.2011
28. Patsenker E, Popov Y, Stickel F, Schneider V, Ledermann M, Sagesser H et al (2009) Pharmacological inhibition of integrin alphavbeta3 aggravates experimental liver fibrosis and suppresses hepatic angiogenesis. Hepatology 50(5):1501–1511. https://doi.org/10.1002/hep.23144
29. Henderson NC, Arnold TD, Katamura Y, Giacomini MM, Rodriguez JD, McCarty JH et al (2013) Targeting of alphav integrin identifies a core molecular pathway that regulates fibrosis in several organs. Nat Med 19(12):1617–1624. https://doi.org/10.1038/nm.3282
30. Bansal R, Nakagawa S, Yazdani S, van Baarlen J, Venkatesh A, Koh AP et al (2017) Integrin alpha 11 in the regulation of the myofibroblast phenotype: implications for fibrotic diseases. Exp Mol Med 49(11):e396. https://doi.org/10.1038/emm.2017.213
31. van Beuge MM, Prakash J, Lacombe M, Post E, Reker-Smit C, Beljaars L et al (2013) Enhanced effectivity of an ALK5-inhibitor after cell-specific delivery to hepatic stellate cells in mice with liver injury. PLoS One 8(2):e56442. https://doi.org/10.1371/journal.pone.0056442
32. van Beuge MM, Prakash J, Lacombe M, Post E, Reker-Smit C, Beljaars L et al (2011) Increased liver uptake and reduced hepatic stellate cell activation with a cell-specific conjugate of the rho-kinase inhibitor Y27632. Pharm Res 28(8):2045–2054. https://doi.org/10.1007/s11095-011-0430-9
33. van Beuge MM, Prakash J, Lacombe M, Gosens R, Post E, Reker-Smit C et al (2011) Reduction of fibrogenesis by selective delivery of a Rho kinase inhibitor to hepatic stellate cells in mice. J Pharmacol Exp Ther 337(3):628–635. https://doi.org/10.1124/jpet.111.179143
34. Moreno M, Gonzalo T, Kok RJ, Sancho-Bru P, van Beuge M, Swart J et al (2010) Reduction of advanced liver fibrosis by short-term targeted delivery of an angiotensin receptor blocker to hepatic stellate cells in rats. Hepatology 51(3):942–952. https://doi.org/10.1002/hep.23419
35. Klein S, Van Beuge MM, Granzow M, Beljaars L, Schierwagen R, Kilic S et al (2012) HSC-specific inhibition of rho-kinase reduces portal pressure in cirrhotic rats without major systemic effects. J Hepatol 57(6):1220–1227. https://doi.org/10.1016/j.jhep.2012.07.033
36. Lee J, Byun J, Shim G, Oh YK (2022) Fibroblast activation protein activated antifibrotic peptide delivery attenuates fibrosis in mouse models of liver fibrosis. Nat Commun 13(1):1516. https://doi.org/10.1038/s41467-022-29186-8
37. Sato Y, Murase K, Kato J, Kobune M, Sato T, Kawano Y et al (2008) Resolution of liver cirrhosis using vitamin A-coupled liposomes to deliver siRNA against a collagen-specific chaperone. Nat Biotechnol 26(4):431–442. https://doi.org/10.1038/nbt1396
38. Huang L, Xie J, Bi Q, Li Z, Liu S, Shen Q et al (2017) Highly selective targeting of hepatic stellate cells for liver fibrosis treatment using a d-enantiomeric peptide ligand of Fn14 identified by mirror-image mRNA display. Mol

Pharm 14(5):1742–1753. https://doi.org/10.1021/acs.molpharmaceut.6b01174

39. Kurniawan DW, Booijink R, Pater L, Wols I, Vrynas A, Storm G et al (2020) Fibroblast growth factor 2 conjugated superparamagnetic iron oxide nanoparticles (FGF2-SPIONs) ameliorate hepatic stellate cells activation in vitro and acute liver injury in vivo. J Control Release 328:640–652. https://doi.org/10.1016/j.jconrel.2020.09.041

40. Nagorniewicz B, Mardhian DF, Booijink R, Storm G, Prakash J, Bansal R (2019) Engineered Relaxin as theranostic nanomedicine to diagnose and ameliorate liver cirrhosis. Nanomedicine 17:106–118. https://doi.org/10.1016/j.nano.2018.12.008

41. Ramachandran P, Dobie R, Wilson-Kanamori JR, Dora EF, Henderson BEP, Luu NT et al (2019) Resolving the fibrotic niche of human liver cirrhosis at single-cell level. Nature 575(7783):512–518. https://doi.org/10.1038/s41586-019-1631-3

Chapter 18

Experimental Workflow for Preclinical Studies of Human Antifibrotic Therapies

Lien Reolizo, Michitaka Matsuda, and Ekihiro Seki

Abstract

Chronic liver diseases accompanied by liver fibrosis have caused significant morbidity and mortality in the world with increasing prevalence. Nonetheless, there are no approved antifibrotic therapies. Although numerous preclinical studies showed satisfactory results in targeting fibrotic pathways, these animal studies have not led to success in humans. In this chapter, we summarize the experimental approaches currently available, including in vitro cell culture models, in vivo animal models, and new experimental tools relevant to humans, and discuss how we translate laboratory results to clinical trials. We will also address the obstacles in transitioning promising therapies from preclinical studies to human antifibrotic treatments.

Key words Liver fibrosis, Animal models, In vitro models, In vivo models, Hepatic stellate cells

1 Introduction

Liver fibrosis is the result of abnormal repair action to prolonged chronic liver injury, which can progress into cirrhosis and hepatocellular carcinoma (HCC). The hepatic fibrotic response is initiated in many etiologies of chronic liver injury, including infections of hepatitis viruses and parasites, alcohol abuse, autoimmune disorders, and the metabolic syndrome with nonalcoholic fatty liver disease. Fibrosis is characterized by the balance of extracellular (ECM) turnover favoring net deposition in the liver [1, 2]. The key underlying mechanism is the activation of hepatic stellate cells (HSCs), which transdifferentiate into myofibroblast phenotype following hepatic epithelial injury and lobular inflammation. During HSC activation, collagen and other ECM components are deposited into the space of Disse, which disrupts the physiological architecture of the liver. Longstanding injury provokes chronic inflammation and accumulation of ECM, leading to progressive substitution of normal liver parenchyma by fibrotic scar tissue. The transition to a fibrotic liver entails activation and regulation

Ralf Weiskirchen and Scott L. Friedman (eds.), *Hepatic Stellate Cells: Methods and Protocols*, Methods in Molecular Biology, vol. 2669, https://doi.org/10.1007/978-1-0716-3207-9_18,

of complex molecular signals and cell-cell communication between liver-resident and infiltrating cells including hepatocytes and non-parenchymal cells [3]. Yet, there are still many difficulties in unraveling the precise molecular events during HSC activation in chronic liver diseases.

The development of antifibrotic therapies relies on a comprehensive understanding of profibrogenic mechanisms in multiple organ systems, as well as disease-specific contexts. Ongoing efforts seek to understand the molecular basis of HSC activation with the hope of developing novel strategies to combat liver fibrosis. The largely disappointing results of clinical phase II and III trials are in sharp contrast to a long pipeline of promising antifibrotic candidate agents in preclinical models. Therefore, mitigating insufficient and/or off-target effects are essential for successful translation of antifibrotic agents and their effective use in clinical practice [2, 3].

To grasp a more complete understanding of disease targets, it is useful to explore the current extensive experimental methods used to study fibrosis progression. In this chapter, we summarize the hepatic fibrosis models currently available for the development of liver antifibrotic therapy and discuss possible reasons for the failure of so many antifibrotic drugs at the clinical level (Fig. 1). We will then explore the future directions of how we can translate our laboratory findings to antifibrotic therapies in humans. A better understanding of the differences between animal models and human pathology and an improved insight into carefully designed trials with appropriate endpoints and dosing need to be considered in order to identify more effective antifibrotics in patients with chronic liver disease.

1.1 Etiological Treatment

To date, there is no effective US Food and Drug Administration-approved pharmacotherapy against liver fibrosis. Liver fibrosis and cirrhosis were considered irreversible disease entities, and spontaneous regression was not expected without treatment of the underlying liver disease [4]. However, the recent development of effective antiviral therapies for viral hepatitis has shown that liver fibrosis can be reversed and inflammation is attenuated. Therefore, the only treatments currently available are to eliminate pathogenic factors, including antiviral drugs for hepatitis virus, and lifestyle modifications for nonalcoholic fatty liver diseases (NAFLD) and alcohol-associated liver disease (ALD). While responses to antiviral therapies can be dramatic, interventions for alcoholic and nonalcoholic liver diseases are less effective because the underlying disease drivers persist. In HCV cirrhosis patients, treating HCV infection may regress some levels of fibrosis but does not improve portal hypertension. Curative therapy still relies on liver transplantation although suitable deceased-donor or transplantable organs are critically scarce and require long-term immunosuppression [2]. Thus, there are significant unmet needs to develop effective therapies for

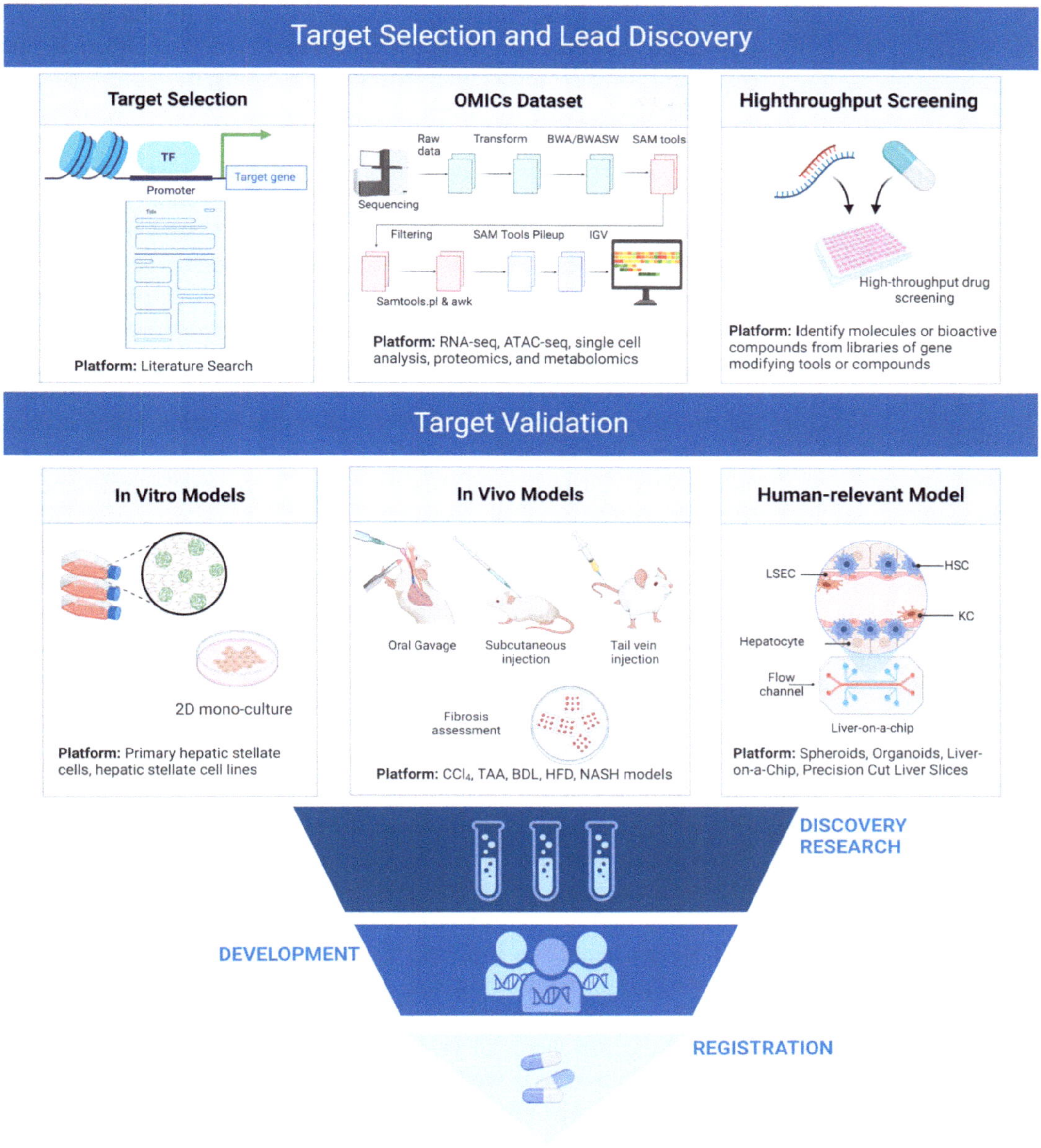

Fig. 1 The scheme depicts the underpinnings of drug discovery process. Translation to human antifibrotic therapies requires several key steps: target selection/identification, target validation, and clinical trials. Target selection and lead discovery use the molecular approach consisting of literature review, OMICS technology, and high-throughput screening to identify a drug-like small molecule or biological therapeutic against liver fibrosis. Following on from this, target validation utilizes in vitro, in vivo, and human-relevant models to confirm the role of the target, and if successful, that will progress into clinical development. Abbreviations: BDL bile duct ligation, CCl_4 carbon tetrachloride, HSC hepatic stellate cells, HFD high-fat diet, KC Kupffer cells, LSECs liver sinusoidal endothelial cells, NASH nonalcoholic steatohepatitis, TF transcription factor, TAA thioacetamide

liver fibrosis. This highlights the importance of clarifying the molecular mechanism of the development of liver fibrosis and searching for new molecular targets for treating it.

1.2 Identification of New Molecular Targets for Drug Discovery

Fibrosis is a crucial constituent in the pathogenesis of a variety of diseases and therefore may share common mechanisms and therapeutic targets. For these reasons, common intervention strategies and medicines may be applicable for fibrosis among different organs. Preclinical research has yielded numerous targets for antifibrotic agents, some of which have entered early-phase clinical studies, but progress has been hampered due to insufficient efficacy to reverse or halt fibrosis in human trials [5].

There are several limitations of the currently available preclinical models to determine effective molecular targets. For instance, there is no perfect animal model relevant to human liver fibrosis, and in vitro cell and tissue culture models have variable relevance to human pathophysiology. Another limitation is the relative lack of sensitive and specific noninvasive biomarkers and imaging modalities to measure the progression or regression of fibrosis, particularly in the liver [4]. These limitations are often associated with the failure to identify and test the right targets for clinical trials, in part because the development of liver fibrosis and cirrhosis takes decades in humans and improving fibrosis may also require many years. The following sections discuss several approaches, including preclinical models, to identify the molecular targets to treat liver fibrosis, which can be the foundation for future trials.

1.2.1 Search Targets from Previous Literature

A crucial step in the drug discovery process is the determination and characterization of a molecular lead that can modify the selected target to generate a desirable pharmacological outcome that is translatable to humans. To identify novel drug targets and uncover potential pathogenic pathways, a literature search can identify previously investigated signaling pathways, experimental models (e.g., in vitro, in vivo), and analytic tools already used to study liver fibrosis, as well as reported efficacy of existing pharmacologic agents [5]. Based on this information, conceptually new hypotheses and experimental approaches not tested previously may be proposed.

An example from our laboratory reported the role of toll-like receptor 4 (TLR4) expression by HSCs and intestinal microbiota in regulating hepatic fibrogenesis [6]. Prior to this study, the expression of TLR4 had been studied using culture HSCs. However, the direct effect of TLR4 in HSC activation and fibrosis in vivo was yet to be delineated, with implications for clarifying the mechanisms of fibrogenesis. Thus, we validated the functional role of TLR4 signaling in vivo using TLR4-mutant mice and TLR4 bone marrow-chimeric mice in three mouse liver fibrosis models – bile duct ligation (BDL) and chronic treatment of carbon tetrachloride

(CCl_4) and thioacetamide (TAA). The study cemented several lines of evidence that HSCs, and not Kupffer cells, are the primary targets that drive fibrogenesis through TLR4 signaling. As the gut microbiome was previously associated in other liver disease, except liver fibrosis, this study was one of the earliest to unveil that gut-derived LPS is an important mediator of hepatic fibrogenesis, which endogenously activates TLR4 in the liver. This has propelled our interest to conduct further studies to comprehensively understand bacterial metabolite-mediated gut-liver interactions in NAFLD fibrosis in hopes of identifying new molecular treatment targets.

Another approach is to search for targets from the studies of other organ fibrosis, for example, lung, kidney, and cardiac fibrosis. For example, the roles of hyaluronan and HAS2 have been well established in lung fibrosis research [7]. However, except as a biomarker, the behaviors of hyaluronan and its regulation by HAS proteins were unknown in liver fibrosis. To address this question, our group utilized hepatic stellate cell-specific HAS2 knockout mice in multiple fibrosis models complemented by promoter analysis of HAS2 [8]. The study also provided evidence that inhibiting HA synthesis is a promising candidate for the treatment of hepatic fibrosis. This indicates that some common pathway signatures may underlie fibrosis in different organs (e.g., lung and liver).

An additional approach is drug repurposing. Drug repurposing can uncover new indications of established drugs that extend beyond their original roles. As of recently, this approach has accounted for ~30% of all drugs issued by the US Food and Drug Administration [9, 10]. A major advantage of drug repurposing is that established drugs selected for a novel indication have already passed the time-consuming pharmacokinetics, pharmacodynamics, and toxicity profiling evaluation and may already be approved for other indications and for which their safety is well established. For instance, type 2 diabetes mellitus (T2DM) is known to share targets related to metabolic syndrome, obesity, and NAFLD. Correcting the metabolic state may improve NAFLD and its associated fibrosis. Thus, the recent NAFLD-fibrosis clinical trials often test drugs developed to treat diabetes. The antidiabetics pioglitazone, metformin, GLP-1 receptor agonists, and sGLT2 inhibitors are good examples [11]. Pioglitazone can improve NAFLD, but the adverse effects, including possible weight gain and increased risk for bladder cancer, have limited their use in patients. Metformin induces weight loss and diabetes improvement, but the NAFLD state does not improve significantly, and thus it is ineffective for NAFLD. While evidence is still incomplete, GLP-1 receptor agonists and sGLT2 inhibitors warrant further investigations in clinical trials for NAFLD patients. Harrison et al. recently reported data from a randomized, double-blind, placebo-controlled, phase IIa study of licogliflozin. Treatment with licogliflozin, a selective and potent inhibitor of both sodium-glucose cotransporter 1 (SGLT1) and SGLT2, lowered ALT levels and hepatic fat content compared to placebo [12]. Thus, the study concluded that the repurposing of antidiabetic sGLT inhibitor can reduce the risk of NASH. In summary, careful literature search and potential repurposing of existing drugs represent a rational experimental workflow that establishes a clear rationale for testing preclinical models and their translation to effective clinical trials.

1.2.2 Secondary Analysis of Publicly Available OMICS Datasets

The recent advancement of OMICS analysis, such as RNA-seq, ATAC-seq, single cell analysis, proteomics, and metabolomics, has increased OMICS datasets that are publicly available through NCBI Gene Expression Omnibus (GEO), SRA, and EBI-ENA databases [13]. With the development of several bioinformatics tools, one strategy is reanalysis of previously published large-scale datasets. In fact, a substantial proportion of these data can be aptly categorized as unprocessed data which have yet to be analyzed. In order to maximize their reusability, a secondary analysis using the same dataset has the potential to identify and evaluate potentially novel molecular targets in a time- and cost-saving manner. One example is a recent work indicating that WISP1 can be a treatment target for liver fibrosis [14]. In this study, authors secondarily analyzed the previously reported liver transcriptomic profiles from the HCV and NAFLD-fibrosis patients [15] (/ENA: ERP109255) by combining gene ontology (GO) analysis and Molecular Signatures Database (MSigDB) and identified that serum response factor (SRF) putative target genes were enriched, which overlapped in cirrhosis from HCV and NAFLD. The same datasets further identified the enriched gene signature for MRTFs, the coactivators of SRF for cytoskeleton gene expression and cell motility. An additional genome-wide expression correlation analysis between extracellular factor genes and the MRTF response signature determined that WISP1 is one of the genes most significantly correlated with the MRTF response signature in cirrhotic livers. Subsequently, functional studies evaluated fibrosis in WISP1 knockout mice as well as through the administration of a potent neutralizing anti-WISP1 antibody. Efficacy of this antibody in mice indicated its potential to treat human fibrotic disease. Thus, a secondary analysis combining multiple publicly available OMICS data with additional bioinformatics analysis can generate new hypotheses which can be tested functionally in mouse models, thereby accelerating the translation of laboratory results to human disease.

1.2.3 High-Throughput Screening Approaches

High-throughput screening (HTS) is a powerful approach to identify target molecules or bioactive compounds from libraries of gene modifying tools or compounds by measuring the readouts (here, HSC activation or liver fibrosis) in vitro or in vivo. This approach has potential to identify compounds or molecules, whose effects or roles in liver fibrosis and HSC activation are previously unknown, in a rapid and cost-effective manner [16–20]. The first approach is typically to use libraries of small molecule compounds to screen for those compounds that can suppress HSC activation. HSC activation can be induced either spontaneously or by stimulation using fibrogenic factors, such as TGF-β or PDGF. HSC activation can be assessed by the expression of αSMA or using a HSC reporter system with GFP or luciferase expression under control of collagen 1a1 promoter. Once the "hit" compounds are identified, the target pathways responsible in regulating HSC activation can be explored. If the "hit" compounds do not have suitable pharmacokinetic properties or have off-target effects, an additional structural analysis can search for or develop other compounds with similar HSC suppressing capability exclusive of the off-target effects. Using this screening approach, an antifungal itraconazole (ITA) [17], tricyclic antidepressants (TCAs) [18], and polyether ionophore nanchangmycin (NCMC) [19] were identified to inactivate HSCs. Further analyses determined ITA inhibited Hedgehog and VEGF pathways; TCAs inhibited sphingomyelinase pathway; and NCMC inhibited activities of FAK, ERK2/1, and HSP27.

The second approach is to screen using libraries of genetic modifying tools, such as siRNA gene silencing or CRISPR-Cas9 gene editing libraries in cultured HSCs [20]. Fibrotic factors (e.g., TGFβ1) can be used to activate the cells based on the expression of αSMA or a GFP or luciferase reporter system. As with small molecule screening, once the "hit" targets are identified, a literature search or structural analysis may discover existing compounds or develop new compounds that regulate the activity of target pathways. Functional studies in vivo can then be evaluated as described below.

Once we have identified a shortlist of potential target genes associated with liver fibrosis, functional impact assessments using in vivo gene silencing or editing approaches in preclinical models should be considered. Vollmann et al. selected 24 genes of interest by filtering 169 common upregulated genes in animal models of liver and kidney fibrosis [16]. They silenced 24 genes in CCl_4-induced liver fibrosis model and found 7 genes whose silencing reduced collagen 1a1 mRNA levels. For five out of seven genes (*Egr2*, *Atp1a2*, *Fkbp10*, *Fstl1*, and *Has2*), silencing led to a 75–100% reduction in liver fibrosis. Importantly, three genes (*Egr2*, *Atp1a2*, and *Fkbp10*) were not linked to liver fibrosis prior to this study, indicating this in vivo silencing approach can effectively identify fibrogenic factors, including those not previously discovered. In summary, high-throughput screening approaches can be effective to rapidly identify new target molecules and treatments. Additional validations using in vivo or human-based systems can bring these efforts closer to clinical validation.

2 Functional Studies Using In Vitro/In Vivo Models of the Liver Fibrosis

Target gene identification along typically provides an incomplete picture of the fibrotic mechanisms. Therefore, validation experiments are the next step using either in vitro or in vivo models. These models are essential tools in the discovery and preclinical stages of drug development and biomarker discovery.

2.1 In Vitro Cell Culture Models of Liver Fibrosis

In vitro models are an important approach to advance our understanding of the molecular pathogenesis. Two-dimensional (2D) monolayer cultures of primary human and rodent HSCs and established HSC cell lines, such as LX-2 cells, are the standard methodologies to study liver fibrosis in vitro. These models can be used for functional assays to identify and test novel molecular targets and are suitable for high-throughput testing and development of candidate antifibrotic agents [21]. However, the current 2D HSC culture models do not fully replicate the complexities in human fibrogenesis or Fibrinolysis in vivo. These limitations include lack of (i) physiological HSC plasticity; (ii) dynamic interactions between other cell types (e.g., hepatocytes, endothelial cells); (iii) growth factors produced by hematopoietic and immune cell types; (iv) pathologic ECM production, assembly, and interactions; and (v) physiological vascular architecture or 3D morphology. Nevertheless, cell culture models are highly accessible with reasonable reproducibility and high cost-effectiveness [22]. They can be used as the front-line methodology to test new hypotheses associated with selected targets through a series of surrogate assays. The application of in vitro HSC culture system has also become a cornerstone of drug development programs to test antifibrotic or fibrinolytic actions of candidate agents. In addition, the 2D assay models of primary human liver cells (e.g., hepatocytes, endothelial cells) are still the gold standard for toxicity screening in drug development.

2.1.1 Primary Hepatic Stellate Cells (HSCs)

Friedman et al. were among the first to pioneer the isolation and characterization of the hepatic stellate cell (HSC), the key cell type responsible for scar production in the liver [23, 24]. Primary HSCs can be isolated from normal rat and mouse livers using in situ perfusion methods with two step collagenase-pronase digestions followed by density gradient centrifugation with Nycodenz [25]. Human HSCs can also be isolated from a piece of resected or explanted livers with similar methods. The density gradient method can purify lipid droplets storing HSCs and is suitable to isolate quiescent HSCs from normal liver tissues. Plated quiescent

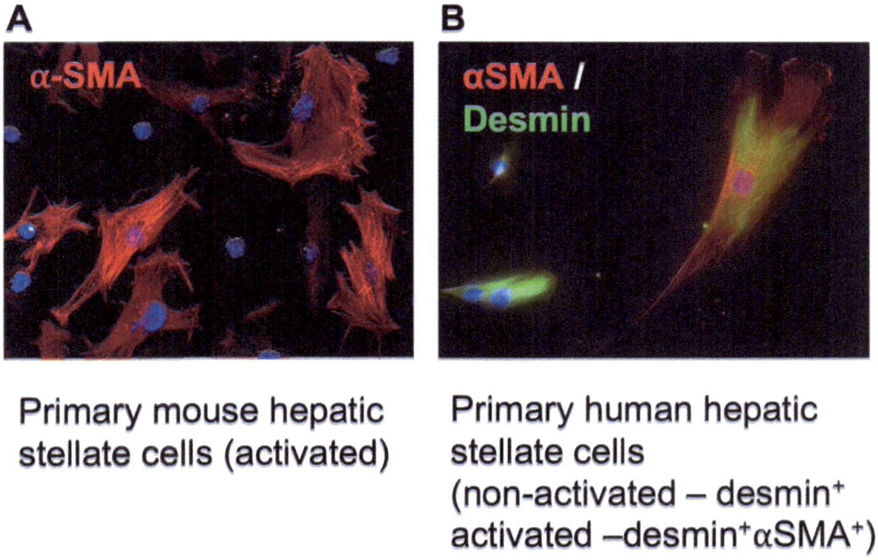

Fig. 2 Primary mouse hepatic stellate cells. (**a**) Cultured activated mouse hepatic stellate cells with overexpression of α smooth muscle actin (αSMA, red fluorescence). Blue, nucleus. (**b**) Human hepatic stellate cells – nonactivated (desmin, green) and activated (desmin, green; αSMA, red)

HSCs will be transdifferentiated into activated myofibroblasts that acquire the ability to produce ECM spontaneously on plastic culture dishes or by treating with TGF-β, a profibrogenic factor, or PDGF, a proliferative factor (Fig. 2). These cells are suitable to identify pathways associated with HSC activation. Gene modifications and pharmacologic inhibition or activation are used for functional assays to study target molecules. Primary HSCs are generally resistant to transfection of plasmids. Instead, our laboratory uses viral vector mediated gene modification approaches, such as adenovirus vectors. Gene silencing using siRNA transfection works well in primary HSCs. To validate the functions of target molecules and pathways, gene-modified or pharmacologically modulated HSCs are used, and cellular activation markers are tracked. Our laboratory uses Desmin and L-rat as general HSC markers, whereas we evaluate col1a1, αSMA, and Timp1 as activation markers (Fig. 2); Gfap, PPAR-γ, and Bambi as quiescent markers; and Gabra3 as markers of HSC inactivation (i.e., previously activated HSCs that have now quiescence) [26, 27]. In vitro activated HSCs reflect, but not perfectly recapitulate, in vivo activation states. If a more representative cell model is required, primary HSCs can be isolated from fibrotic livers, including animal models of fibrosis (e.g., BDL, CCl_4 models) and cirrhotic human livers. In vivo activated HSCs partially lose lipid droplets in BDL and CCl_4 rodent models. In contrast, the loss of lipid droplets in HSCs from alcohol-associated liver injury model is more rapid, making it impossible to use a density gradient isolation method to recover the cells. Similarly in NAFLD models, lipid accumulated hepatocytes can confound the purification of HSCs during density gradient centrifugation. In these circumstances, a FACS sorting approach using a transgenic mouse model expressing GFP under a control of collagen promoter can be used for isolation of activated HSCs. Limitations to the use of primary HSCs include the nonphysiologic mode of activation in primary culture and the lack of complexity relevant to human liver anatomy and physiology. To block spontaneous activation induced by culture on a hard substratum, various culturing strategies have been proposed, including the use of collagen- or Matrigel®-coated model [28] or suspension culture on a nonadherent surface. These approaches can restore HSCs to a more quiescent or mild activation state. Nonetheless, the 2D HSC monolayer culture model is simple and easy. Proper uses of primary HSC culture models are still a powerful tool to test novel hypotheses before going to the next step.

2.1.2 Hepatic Stellate Cell Line Models

LX-2 cells are one of the most commonly used human HSC lines. The HSC lines are maintained as a fully activated state and therefore not ideal to study the early stages of HSC activation. However, the cells can still respond to fibrotic stimuli, such as TGF-β and PDGF, to activate their downstream signaling pathways, such as SMADs and MAPKs, and promote ECM production and HSC proliferation [29]. The proper uses of HSC lines are still important to elucidate HSC biology and test novel hypotheses in vitro. As mentioned earlier, primary HSCs are resistant to transfection of plasmids. In contrast, LX-2 cells have higher efficacy of plasmid transfection. LX-2 cells are suitable for gene modifications using plasmid transfection and for reporter assays using plasmid constructs [30]. Previously, our laboratory used LX-2 cells for Bambi and HAS2 promoter analyses which are regulated by TLR4 signaling and TGF-β signaling, respectively [8, 31]; LX-2 cells worked well for this purpose [8, 31]. Another potential use of this and related human cell lines is to seek molecular mechanism or drugs that revert activated HSCs into their quiescent status. LX-2 cells are typically fully activated and do not contain lipid droplets in their cytoplasm. Genetic or pharmacologic interventions that suppress HSC activation or even revert the cells to a quiescent status with accumulation of lipid droplets could provide important new targets to downregulate HSCs therapeutically. There are several other HSC lines including HSC-T6, LI90, and GRX cells [29], each one having some limitations including unphysiological HSC activation states and atypical responses to fibrogenic stimuli. Thus, the choice of cell line must optimally match the intended use and model the behavior of HSCs in vivo as much as possible.

2.2 Experimental Rodent Models to Validate Functions of the Target(s) of Interest

After in vitro validation of lead targets, pursuing animal models is the next step to elucidate the intricate mechanisms governing liver fibrosis and accelerate the translation into effective clinical therapies. Hypothesized mechanisms may then be tested with specific interventional studies that may lead to improved understanding of the disease. However, in liver fibrosis research, there is no single perfect animal model recapitulating all features of human liver fibrosis. Therefore, replication of key findings or responses across multiple animal models can strengthen the evidence for a global antifibrotic activity of candidate drugs. On the other hand, a lead molecule sometimes elicits divergent actions in different models. Therefore, etiology-dependent fibrotic mechanisms should also be considered with strong evidence of efficacy in preclinical models before considering clinical trials. To date, we mainly use preclinical liver fibrosis models induced by chronic hepatotoxin exposure (e.g., CCl_4), BDL, and high-fat/high caloric diets. We also want to note that rat fibrosis models can lead to cirrhosis eventually, although mouse models are unable to develop pre-cirrhotic condition.

2.2.1 Liver Fibrosis Models Induced by Chronic Hepatotoxin Exposure

In liver fibrosis caused by chronic HBV or HCV infection, chronic hepatocyte injury is the primary mechanism that drives chronic inflammation and fibrosis. Unfortunately, relevant preclinical fibrosis models for HBV or HCV are not widely accessible [1, 2]. Hepatotoxin-induced liver fibrosis models can be considered as the potential replacement model for HBV-/HCV-induced fibrosis, depending on the pathway being explored. While likely underestimated, chronic environmental toxin exposure may be an overlooked accelerator of liver fibrosis induced by other etiologies (HBV, HCV, ALD, NAFLD, and cholestasis). Also, hepatotoxin-induced fibrosis models are good to study the mechanism of fibrosis regression or reversal after halting hepatotoxin exposure to animals [32].

Carbon Tetrachloride

Carbon tetrachloride (CCl_4) is the most commonly used hepatotoxin-induced liver fibrosis model in the laboratory. Animals can receive CCl_4 through IP, oral gavage, or inhalation. Pericentral fibrosis and central-central bridging fibrosis is observed after injection of twice per week for 6–12 weeks. Once discontinuing the injection, liver fibrosis rapidly regresses. Therefore, this model is useful to study the mechanisms of both fibrosis due to HSC activation, as well as regression or reversal of fibrosis with HSC inactivation [33]. Limitations of this model the reliance on repeating hepatocyte injury, ECM deposition, resolution, and liver regeneration (but not continuous progression), and the injury and fibrosis location are primarily limited in peri-central vein area, leading to central-central bridging fibrosis [32].

Thioacetamide

Thioacetamide (TAA) model is not as common as CCl_4. Compared to the CCl_4 model, TAA induces a more progressive pattern of fibrosis and prolonged collagen deposition even after ceasing toxin treatment. Despite being good for studying the mechanism of progressive liver fibrosis, investigating the reversal of liver fibrosis takes longer after discontinuing TAA treatment [34]. The TAA model displays portal-portal and portal-central bridging fibrosis after 12–24 weeks of treatment in drinking water or IP injection once or twice per week. The model is more rapid in rats when given the agent intraperitoneally, typically developing cirrhosis in 6–8 weeks.

2.2.2 Cholestatic Liver Fibrosis Model

Cholestasis-induced liver fibrosis is among the most reliable animal models to mimic human liver fibrosis by causing periportal ECM deposition and portal-portal bridging fibrosis. Whereas hepatotoxic liver fibrosis largely relies on the activation of HSCs, in cholestasis, both HSCs and portal fibroblasts (PFs) are main sources of fibrotic liver myofibroblasts in establishing biliary fibrosis [35]. BDL is a common procedure to induce cholestasis-induced portal fibrosis. This model is not suitable for studying fibrosis resolution unless the biliary occlusion can be reversed surgically or through use of a reversible ligature around the bile duct. Typically fibrosis develops 2–3 weeks after surgery but can be mouse strain-dependent. Because BDL-induced liver fibrosis is induced using a surgical procedure, some laboratories may have limitations to use this model, and an additional surgical training is required. Alternatively, chronic feeding of the porphyrinogen 3,5-diethoxycarbonyl-1,4-dihydrocollidine, termed "DDC," is an alternative in vivo model for cholestatic liver fibrosis [36]. The DDC-enriched diet increases the secretion of hepatotoxic protoporphyrins, along with the formation of protoporphyrin plugs leading to intermittent bile duct blockage, activation of biliary epithelial cells liver injury together with ductular reaction. Consequently, a 0.1% DDC-containing diet feeding induces a rapid onset liver injury, and then cholestasis and periportal liver fibrosis typically develop within 2–4 weeks [37, 38]. Another commonly used periportal fibrosis model is the $Mdr2^{-/-}$ mouse model that resembles the pathophysiology of human primary sclerosing cholangitis (PSC) [39]. This mouse model lacks the multidrug resistance associated protein 2 (Mdr2, encoded by *Abcb4* in mice and *MDR3* in humans), a transporter that plays a role in phospholipid excretion and instigates biliary epithelial injury and sterile inflammation and ultimately leads to fibrosis [39] by disruption of tight junctions and basement membranes of bile ducts and bile leakage to the portal tract [38]. An advantage of the DDC model and $Mdr2^{-/-}$ mice compared to the BDL model is they don't require surgical expertise. However, it is still limited by breeding issues such that the $Mdr2^{-/-}$ model contains FVB/N background which differs from the most often used laboratory mouse strain – C57BL/6 and Balb/c. This strategy takes a long time to develop, and crossing of different inbred strains of mice increases mouse genetic variability.

2.2.3 NASH-Associated Fibrosis

An ideal mouse NAFLD-fibrosis model would combine key metabolic features of human NAFLD, including obesity and insulin resistance with hepatic steatosis, inflammation, and development of fibrosis. Nevertheless, a perfect model does not exist, and no single model ideally captures all features of human NAFLD, but they are improving and include those listed below.

High-Fat Diet (HFD)-Based Model

Long-term HFD feeding (6–12 months) reproducibly produces hepatic steatosis, obesity, and insulin resistance with moderate elevation of hepatocyte injury markers AST and ALT, but inflammation and fibrosis are very mild (Fig. 3a). For studying hepatic steatosis, obesity, and insulin resistance, the HFD model is informative, but it is not recommended for studying NASH-fibrosis [40]. To recapitulate NASH fibrosis, additional supplementations are often used, including moderate (0.2%) or high (1–2%) cholesterol and high sugar (glucose and fructose) (a.k.a. Western/Fast food diet or the American Lifestyle-Induced Obesity Syndrome

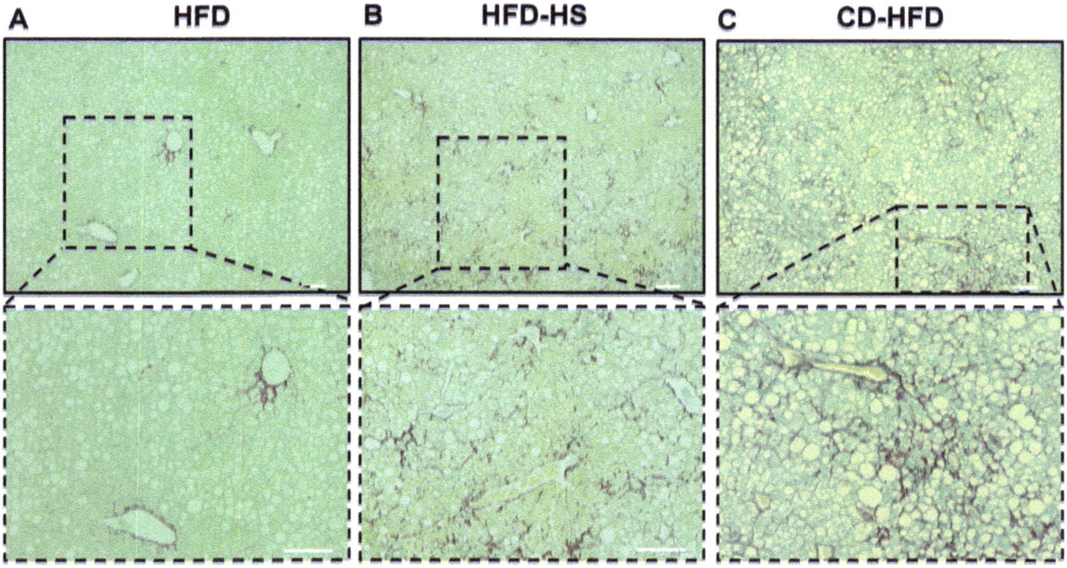

Fig. 3 Histological characterization of NASH liver fibrosis models. To visualize the most widely used experimental models, liver sections were prepared from mice subjected to (**a**) HFD, (**b**) HFD-HS, and (**c**) CD-HFD feeding, respectively. Sirius Red staining shows very mild fibrosis (**a**), moderate fibrosis (**b**), and massive pericellular fibrosis (**c**) in HFD, HFD-HS, and CD-HFD fed mice, respectively. Scale bar: 100 μm (top), 100 μm (bottom). Abbreviations: HFD high-fat diet, HFD-HS high-fat diet-high sugar, CD-HFD choline-deficient HFD

(ALIOS) Diet), or high sugar (glucose and fructose) only. Supplementation with palmitate, a key lipotoxic saturated free fatty acid, in high fat, cholesterol, and fructose can be used as "FPC" diet [41, 42]. Because high cholesterol supplementations are not relevant to human diets, our lab preferentially uses HFD plus high sugar (glucose + fructose) water. These modified HFD-based diets display almost the full course of NAFLD liver phenotype (hepatocyte injury, steatosis, inflammation, and fibrosis), with obesity and insulin resistance [43]. Of note, hepatocyte ballooning, a hallmark of hepatocyte injury in human NASH, is minimally or slightly observed in all mouse NAFLD models [44]. While rodent models still have limitations to recapitulate all phenotypes of human NASH-fibrosis and are time-consuming when feeding for 6--12 months, we favorably use the 6 months of HFD plus high sugar (glucose + fructose) water protocol to study NAFLD with fibrosis (Fig. 3b).

Choline-Deficient Diet-Based Model

In the past, methionine-choline-deficient (MCD) diet was a gold standard model to study NASH. However, because of its unphysiological systemic phenotype (body weight loss, lack of insulin resistance), few laboratories continue to use it [44]. Instead, the choline-deficient amino acid-defined (CDAA) diet model may be used to study NASH-fibrosis. This diet contains methionine that can be converted to choline to produce low amounts of choline, which makes the rodents survive longer than the MCD model, with better body weight gain and mild insulin resistance and evident liver fibrosis after 6 months [45, 46]. More recently, we have favored the choline-deficient HFD (CD-HFD) model as a NASH-fibrosis model (Fig. 3c). This diet contains high fat (60% or 45% dietary fat) but is choline-deficient [8]. This model nicely produces hepatic injury, steatosis, inflammation, and fibrosis (pericellular and bridging fibrosis) and obvious weight gain and insulin resistance [8]. It is known that systemic choline level is decreased in obese patients, which is associated with the conversion of choline to methylamine, reducing plasma levels of phosphatidylcholine by an altered gut microbiome in patients with metabolic syndrome [47]. Currently, we believe this diet model is the most representative rodent NASH-fibrosis model resembling human NASH-fibrosis. Therefore, we prefer to use CD-HFD as a NASH-fibrosis model. Some laboratories use choline-deficient amino acid-defined HFD (CDA-HFD). The CDA-HFD is a choline-deficient and low methionine (0.1%) diet, and the rodents do not gain weight and show insulin resistance [48]. This model differs from the CD-HFD model, which investigators need to take into account when choosing a suitable model for their research.

A Hybrid Model with the Combination of Western Diet + CCl_4

Tsuchida et al. reported the Western diet (21.1% fat, 41% sucrose, 1.25% cholesterol, and glucose-fructose solution) with CCl_4 can rapidly develop stage 3 bridging fibrosis with HSC activation and ductular reaction at 12 weeks and stage 4 cirrhosis and HCC at 24 weeks [49]. Importantly, transcriptomic profiles of this model exhibited a similar transcriptomic profile with human NASH compared with other 18 rodent NASH models [49]. Having said that, the model is limited by no insulin resistance and downregulated cholesterol metabolism pathways. Nonetheless, many researchers agree with the human-relevant transcriptome in this model, and the use of this rodent NASH models may maximize the translational implications of the targets of interest being tested. Because it can recapitulate the progressive stages of fibrosis, this model may be useful for testing new clinical imaging tools, such as new contrast agents for MR imaging, and for testing treatment agents by evaluating liver fibrosis improvement using noninvasive imaging tools.

Lastly, the ob/ob leptin-deficient mouse model develops obesity, insulin resistance, and fatty liver driven by hyperphagia. However, because HSC activation and ECM production require leptin signaling, ob/ob mice are therefore relatively resistant to liver fibrosis [50]. Thus, this model is highly suitable to study obesity, diabetes, and hepatic steatosis, but it is not recommended for studying NASH-fibrosis.

In conclusion, in vivo rodent models of liver fibrosis demonstrate both shared and distinct pathogenic features and gene regulation compared to human liver fibrosis. It is important to carefully select fibrosis models with characteristics most suitable for the study objective.

2.3 Validation of Experimental Findings in Additional Human-Relevant Models of Liver Fibrosis to Further Refine Druggable Targets

Currently, antifibrotic actions have been validated for numerous drugs at preclinical stages using cell culture and animal models. However, many clinical trials for testing antifibrotics remain unsatisfactory [2, 3]. One concern is the gap in liver fibrosis pathophysiology and drug metabolism between rodents and humans. For example, there may be reduced activities of hepatic cytochrome P450 enzymes in rodents and higher sensitivity of human immune cells and HSCs to fibrotic factors and LPS compared to rodent cells. Additionally, the composition of the gut microbiome may differ (gram-negative dominant in humans vs. gram-positive dominant in mice): there are different circadian rhythms (active in daytime in humans vs. active in nighttime in rodents), different durations of injury in rodents (years in humans vs. weeks to months in rodents), different life expectancies (80 years in humans vs. ~2 years in rodents), divergent sizes of body and livers, and different heart rates. Importantly, mice are typically inbred and therefore genetically identical unlike the highly varied genetic backgrounds of humans [32].

The development of 3D multicellular culture system is increasingly used as organoids that recapitulate relevant physiological hepatic conditions that may overcome some of the limitations of animal models or 2D cell culture systems [51]. For example, when hepatocytes and HSCs are cultured on a 2D surface, they adopt a flat and elongated shape that is not a morphological characteristic for these cell types in normal liver but rather mimic the response to injury with loss of hepatocyte function and activation of HSCs. In contrast, when cultured in a 3D microenvironment, hepatocytes retain their morphological shape and metabolic functions, whereas HSCs maintain their natural star-shaped morphology and display greater sensitivity to fibrotic stimuli and therapeutic compounds [51]. This section will discuss human-relevant preclinical in vitro liver models that we recommend for additional validation to select appropriate drugs prior to clinical trials, thereby saving time and cost.

2.3.1 Spheroid-Based Mini Human Liver Tissues

Primary hepatocytes can form spheroids using (i) ultralow attachment or cell-repellent plates and (ii) gravitational aggregation in hanging drop cultures. Unlike the conventional 2D cultures, an advantage of 3D hepatic spheroids is the stability of hepatocyte metabolic functions. The spheroid system can maintain primary human hepatocyte viability and functionality, including ATP levels, albumin secretion, and stable cytochrome p450 enzyme activity up to 1 month [52]. By coculturing hepatocyte spheroids with primary human HSCs and adding fibrotic stimuli, such as TGF-β, PDGF, lipids, and alcohol, this mini-liver model has value in studying liver fibrosis, NAFLD, and ALD with human-relevant metabolic and fibrotic features. Additional coculturing with human endothelial cells and immune cells, for example, resident macrophages (Kupffer cells), may further mimic in vivo human liver as an ex vivo model. However, limitations include unphysiological cell-cell contact between hepatocytes and other cells and an uneven supply of nutrients, media, and oxygen throughout the spheroid. However, this model is easy and convenient and has a reasonable cost, and its metabolic function is differentiated. We recommend this model as the first step in testing drug efficacy as well as toxicity in a relevant human metabolic state following in vivo validation in animals but prior to clinical trials. Lastly, we do not recommend using cell lines in these spheroids, as their functional profiles may differ greatly from primary cells and/or cells may develop nonphysiologic interactions with other cell types within the spheroid.

2.3.2 Organoids

The organoid system is an exciting, emerging tool to study liver disease ex vivo using iPSC (induced pluripotent stem cell) technology. The established protocol can generate hepatocytes from iPSCs using a combination of growth factors or small molecules [53]. More recently, Ouchi et al. generated an iPSC-derived human liver organoids by co-differentiating human-derived iPSCs using a Matrigel scaffold and treatment with retinoid acid to epithelial and mesenchymal lineage cells [54]. The iPSC-derived organoids contain not only hepatocyte-like cells but also Kupffer cell-like and HSC-like cells. Due to the maintenance of a high metabolic state, including cytochrome p450 activities, the iPSC-derived liver organoids nicely develop fatty liver and fibrosis-like phenotype by demonstrating hepatocyte lipid accumulation, HSC activation, and ECM production in response to fatty acids. This indicates iPSC-derived liver organoids can be employed as an ex vivo NAFLD-fibrosis model. Furthermore, the efficacy of treatments for NAFLD has also been tested using FGF19, a candidate treatment for NAFLD fibrosis. FGF19 treatment reduces fatty liver and fibrosis-like phenotypes in iPSC-derived liver organoids, indicating that this model can validate drug treatment efficacy as well as toxicity [54]. The limitations of this model are similar to the spheroid models but further include the potential of containing undifferentiated cells, the cost for differentiation culture media, and the handling of iPSCs. Nevertheless, this iPSC-derived organoids may be good for studying how genetic susceptibility contributes to NAFLD-fibrosis development, for example, using the iPSCs derived from PNPLA3 I148M patients to assess the treatment efficacy using a precision medicine approach.

2.3.3 Liver-on-a-Chip

The organ-on-a-chip model uses a microfluidic chip equivalent to a size of an AA battery that is composed of tiny top and bottom fluidic channels that are separated by an ECM-coated, porous membrane (Fig. 4). The liver-on-a-chip model is comprised of hepatocytes in the top channel, organ- and species-specific vascular endothelial cells, and other nonparenchymal cells (e.g., HSCs and Kupffer cells) in the bottom channel. Similar to liver spheroids and

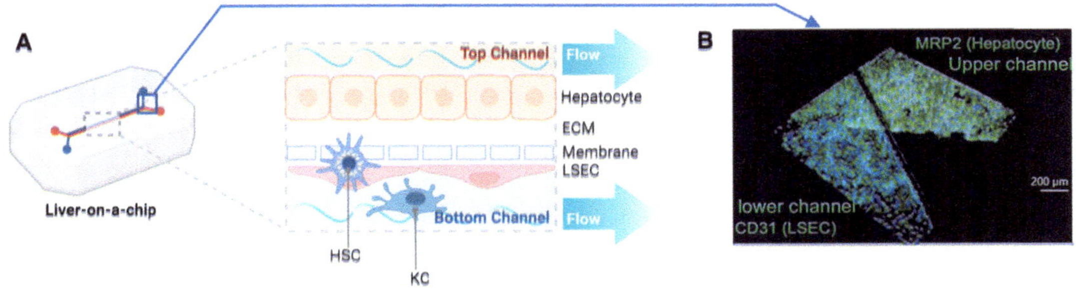

Fig. 4 Liver-on-a-chip. (**a**) Primary hepatocytes are grown in the upper parenchymal channel (red) within an ECM sandwich, on top of an ECM-coated, porous membrane that separates the two parallel microchannels. Relevant species-specific LSECs with or without liver KC or HSC are cultured on the opposite side of the membrane in the lower vascular channel (blue). (**b**) Immunofluorescence picture – MRP2 staining indicates hepatocytes in the upper chamber; CD31 staining indicates LSECs in the lower chamber. Abbreviations: ECM extracellular matrix, HSCs hepatic stellate cells, KC Kupffer cells, LSECs liver sinusoidal endothelial cells

organoids, liver-on-a-chip has a better metabolic state, cell sensitivity to substances, and physiological cell plasticity with a long culturing capacity of up to 4 weeks. The advantage of this model is the use of human-originated primary cells with continuous physiological media flow that maintains proper oxygen and nutrient concentrations and washes out metabolites and cell debris. The device also maintains a physiologic level of shear stress and a relevant mechanical microenvironment, with an appropriate sinusoidal architecture [55]. A noteworthy advantage of this model is using organ- and species-specific vascular endothelial cells that are crucial to maintain the canonical functions of hepatocyte metabolism, including albumin production and activities of cytochrome P450 and alcohol-metabolizing enzymes. In addition to well-described drug toxicity assays [56], liver-on-a-chip responses to ethanol exposure have been tested [57]. This model can serve as a nice validation tool relevant to human NAFLD and ALD.

2.3.4 Precision-Cut Liver Slices

Precision-cut liver slices (PCLS) model is considered especially useful for the study of induction and reversion of hepatic fibrogenesis [58]. Moreover, PCLS have emerged as a versatile ex vivo tool because they can retain the 3D structure, physiological ECM composition, and the native complex cell-cell interactions of the liver [59]. The use of fresh surgical or biopsy specimens from patients with liver fibrosis, cirrhosis, NAFLD, and ALD is relevant to test antifibrotic effects of new compounds within conditions relevant to human diseases [48]. Limitations are relatively short functional life span of no greater than 72–96 h, which is attributed to hypoxia, downregulation of hepatocyte functions, and progressive HSC activation during slice culture. Compared to iPSC-derived organoids and liver-on-a-chip models, the PCLS model is relatively simple and still a relevant human ex vivo model for testing candidate drugs for fibrosis but requires regular access to fresh human liver tissue.

3 Translation of Preclinical Studies to Clinical Trials and Current Limitations

The clinical failures of drugs to date have led to a reassessment of how they are tested during the early stages of development. Screening for potential druggable targets through the strategies described at Subheading 1.2 is essential, as the current knowledge about etiology, epigenetic, molecular, and cellular events during fibrosis is advancing. Considering the importance of large-scale datasets or libraries in finding drugs that may attenuate HSC activation, their use for in-depth study will be impactful for target identification. Extensive target validation at an early stage leads to increased likelihood of a successful clinical trial. In vitro models have several ethical and economic advantages relative to in vivo. On the other hand, animal models with reliable physiological processes and disease features are widely used in drug development and have facilitated the clinical translation of many drugs. Once a target has reached an acceptable level of validation, another validation step during the late stages of preclinical studies using human-relevant models is essential. For example, compared to conventional 2D cultures, microfluidics-based system liver-on-a-chip is a developing tool that may offer a more relevant means of antifibrotic drug testing to supplement standard cell culture and animal testing. This technology may lead to better predictions of human hepatotoxicity (e.g., drug-induced liver injury (DILI)) to determine clinical risk, thus providing data to accelerate drug approvals. The translational goal of liver-on-a-chip is to better recapitulate the patient in vitro to deliver precision medicine approach to treatment.

To propose and start new clinical trials, strong evidence of drug efficacy for HSC inactivation and liver fibrosis reduction or halting must be shown in preclinical studies. We have reviewed the advantages and limitations of in vitro cell culture models and in vivo rodent animal models. Proper selection of these models can determine the efficacy and success of candidate drugs for liver fibrosis. We also seek to detect hepatotoxicity and unfavorable off-target effects during preclinical testing. The drug efficacy for liver fibrosis and off-target effects may also be influenced by administration routes and doses used. Therefore, the following points should be evaluated: (1) the drug's safety in doses equivalent to approximated human exposures, (2) pharmacodynamics (i.e., mechanisms of action and the relationship between drug levels and clinical response), and (3) pharmacokinetics (i.e., drug absorption, distribution, metabolism, excretion, and potential drug-drug interactions) through relevant administration routes similar with clinical trials. Preclinical animal studies often use IP routes because of its convenience. However, if clinical trials consider non-IV routes, such as PO, oral gavage must be used for animal studies. Nonetheless, IV and IP routes may not generate equivalent drug safety, doses, pharmacokinetics, and pharmacodynamics.

We anticipate that additional human-relevant ex vivo models as described in Subheading 2.3 will streamline drug development.

4 Conclusions and Perspectives

In this chapter, we have outlined the approaches used to identify targets and discussed the key steps involved in target validation and the benefits and challenges of using the strategies mentioned to validate targets. This includes the application of a range of techniques that aim to deliver greater therapeutic efficacy. Identifying a good antifibrotic drug target needs to be relevant to the disease phenotype and should be amenable to therapeutic modulation during experimental validation.

In conclusion, the complex mechanisms underlying hepatic fibrosis may delay the development of safe and effective antifibrotic therapies. A better understanding of currently available models will accelerate progress by noting their specific properties, advantages, and limitations. Further exploration of these combination models will undoubtedly pave the way for improved modeling of human hepatic fibrogenesis and provide the possibility of more accurate mechanistic and therapeutic studies and faster drug development.

Financial Support

This work is supported by NIH grants R01DK085252.

References

1. Acharya P, Chouhan K, Weiskirchen S, Weiskirchen R (2021) Cellular mechanisms of liver fibrosis. Front Pharmacol 12:671640
2. Friedman SL, Pinzani M (2022) Hepatic fibrosis 2022: unmet needs and a blueprint for the future. Hepatology 75(2):473–488
3. Kisseleva T, Brenner D (2021) Molecular and cellular mechanisms of liver fibrosis and its regression. Nat Rev Gastroenterol Hepatol 18(3):151–166
4. Shan L, Wang F, Zhai D, Meng X, Liu J, Lv X (2022) New drugs for hepatic fibrosis. Front Pharmacol 13:874408
5. Tan Z, Sun H, Xue T, Gan C, Liu H, Xie Y et al (2021) Liver fibrosis: therapeutic targets and advances in drug therapy. Front Cell Dev Biol 9:2622
6. Seki E, De Minicis S, Österreicher CH, Kluwe J, Osawa Y, Brenner DA, Schwabe RF (2007) TLR4 enhances TGF-β signaling and hepatic fibrosis. Nat Med 13(11):1324–1332
7. Li Y, Jiang D, Liang J, Meltzer EB, Gray A, Miura R et al (2011) Severe lung fibrosis requires an invasive fibroblast phenotype regulated by hyaluronan and CD44. J Exp Med 208(7):1459–1471
8. Yang YM, Noureddin M, Liu C, Ohashi K, Kim SY, Ramnath D et al (2019) Hyaluronan synthase 2–mediated hyaluronan production mediates Notch1 activation and liver fibrosis. Sci Transl Med 11(496):eaat9284
9. Hughes B (2010) 2009 FDA drug approvals: in a year in which regulators and drug companies gained familiarity with risk management strategies, the number of new drug approvals was similar to that in 2008. Nat Rev Drug Discov 9(2):89–93
10. Jin G, Wong ST (2014) Toward better drug repositioning: prioritizing and integrating existing methods into efficient pipelines. Drug Discov Today 19(5):637–644
11. Blazina I, Selph S (2019) Diabetes drugs for nonalcoholic fatty liver disease: a systematic review. Syst Rev 8(1):1–13
12. Harrison SA, Manghi FP, Smith WB, Alpenidze D, Aizenberg D, Klarenbeek N et al (2022) Licogliflozin for nonalcoholic steatohepatitis: a randomized, double-blind, placebo-controlled, phase 2a study. Nat Med 28(7):1432–1438
13. Paananen J, Fortino V (2020) An omics perspective on drug target discovery platforms. Brief Bioinform 21(6):1937–1953
14. Xi Y, LaCanna R, Ma HY, N'Diaye EN, Gierke S, Caplazi P et al (2022) A WISP1 antibody inhibits MRTF signaling to prevent the progression of established liver fibrosis. Cell Metab 34(9):1377–1393
15. Ramnath D, Irvine KM, Lukowski SW, Horsfall LU, Loh Z, Clouston AD et al (2018) Hepatic expression profiling identifies steatosis-independent and steatosis-driven advanced fibrosis genes. JCI Insight 3(14): e120274
16. Vollmann EH, Cao L, Amatucci A, Reynolds T, Hamann S, Dalkilic-Liddle I et al (2017) Identification of novel fibrosis modifiers by in vivo siRNA silencing. Mol Ther Nucleic Acids 7: 314–323

17. Bollong MJ, Yang B, Vergani N, Beyer BA, Chin EN, Zambaldo C et al (2017) Small molecule-mediated inhibition of myofibroblast transdifferentiation for the treatment of fibrosis. Proc Natl Acad Sci 114(18):4679–4684
18. Chen JY, Newcomb B, Zhou C, Pondick JV, Ghoshal S, York SR et al (2017) Tricyclic antidepressants promote ceramide accumulation to regulate collagen production in human hepatic stellate cells. Sci Rep 7(1):1–13
19. Li W, Chen JY, Sun C, Sparks RP, Pantano L, Rahman RU et al (2022) Nanchangmycin regulates FYN, PTK2, and MAPK1/3 to control the fibrotic activity of human hepatic stellate cells. elife 11:e74513
20. Yu S, Ericson M, Fanjul A, Erion DM, Paraskevopoulou M, Smith EN et al (2022) Genome-wide CRISPR screening to identify drivers of TGF-β-induced liver fibrosis in human hepatic stellate cells. ACS Chem Biol 17(4):918–929
21. Weiskirchen R, Weimer J, Meurer SK, Kron A, Seipel B, Vater I et al (2013) Genetic characteristics of the human hepatic stellate cell line LX-2. PLoS One 8(10):e75692
22. Chiabotto G, Ceccotti E, Bruno S (2022) Narrative review of in vitro experimental models of hepatic fibrogenesis, vol 5
23. Friedman SL, Roll FJ, Boyles J, Bissell D (1985) Hepatic lipocytes: the principal collagen-producing cells of normal rat liver. Proc Natl Acad Sci 82(24):8681–8685
24. Friedman SL (2008) Hepatic stellate cells: protean, multifunctional, and enigmatic cells of the liver. Physiol Rev 88(1):125–172
25. Maschmeyer P, Flach M, Winau F (2011) Seven steps to stellate cells. J Vis Exp JoVE 51:e2710
26. Rosenthal SB, Liu X, Ganguly S, Dhar D, Pasillas MP, Ricciardelli E et al (2021) Heterogeneity of HSC s in a mouse model of NASH. Hepatology 74(2):667–685
27. Liu X, Xu J, Rosenthal S, Zhang LJ, McCubbin R, Meshgin N et al (2020) Identification of lineage-specific transcription factors that prevent activation of hepatic stellate cells and promote fibrosis resolution. Gastroenterology 158(6):1728–1744
28. Gaça MD, Zhou X, Issa R, Kiriella K, Iredale JP, Benyon RC (2003) Basement membrane-like matrix inhibits proliferation and collagen synthesis by activated rat hepatic stellate cells: evidence for matrix-dependent deactivation of stellate cells. Matrix Biol 22(3):229–239
29. Herrmann J, Gressner AM, Weiskirchen R (2007) Immortal hepatic stellate cell lines: useful tools to study hepatic stellate cell biology and function? J Cell Mol Med 11(4):704–722
30. Xu L, Hui AY, Albanis E, Arthur MJ, O'byrne SM, Blaner WS et al (2005) Human hepatic stellate cell lines, LX-1 and LX-2: new tools for analysis of hepatic fibrosis. Gut 54(1): 142–151
31. Liu C, Chen X, Yang L, Kisseleva T, Brenner DA, Seki E (2014) Transcriptional repression of the transforming growth factor β (TGF-β) Pseudoreceptor BMP and activin membrane-bound inhibitor (BAMBI) by nuclear factor κB (NF-κB) p50 enhances TGF-β signaling in hepatic stellate cells. J Biol Chem 289(10): 7082–7091
32. Delire B, Stärkel P, Leclercq I (2015) Animal models for fibrotic liver diseases: what we have, what we need, and what is under development. J Clin Transl Hepatol 3(1):53
33. Iredale JP, Benyon RC, Pickering J, McCullen M, Northrop M, Pawley S et al (1998) Mechanisms of spontaneous resolution of rat liver fibrosis. Hepatic stellate cell apoptosis and reduced hepatic expression of metalloproteinase inhibitors. J Clin Invest 102(3): 538–549
34. Reif S, Aeed H, Shilo Y, Reich R, Kloog Y, Kweon YO, Bruck R (2004) Treatment of thioacetamide-induced liver cirrhosis by the Ras antagonist, farnesylthiosalicylic acid. J Hepatol 41(2):235–241
35. Dranoff JA, Wells RG (2010) Portal fibroblasts: underappreciated mediators of biliary fibrosis. Hepatology 51(4):1438–1444
36. Pose E, Sancho-Bru P, Coll M (2019) 3, 5-diethoxycarbonyl-1, 4-dihydrocollidine diet: a rodent model in cholestasis research. In: *Experimental cholestasis research*. Humana, New York, pp 249–257
37. Meerman LEO, Koopen NR, Bloks V, van Goor H, Havinga R, Wolthers BG et al (1999) Biliary fibrosis associated with altered bile composition in a mouse model of erythropoietic protoporphyria. Gastroenterology 117(3):696–705
38. Fickert P, Stöger U, Fuchsbichler A, Moustafa T, Marschall HU, Weiglein AH et al (2007) A new xenobiotic-induced mouse model of sclerosing cholangitis and biliary fibrosis. Am J Pathol 171(2):525–536
39. Iwaisako K, Jiang C, Zhang M, Cong M, Moore-Morris TJ, Park TJ et al (2014) Origin of myofibroblasts in the fibrotic liver in mice. Proc Natl Acad Sci 111(32):E3297–E3305
40. Nakamura A, Terauchi Y (2013) Lessons from mouse models of high-fat diet-induced NAFLD. Int J Mol Sci 14(11):21240–21257

41. Ipsen DH, Lykkesfeldt J, Tveden-Nyborg P (2020) Animal models of fibrosis in nonalcoholic steatohepatitis: do they reflect human disease? Adv Nutr 11(6):1696–1711
42. Eng JM, Estall JL (2021) Diet-induced models of non-alcoholic fatty liver disease: food for thought on sugar, fat, and cholesterol. Cell 10(7):1805
43. Liang H, Jiang F, Cheng R, Luo Y, Wang J, Luo Z et al (2021) A high-fat diet and high-fat and high-cholesterol diet may affect glucose and lipid metabolism differentially through gut microbiota in mice. Exp Anim 70(1): 73–83
44. Stephenson K, Kennedy L, Hargrove L, Demieville J, Thomson J, Alpini G, Francis H (2018) Updates on dietary models of nonalcoholic fatty liver disease: current studies and insights. Gene Expr 18(1):5
45. Kodama Y, Kisseleva T, Iwaisako K, Miura K, Taura K, De Minicis S et al (2009) c-Jun N-terminal kinase-1 from hematopoietic cells mediates progression from hepatic steatosis to steatohepatitis and fibrosis in mice. Gastroenterology 137(4):1467–1477
46. Miura K, Kodama Y, Inokuchi S, Schnabl B, Aoyama T, Ohnishi H et al (2010) Toll-like receptor 9 promotes steatohepatitis by induction of interleukin-1β in mice. Gastroenterology 139(1):323–334
47. Schnabl B, Brenner DA (2014) Interactions between the intestinal microbiome and liver diseases. Gastroenterology 146(6):1513–1524
48. Alsamman S, Christenson SA, Yu A, Ayad NME, Mooring MS, Segal JM et al (2020) Targeting acid ceramide inhibits YAP/TAZ signaling to reduce fibrosis in mice. Sci Transl Med 12(557):eaay9798
49. Tsuchida T, Lee YA, Fujiwara N, Ybanez M, Allen B, Martins S et al (2018) A simple diet- and chemical-induced murine NASH model with rapid progression of steatohepatitis, fibrosis and liver cancer. J Hepatol 69(2):385–395
50. Ikejima K, Takei Y, Honda H, Hirose M, Yoshikawa M, Zhang YJ et al (2002) Leptin receptor-mediated signaling regulates hepatic fibrogenesis and remodeling of extracellular matrix in the rat. Gastroenterology 122(5): 1399–1410
51. Saxton SH, Stevens KR (2022) 2D and 3D liver models. J Hepatol S0168-8278(22): 02920–02928
52. Bell CC, Hendriks DF, Moro SM, Ellis E, Walsh J, Renblom A et al (2016) Characterization of primary human hepatocyte spheroids as a model system for drug-induced liver injury, liver function and disease. Sci Rep 6(1):1–13
53. Takebe T, Sekine K, Enomura M, Koike H, Kimura M, Ogaeri T et al (2013) Vascularized and functional human liver from an iPSC-derived organ bud transplant. Nature 499(7459):481–484
54. Ouchi R, Togo S, Kimura M, Shinozawa T, Koido M, Koike H et al (2019) Modeling steatohepatitis in humans with pluripotent stem cell-derived organoids. Cell Metab 30(2): 374–384
55. Hayward KL, Kouthouridis S, Zhang B (2020) Organ-on-a-chip systems for modeling pathological tissue morphogenesis associated with fibrosis and cancer. ACS Biomater Sci Eng 7(7):2900–2925
56. Jang KJ, Otieno MA, Ronxhi J, Lim HK, Ewart L, Kodella KR et al (2019) Reproducing human and cross-species drug toxicities using a Liver-Chip. Sci Transl Med 11(517):eaax5516
57. Nawroth JC, Petropolis DB, Manatakis DV, Maulana TI, Burchett G, Schlünder K et al (2021) Modeling alcohol-associated liver disease in a human Liver-Chip. Cell Rep 36(3): 109393
58. Pearen MA, Lim HK, Gratte FD, Fernandez-Rojo MA, Nawaratna SK, Gobert GN et al (2020) Murine precision-cut liver slices as an ex vivo model of liver biology. J Vis Exp (JoVE) 157:e60992
59. Westra IM, Oosterhuis D, Groothuis GM, Olinga P (2014) Precision-cut liver slices as a model for the early onset of liver fibrosis to test antifibrotic drugs. Toxicol Appl Pharmacol 274(2):328–338

Index

A

B

C

D

E

F

Ralf Weiskirchen and Scott L. Friedman (eds.), *Hepatic Stellate Cells: Methods and Protocols*, Methods in Molecular Biology, vol. 2669, https://doi.org/10.1007/978-1-0716-3207-9,

MIX
Papier aus verantwortungsvollen Quellen
Paper from responsible sources
FSC® C105338

If you have any concerns about our products,
you can contact us on
ProductSafety@springernature.com

In case Publisher is established outside the EU,
the EU authorized representative is:
Springer Nature Customer Service Center GmbH
Europaplatz 3, 69115 Heidelberg, Germany

Printed by Libri Plureos GmbH
in Hamburg, Germany